AF360733

Particles Separation in Microfluidic Devices

Particles Separation in Microfluidic Devices

Editors

Takasi Nisisako
Naotomo Tottori

MDPI • Basel • Beijing • Wuhan • Barcelona • Belgrade • Manchester • Tokyo • Cluj • Tianjin

Editors
Takasi Nisisako
Tokyo Institute of Technology
Japan

Naotomo Tottori
Kyushu University
Japan

Editorial Office
MDPI
St. Alban-Anlage 66
4052 Basel, Switzerland

This is a reprint of articles from the Special Issue published online in the open access journal *Micromachines* (ISSN 2072-666X) (available at: https://www.mdpi.com/journal/micromachines/special_issues/Particles_Separation).

For citation purposes, cite each article independently as indicated on the article page online and as indicated below:

LastName, A.A.; LastName, B.B.; LastName, C.C. Article Title. *Journal Name* **Year**, *Article Number*, Page Range.

ISBN 978-3-03936-694-1 (Hbk)
ISBN 978-3-03936-695-8 (PDF)

Contents

About the Editors . . . . **vii**

Naotomo Tottori and Takasi Nisisako
Editorial for the Special Issue on Particles Separation in Microfluidic Devices
Reprinted from: *Micromachines* **2020**, *11*, 602, doi:10.3390/mi11060602 . . . . **1**

Fadi Alnaimat, Bobby Mathew and Ali Hilal-Alnaqbi
Modeling a Dielectrophoretic Microfluidic Device with Vertical Interdigitated Transducer
Electrodes for Separation of Microparticles Based on Size
Reprinted from: *Micromachines* **2020**, *11*, 563, doi:10.3390/mi11060563 . . . . **5**

Amanda Bogseth, Jian Zhou and Ian Papautsky
Evaluation of Performance and Tunability of a Co-Flow Inertial Microfluidic Device
Reprinted from: *Micromachines* **2020**, *11*, 287, doi:10.3390/mi11030287 . . . . **19**

**Fleming Dackson Gudagunti, Logeeshan Velmanickam, Dharmakeerthi Nawarathna and
Ivan T. Lima Jr.**
Nucleotide Identification in DNA Using Dielectrophoresis Spectroscopy
Reprinted from: *Micromachines* **2020**, *11*, 39, doi:10.3390/mi11010039 . . . . **35**

Christopher Sobecki, Jie Zhang and Cheng Wang
Numerical Study of Paramagnetic Elliptical Microparticles in Curved Channels and Uniform
Magnetic Fields
Reprinted from: *Micromachines* **2020**, *11*, 37, doi:10.3390/mi11010037 . . . . **47**

**Jonathan Kottmeier, Maike Wullenweber, Sebastian Blahout, Jeanette Hussong,
Ingo Kampen, Arno Kwade and Andreas Dietzel**
Accelerated Particle Separation in a DLD Device at Re > 1 Investigated by Means of μPIV
Reprinted from: *Micromachines* **2019**, *10*, 768, doi:10.3390/mi10110768 . . . . **71**

Salini Krishna, Fadi Alnaimat and Bobby Mathew
Nozzle-Shaped Electrode Configuration for Dielectrophoretic 3D-Focusing of Microparticles
Reprinted from: *Micromachines* **2019**, *10*, 585, doi:10.3390/mi10090585 . . . . **89**

**Charles P. Clark, Vahid Farmehini, Liam Spiers, M. Shane Woolf, Nathan S. Swami and
James P. Landers**
Real Time Electronic Feedback for Improved Acoustic Trapping of Micron-Scale Particles
Reprinted from: *Micromachines* **2019**, *10*, 489, doi:10.3390/mi10070489 . . . . **105**

Gangadhar Eluru, Pavan Nagendra and Sai Siva Gorthi
Microfluidic In-Flow Decantation Technique Using Stepped Pillar Arrays and Hydraulic
Resistance Tuners
Reprinted from: *Micromachines* **2019**, *10*, 471, doi:10.3390/mi10070471 . . . . **119**

Takuma Yanai, Takatomo Ouchi, Masumi Yamada and Minoru Seki
Hydrodynamic Microparticle Separation Mechanism Using Three-Dimensional Flow Profiles in
Dual-Depth and Asymmetric Lattice-Shaped Microchannel Networks
Reprinted from: *Micromachines* **2019**, *10*, 425, doi:10.3390/mi10060425 . . . . **137**

Yanying Jiao, Yongqing He and Feng Jiao
Two-dimensional Simulation of Motion of Red Blood Cells with Deterministic Lateral
Displacement Devices
Reprinted from: *Micromachines* **2019**, *10*, 393, doi:10.3390/mi10060393 . . . . **149**

Haeli Kang, Jinho Kim, Hyungseok Cho and Ki-Ho Han
Evaluation of Positive and Negative Methods for Isolation of Circulating Tumor Cells by
Lateral Magnetophoresis
Reprinted from: *Micromachines* **2019**, *10*, 386, doi:10.3390/mi10060386 . . . . **165**

Annalisa Volpe, Caterina Gaudiuso and Antonio Ancona
Sorting of Particles Using Inertial Focusing and Laminar Vortex Technology: A Review
Reprinted from: *Micromachines* **2019**, , 594, doi:10.3390/mi10090594 . . . . **175**

**Chenlin Zhang, Bingjie Xu, Chaoyang Gong, Jingtang Luo, Quanming Zhang
and Yuan Gong**
Fiber Optofluidic Technology Based on Optical Force and Photothermal Effects
Reprinted from: *Micromachines* **2019**, *10*, 499, doi:10.3390/mi10080499 . . . . **195**

About the Editors

Takasi Nisisako received a Ph.D. in mechanical engineering from the University of Tokyo, Japan, in 2005. Presently, he is an associate professor working at the Institute of Innovative Research, Tokyo Institute of Technology, Japan. His current research interests mainly include microfluidic systems for analytical and/or manufacturing applications.

Naotomo Tottori received a Ph.D in mechanical engineering from the Tokyo Institute of Technology in 2018. In 2018–2020, he was a specially appointed assistant professor working at the Institute of Innovative Research, Tokyo Institute of Technology, Japan. Presently, he is working as an assistant professor with the Department of Mechanical Engineering, Faculty of Engineering, Kyushu University, Japan. His current research interests mainly include microfluidic systems for particles processing.

Editorial

Editorial for the Special Issue on Particles Separation in Microfluidic Devices

Naotomo Tottori [1] and Takasi Nisisako [2],*

[1] Department of Mechanical Engineering, Faculty of Engineering, Kyushu University, W4-729, 744, Motooka, Nishi-ku, Fukuoka 819-0395, Japan; tottori@mech.kyushu-u.ac.jp

[2] Institute of Innovative Research, Tokyo Institute of Technology, R2-9, 4259 Nagatsuta-cho, Midori-ku, Yokohama, Kanagawa 226-8503, Japan

* Correspondence: nisisako.t.aa@m.titech.ac.jp

Received: 20 June 2020; Accepted: 20 June 2020; Published: 22 June 2020

The separation and sorting of micro- and nano-sized particles is an important step in chemical, biological, and medical analyses. In the past two decades, micro- and nanofluidic platforms have been increasingly applied for the separation, fractionation, sorting, and purification of all classes of particles based on their physical and chemical properties because of their advantages of minimal consumption of sample and reagent, ease of use, and enabling of the integration of multicomponent for comprehensive analysis. The separation techniques using micro- and nanofluidic devices are classified into passive methods using geometries and hydrodynamic effects at micro/nanoscale, and active methods using external fields such as electric, magnetic, optical, and acoustic forces.

This Special Issue collects some state-of-the-art developments in active and passive microfluidic separation, isolation, and manipulation for a wide range of particles. In this Special Issue, 11 research papers, and two review articles are published. Five papers [1–5] and a review article [6] present (1) passive microfluidic techniques using inertial focusing [1,6], deterministic lateral displacement (DLD) [2,3], and hydrodynamic methods [4,5]. The remaining papers [7–12] and a review article [13] cover (2) active microfluidic techniques using electric [7–9], acoustic [10], magnetic [11,12], and optical forces [13].

(1) Passive microfluidic technique: Bogseth et al. proposed a co-flow inertial microfluidic device that is tunable in multiple ways for adaptation to different application requirements [1]. They evaluated flow rate, flow rate ratio, and output resistance ratio to flexibly tune the cutoff size of the device and separation performance even after the devices are fabricated. Kottmeier et al. experimentally observed an asymmetric flow field pattern caused by vortices behind DLD mircopost at high Reynolds number (Re > 1) using microparticle image velocimetry and compared this experimental result with CFD simulations [2]. Jiao et al. reported a numerical simulation of the motion of red blood cells (RBCs) flowing through DLD devices with different pillar shapes and gap configurations [3]. Eluru et al. proposed a microfluidic in-flow decantation technique that enables continuous separation of particles from fluid [4]. They achieved clog-free separation during the operation for at least an hour and could obtain purities close to 100% and yields as high as 14%. Yanai et al. demonstrated a new hydrodynamic mechanism of microparticle separation using dual-depth, lattice-patterned asymmetric microchannel networks [5]. By precisely observing the motion of model particles in the microchannel, they revealed that the 3D laminar flow profile affects the size-selective particle separation. They also demonstrated that the input position of particles in both x and z directions could improve the separation performance significantly. In addition to these research articles for passive techniques, Volpe et al. wrote a comprehensive review of microfluidic particles sorting using inertial focusing and laminar vortex technology [6].

(2) Active microfluidic technique: Krishna et al. presented an experimentally validated mathematical model of a microfluidic device with nozzle-shaped electrode configuration for dielectrophoretic 3D-focusing of particles [7]. They investigated the effect of operating/geometric parameters on the

3D-focusing efficiency of the device through the proposed mathematical model. Alnaimat et al. conceptualized and mathematically modeled a dielectrophoretic microfluidic device with two sets of interdigitated transducer vertical electrodes for separation of a binary heterogeneous mixture of particles based on size [8]. The proposed model is used for a parametric study to investigate the effect of parameters on the performance of the microfluidic device. Gudagunti et al. used negative dielectrophoresis (DEP) spectroscopy as an effective transduction mechanism of a biosensor to accurately detect single nucleotide polymorphism (SNP) in a short DNA strand [9]. Clark et al. demonstrated real-time monitoring of voltage measurements and immediate, corresponding adjustments to acoustic trapping frequency to improve their acoustic differential extraction [10]. Kang et al. introduced positive and negative methods for isolating circulating tumor cells (CTCs) by lateral magnetophoresis [11]. They compared the CTCs recovery rates, WBC depletion rates, and CTC purities between the positive and negative methods to discuss their strengths and weaknesses points for CTC-based diagnostics, prognostics, and therapeutics for cancer. Sobecki et al. reported numerical simulation of the dynamics of a paramagnetic elliptical particle in a low Reynolds number Poiseuille flow in a curved channel and under a uniform magnetic field [12]. In addition to these research articles for active techniques, Zhang et al. presented a comprehensive review of the latest progress in fiber optofluidics (FOF) based on two major opto-physical effects, namely optical force and the photothermal effect, in manipulation and sensing applications [13].

We would like to thank all authors for submitting their papers to this Special Issue. We would also like to acknowledge all the reviewers for dedicating their time and timely reviews to improve the quality of this Special Issue.

Conflicts of Interest: The authors declare no conflict of interest.

References

1. Bogseth, A.; Zhou, J.; Papautsky, I. Evaluation of Performance and Tunability of a Co-Flow Inertial Microfluidic Device. *Micromachines* **2020**, *11*, 287. [CrossRef]
2. Kottmeier, J.; Wullenweber, M.; Blahout, S.; Hussong, J.; Kampen, I.; Kwade, A.; Dietzel, A. Accelerated Particle Separation in a DLD Device at Re > 1 Investigated by Means of μPIV. *Micromachines* **2019**, *10*, 768. [CrossRef] [PubMed]
3. Jiao, Y.; He, Y.; Jiao, F. Two-dimensional Simulation of Motion of Red Blood Cells with Deterministic Lateral Displacement Devices. *Micromachines* **2019**, *10*, 393. [CrossRef] [PubMed]
4. Eluru, G.; Nagendra, P.; Gorthi, S.S. Microfluidic In-Flow Decantation Technique Using Stepped Pillar Arrays and Hydraulic Resistance Tuners. *Micromachines* **2019**, *10*, 471. [CrossRef] [PubMed]
5. Yanai, T.; Ouchi, T.; Yamada, M.; Seki, M. Hydrodynamic Microparticle Separation Mechanism Using Three-Dimensional Flow Profiles in Dual-Depth and Asymmetric Lattice-Shaped Microchannel Networks. *Micromachines* **2019**, *10*, 425. [CrossRef] [PubMed]
6. Volpe, A.; Gaudiuso, C.; Ancona, A. Sorting of Particles Using Inertial Focusing and Laminar Vortex Technology: A Review. *Micromachines* **2019**, *10*, 594. [CrossRef] [PubMed]
7. Krishna, S.; Alnaimat, F.; Mathew, B. Nozzle-Shaped Electrode Configuration for Dielectrophoretic 3D-Focusing of Microparticles. *Micromachines* **2019**, *10*, 585. [CrossRef] [PubMed]
8. Alnaimat, F.; Mathew, B.; Hilal-Alnaqbi, A. Modeling a Dielectrophoretic Microfluidic Device with Vertical Interdigitated Transducer Electrodes for Separation of Microparticles Based on Size. *Micromachines* **2020**, *11*, 563. [CrossRef] [PubMed]
9. Gudagunti, F.D.; Velmanickam, L.; Nawarathna, D.; Lima, I.T., Jr. Nucleotide Identification in DNA Using Dielectrophoresis Spectroscopy. *Micromachines* **2019**, *11*, 39. [CrossRef] [PubMed]
10. Clark, C.P.; Farmehini, V.; Spiers, L.; Woolf, M.S.; Swami, N.S.; Landers, J.P. Real Time Electronic Feedback for Improved Acoustic Trapping of Micron-Scale Particles. *Micromachines* **2019**, *10*, 489. [CrossRef] [PubMed]
11. Kang, H.; Kim, J.; Cho, H.; Han, K.-H. Evaluation of Positive and Negative Methods for Isolation of Circulating Tumor Cells by Lateral Magnetophoresis. *Micromachines* **2019**, *10*, 386. [CrossRef] [PubMed]

12. Sobecki, C.; Zhang, J.; Wang, C. Numerical Study of Paramagnetic Elliptical Microparticles in Curved Channels and Uniform Magnetic Fields. *Micromachines* **2019**, *11*, 37. [CrossRef] [PubMed]
13. Zhang, C.; Xu, B.; Gong, C.; Luo, J.; Zhang, Q.; Gong, Y. Fiber Optofluidic Technology Based on Optical Force and Photothermal Effects. *Micromachines* **2019**, *10*, 499. [CrossRef] [PubMed]

Article

Modeling a Dielectrophoretic Microfluidic Device with Vertical Interdigitated Transducer Electrodes for Separation of Microparticles Based on Size

Fadi Alnaimat [1], Bobby Mathew [1] and Ali Hilal-Alnaqbi [2],*

[1]   Mechanical Engineering Department, United Arab Emirates University, Al Ain P. O. Box 15551, UAE; falnaimat@uaeu.ac.ae (F.A.); bmathew@uaeu.ac.ae (B.M.)

[2]   Abu Dhabi Polytechnic, MBZ campus, Abu Dhabi P. O. Box. 111499, UAE

*   Correspondence: ali.alnaqbi@adpoly.ac.ae; Tel.: +971-2-695-1070

Received: 12 March 2020; Accepted: 28 May 2020; Published: 31 May 2020

Abstract: This article conceptualizes and mathematically models a dielectrophoretic microfluidic device with two sets of interdigitated transducer vertical electrodes for separation of a binary heterogeneous mixture of particles based on size; each set of electrodes is located on the sidewalls and independently controllable. To achieve separation in the proposed microfluidic device, the small microparticles are subjected to positive dielectrophoresis and the big microparticles do not experience dielectrophoresis. The mathematical model consists of equations describing the motion of each microparticle, fluid flow profile, and electric voltage and field profiles, and they are solved numerically. The equations of motion take into account the influence of phenomena, such as inertia, drag, dielectrophoresis, gravity, and buoyancy. The model is used for a parametric study to understand the influence of parameters on the performance of the microfluidic device. The parameters studied include applied electric voltages, electrode dimensions, volumetric flow rate, and number of electrodes. The separation efficiency of the big and small microparticles is found to be independent of and dependent on all parameters, respectively. On the other hand, the separation purity of the big and small microparticles is found to be dependent on and independent of all parameters, respectively. The mathematical model is useful in designing the proposed microfluidic device with the desired level of separation efficiency and separation purity.

Keywords: Interdigitated transducer electrodes; dielectrophoresis; microfluidics; microchannel; separation

1. Introduction

Devices employing flow passages with hydraulic diameters smaller than 1 mm are referred to as microfluidic devices [1]. There are several advantages in employing microfluidic devices; these include low sample and reagent requirement, low power consumption, small footprint, and portability. One of the applications for which microfluidic devices are commonly employed is the separation of a heterogeneous sample into multiple homogeneous samples [2,3]; the basis of separation can be either size or type. Separation of a binary heterogeneous mixture into two homogeneous samples based on size requires every microparticle to be acted upon by an actuation force. There are several options for actuation forces that can be employed in microfluidic devices. One of the most commonly used actuation forces is that associated with dielectrophoresis (DEP) [4–8]. DEP is the phenomenon where in dielectric, but polarizable, microparticles suspended in a dielectric medium undergo translation when subjected to an electric field; the electric field needs to be non-uniform for DEP to exist [4–8]. DEP is specifically termed as positive DEP (pDEP) and negative DEP (nDEP) when the translation of the microparticles is towards the highest and lowest gradient of the electric field, respectively [4–8]. The force associated with DEP depends on several factors such as radius of the microparticles, permittivity of the medium, Clausius–Mossotti factor (f_{CM}), and the magnitude and degree of non-uniformity of the electric field [4–8]. The f_{CM} is dependent on the permittivity and conductivity of the microparticle and medium as well as the operating frequency of the electric signal; microparticles experience pDEP and nDEP when real part of Clausius–Mossotti factor, $Re(f_{CM})$, is positive and negative, respectively. The magnitude and degree of non-uniformity of the electric field depends on the dimensions and shape of the microchannel and the electrodes as well as the electrode configuration. Researchers have proposed electrode configurations for purposes of separation of microparticles based on size [4–8]. This article proposes a microfluidic device (Figure 1) for the separation of microparticles based on size using DEP. The device has one inlet and one outlet and consists of multiple interdigitated transducer (IDT) electrodes located on either sides of the microchannel. The set of electrodes on each side is independently controllable. In this device, separation is achieved by subjecting microparticles of a specific size to pDEP while keeping the microparticles of the alternative size unaffected. The microparticles that are subjected to pDEP will be attracted and captured on the electrodes while the microparticles that are not influenced by DEP will pass through the region of the electrodes unaffected and this leads to separation of microparticles based on size. The microparticles that are unaffected by DEP will be collected at the outlet as the sample is being processed, Figure 1b1; however, the microparticles captured by the electrodes will be collected from the same outlet, by switching off the electric power and flushing the microfluidic device with buffer solution, once the entire sample is processed Figure 1b2. The proposed device has the merit that it can handle high throughput in comparison with most devices proposed in literature [4–8]; this is primarily because the electric field does not decay along the height of the device. The proposed device also has the merit that it does not require focusing prior to separation of the heterogeneous sample. Three-dimensional microfabrication techniques required for realizing the conceptualized device are becoming common as can be observed from several articles in literature [5,9–12].

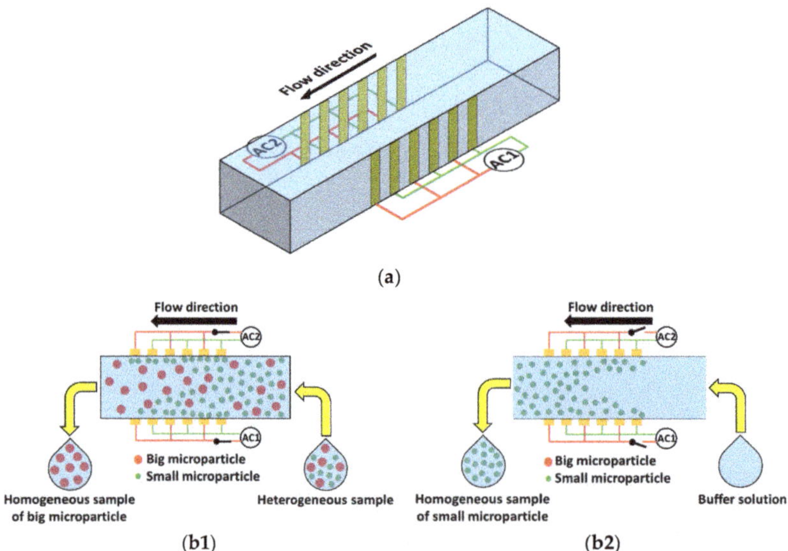

Figure 1. Schematic of the (**a**) proposed microfluidic device (perspective view) and (**b**) working of the device; (**b1**) capture of small microparticles on electrodes and collection of big microparticles at device exit during sample processing (top view) and (**b2**) collection of small microparticles at device exit after their release from electrodes while flushing device with buffer solution (top view).

Among all the different parameters influencing DEP force, the polarity of $Re(f_{CM})$ can be easily altered by varying the operating signal. Thus, the operating frequency of the proposed device should be such that microparticles of a particular size will experience pDEP while microparticles of other size will experience negligible DEP. Figure 2 shows the variation of $Re(f_{CM})$ with operating frequency for polystyrene ($\varepsilon_{ps} = 2.55$ and $K_{s,ps} = 2.85$ nS) and silica microparticles ($\varepsilon_s = 3.8$ and $K_{s,s} = 0.82$ nS) [13]. It can be noticed that $Re(f_{CM})$ is dependent on the operating frequency. It can be noticed that for both types of microparticles, irrespective of their radii, the $Re(f_{CM})$ is positive and negative at low and high operating frequencies, respectively. Furthermore, it can be noticed from Figure 2 that there exists a unique operating frequency, for a mixture of two different sized microparticles, at which the small microparticle experiences pDEP while the big microparticle experiences zero DEP and the proposed device should be operated at this frequency for achieving separation based on size. Frequency at which a microparticle experiences zero DEP force is called cross-over frequency; thus, the proposed device needs to be operated at the cross-over frequency of the big microparticle. Figure 2 has been developed using Equations (1) and (2); cross-over frequency (N_{cr}) of a microparticle can be calculated using Equation (3).

$$Re[f_{CM}] = \frac{(\varepsilon_e + 2\varepsilon_m)(\varepsilon_e - \varepsilon_m) + \frac{(\sigma_e + 2\sigma_m)(\sigma_e - \sigma_m)}{\omega^2}}{(\varepsilon_e + 2\varepsilon_m)^2 + \frac{(\sigma_e + 2\sigma_m)^2}{\omega^2}} \tag{1}$$

$$\sigma_e = \sigma_{bulk,e} + 2\frac{K_{s,e}}{r_e} \tag{2}$$

$$N_{cr} = \frac{\omega}{2\pi} = \frac{1}{2\pi}\sqrt{\frac{(\sigma_e + 2\sigma_m)(\sigma_m - \sigma_e)}{(\varepsilon_e + 2\varepsilon_m)(\varepsilon_e - \varepsilon_m)}} \tag{3}$$

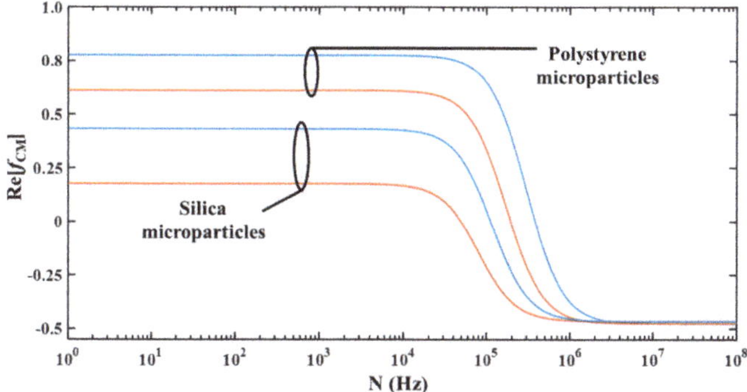

Figure 2. Variation of $Re(f_{CM})$ with operating frequency for 2.5 μm and 5 μm microparticles (■: 2.5 μm, ■ 5 μm and $\sigma_m = 0.0001$ S/m).

Çetin et al. [14] modeled and constructed a DEP-based microfluidic device for separation of microparticles based on size. The device employs a pair of opposing vertical electrodes; one of the electrodes is finite sized while the other electrode is very long in comparison. The mixture with different sized microparticles are focused near the sidewall with the small electrode using sheath flow. In the vicinity of the small electrode, each microparticle is subjected to nDEP force and as it is proportional to the size of the microparticle, the bigger microparticles are pushed further into the microchannel than smaller microparticles. This splits the mixture into two samples with each having microparticles of a particular size. Çetin et al. [14] modeled the trajectory of microparticles in the device; the model consisted of Stokes equation, the equation of electric potential, and equations of motion that considered the influence of forces such as inertia, drag, and DEP. The working of the device is demonstrated by separation a mixture of 5 and 10 μm latex microparticles into homogeneous samples of 5 and 10 μm [14].

Wang et al. [15] constructed and tested a DEP-based microfluidic device, with one set of vertical IDT electrodes located on each of the sidewalls, for separation based on type. The voltage and frequency of operation of one set of IDT electrodes is different from the other. In one instance, this caused one set of IDT electrodes to simultaneously subject the first type of microparticle to weak nDEP and the second type of microparticle to strong nDEP while the other set of IDT electrodes simultaneously subjected first and second types of microparticles to strong nDEP and weak nDEP, respectively. This difference in DEP forces experienced by microparticles allowed for separating a heterogeneous mixture of microparticles and cells into two homogenous samples based on type. Çetin and Li [16] modeled a DEP-based microfluidic device for separation of microparticles based on size. The device consists of two vertical electrodes with one electrode placed upstream of a curved section, of the microchannel, while placing the second electrode downstream of the same. All big microparticles are subjected to pDEP which causes them to move towards the inner wall of the curved section while the small microparticles are subjected to nDEP which causes them to move towards the outer wall of the same thereby leading to desired separation based on size. Kang et al. [17] developed a microfluidic DEP-based device for the separation of microparticles. The device uses a pair of vertical electrodes with one electrode placed, on one of the sidewalls, upstream and the other electrode placed, on the opposing sidewall, downstream of a constriction in the microchannel; moreover, the constriction is closer to the upstream electrode. The incoming microparticles are focused using sheath flow prior to reaching the constriction and as the microparticles pass through the constriction, they are pushed against one of its sidewalls by nDEP force. Subsequently, streamlines passing through the center of the microparticle carry them out of the constriction and as these streamlines are different, the desired separation based on size is achieved. Faraghat et al. [18] developed a DEP-based filter for type-based separation of cells. The filter consists of multiple layers of electrode sandwiched between insulating layers through which several through holes are realized. An electric field is set up between two neighboring electrode layers. With this device, it is possible to subject entities to either pDEP or nDEP; entities subjected to pDEP are attracted and captured on the walls of the through holes while those entities experiencing nDEP are focused at the center of the through holes thereby achieving he desired separation. Mathew et al. [19–21] and Alazzam et al. [22] modeled several microfluidic devices employing spatially varying electric field for realizing field flow fractionation to achieve type based separation of microparticles. In this device the microparticles are subjected to nDEP and sedimentation forces in the vertical direction and this leads to levitation of the microparticles. The levitation height is dependent on the permittivity and density of the microparticle thereby allowing for separation of microparticles. Mathew et al. [19] employed multiple finite sized IDT electrodes located on the bottom surface of the microchannel while Mathew et al. [20] conceptualized a device with multiple finite sized and continuous electrodes on the bottom and top surfaces, respectively. The microfluidic device of Mathew et al. [21] and Alazzam et al. [22] consisted of multiple finite sized electrodes located on its top and bottom surfaces; in Mathew et al. [21], the electrodes on both surfaces are aligned while in Alazzam et al. [22], the electrodes on one surface is aligned with the electrode gap on the opposite surface. Alnaimat et al. [13] developed the mathematical model of a microfluidic device, employing multiple finite sized planar IDT electrodes located on the bottom surface, for separation of microparticles based on type. In this device, one type of microparticle is subjected to pDEP while the other type of microparticle is subjected to nDEP; the microparticles experiencing pDEP are attracted and captured on the electrodes while the microparticles subjected to nDEP are levitated above the electrodes and this achieves separation based on type.

The microfluidic device conceptualized in this document can find application in the area of disease diagnosis, especially that requiring investigation of blood [23,24]. Diagnosis of certain illness depends on identifying foreign entities or rare cells in blood samples [23,24]. For these illnesses, the proposed microfluidic device can be used for identifying foreign entities or rare cells as long as there sizes are different from that of regular cells.

This work is the first to model the microfluidic device shown in Figure 1 while working under the proposed scheme with the aim of separation of microparticles based on size. The model takes into account the influence of all forces associated with the movement of microparticles in microfluidic devices; the forces include that associated with inertia, drag, gravity, buoyancy, and DEP. The inclusion of forces associated with inertia and drag allows for determining the time and length required for achieving a desired performance metric in the proposed device.

2. Mathematical Modeling

The mathematical model of the microfluidic device conceptualized in the previous section of this document is detailed in this section. The mathematical model of the microfluidic device consists of equations of motion, equation of fluid flow, equations of electric voltage, and field. In microfluidic devices the flow is very small and one-dimensional, i.e., flow has velocity only in axial direction. Equation (4) represents the equations of the motion while Equation (5) is the equation of fluid flow (one dimensional). Equations (6) and (7) describe the electric voltage and electric field inside the microchannel, respectively.

$$m_e \frac{d}{dt^2}\left(x_e \boldsymbol{i} + y_e \boldsymbol{j} + z_e \boldsymbol{k}\right) = \sum \left(F_{e,\bar{x}} \boldsymbol{i} + F_{e,\bar{y}} \boldsymbol{j} + F_{e,\bar{z}} \boldsymbol{k}\right) \tag{4}$$

$$\left(\frac{\partial^2}{\partial \bar{y}^2} + \frac{\partial^2}{\partial \bar{z}^2}\right) u_{m,\bar{x}} = \frac{1}{\mu_m} \frac{d}{dx} P \tag{5}$$

$$\left(\frac{\partial^2}{\partial \bar{x}^2} + \frac{\partial^2}{\partial \bar{y}^2} + \frac{\partial^2}{\partial \bar{z}^2}\right) V_{RMS} = 0 \tag{6}$$

$$E_{\bar{x}} \boldsymbol{i} + E_{\bar{y}} \boldsymbol{j} + E_{\bar{z}} \boldsymbol{k} = -\left(\frac{\partial}{\partial \bar{x}} \boldsymbol{i} + \frac{\partial}{\partial \bar{y}} \boldsymbol{j} + \frac{\partial}{\partial \bar{z}} \boldsymbol{k}\right) V_{RMS} \tag{7}$$

The equation of motion, Equation (4), accounts for forces such as that associated with gravity, drag, buoyancy, and DEP. The force associated with gravity and buoyancy are provided in Equations (8) and (9), respectively. The force associated with gravity and buoyancy have only one component and it is in the vertical direction. The force related to drag that is acting on the microparticle is shown in Equation (10) while that related to DEP is presented in Equation (11). The solution of equation of fluid flow, Equation (5), is required for determining drag which in turn is required for calculating the trajectory of the microparticles and is provided in Equation (12) [7].

$$F_{g,\bar{x}} \boldsymbol{i} + F_{g,\bar{y}} \boldsymbol{j} + F_{g,\bar{z}} \boldsymbol{k} = -\frac{4}{3} \pi r_e^3 \rho_e g \boldsymbol{k} \tag{8}$$

$$F_{b,\bar{x}} \boldsymbol{i} + F_{b,\bar{y}} \boldsymbol{j} + F_{b,\bar{y}} \boldsymbol{k} = \frac{4}{3} \pi r_e^3 \rho_m g \boldsymbol{k} \tag{9}$$

$$F_{drag,\bar{z}} \boldsymbol{i} + F_{drag,\bar{y}} \boldsymbol{j} + F_{drag,\bar{z}} \boldsymbol{k} = 6\pi \mu_m r_e \left[\left(u_m|_{X_e} - \frac{d}{dt} x_e\right) \boldsymbol{i} - \frac{d}{dt} y_e \boldsymbol{j} - \frac{d}{dt} z_e \boldsymbol{k}\right] \tag{10}$$

$$F_{DEP,\bar{x}} \boldsymbol{i} + F_{DEP,\bar{y}} \boldsymbol{j} + F_{DEP,\bar{z}} \boldsymbol{k} = 2\pi \varepsilon_m \varepsilon_0 r_e^3 Re[f_{CM}] \left(\frac{\partial E_{RMS}^2}{\partial x}\bigg|_{X_e} \boldsymbol{i} + \frac{\partial E_{RMS}^2}{\partial y}\bigg|_{X_e} \boldsymbol{j} + \frac{\partial E_{RMS}^2}{\partial z}\bigg|_{X_e} \boldsymbol{k}\right) \tag{11}$$

$$u_{m,\bar{x}}|_{X_e} = \frac{48 Q_m \sum\limits_{i=1,3,5}^{\infty} \left(\frac{(-1)^{\left(\frac{i-1}{2}\right)}}{i^3}\right) \left\{1 - \frac{\cosh\left[i \frac{\pi}{W_{ch}} \left(\frac{H_{ch}}{2} - z_e\right)\right]}{\cosh\left(i \frac{\pi}{2} \frac{H_{ch}}{W_{ch}}\right)}\right\} \cos\left[i \frac{\pi}{W_{ch}} \left(\frac{W_{ch}}{2} - y_e\right)\right]}{\pi^3 W_{ch} H_{ch} \left[1 - \frac{192}{\pi^5} \frac{W_{ch}}{H_{ch}} \sum\limits_{i=1,3,5}^{\infty} \frac{\tanh\left(i \frac{\pi}{2} \frac{H_{ch}}{W_{ch}}\right)}{i^5}\right]} \tag{12}$$

Equation (4) is solved using Finite Difference Method (FDM). For this, the differential terms are replaced by difference terms. The second order differential terms are replaced by second order central difference term. The time step is maintained at 10^{-5} s. A MATLAB program was developed for solving

Equation (4). For solving Equation (4), there are two initial conditions as depicted in Equation (13) and Equation (14). The initial displacements of the microparticle are presented in Equation (13) while the initial velocities of the microparticle are shown in Equation (14). The microparticle can start from any location across from the cross-section of the microchannel and this represents the initial displacement of the microchannel. The initial velocities of the microparticle are same as that of the fluid at the initial location of the microparticle.

$$\left[x_e|^{t=0}, y_e|^{t=0}, z_e|^{t=0} \right]^T = [0, W_0, H_0]^T \tag{13}$$

$$\frac{d}{dt}\left[x_e|^{t=0}, y_e|^{t=0}, z_e|^{t=0} \right]^T = \left[u_{m,\bar{x}}|_{X_e}, 0, 0 \right]^T \tag{14}$$

The electric field required for solving Equation (4) is obtained by solving Equation (7). The electric field depends on the electric potential and thus it needs to be determined throughout the microchannel and for this Equation (6) needs to be solved. Equation (6) is solved using FDM as well after replacing second order differential terms by second order central difference schemes; the boundary conditions associated with Equation (6) include known voltages on the electrodes while the remaining boundaries are assumed to be insulated. It needs to be stressed here that solving Equation (6) for the entire microchannel will be computationally taxing and time consuming. To overcome this, Eqution (6) is solved only in a repeating unit of the microchannel and later information on electric voltage inside the repeating unit is mapped on the entire microchannel; similar apporach is taken with regards to the electric field and DEP force. The repeating unit is schematically shown in Figure 3; each repeating unit contains one electrode pair on either side of the microchannel.The internode distance for implementing FDM is maintained at 1 μm. The several linear equations generated by application of FDM are solved using Gauss-Seidel method. Electric field is calculated after replacing the first order differential terms of Equation (7) by second order forward/central/backward difference schemes. A MATLAB program was developed for solving Equation (6).

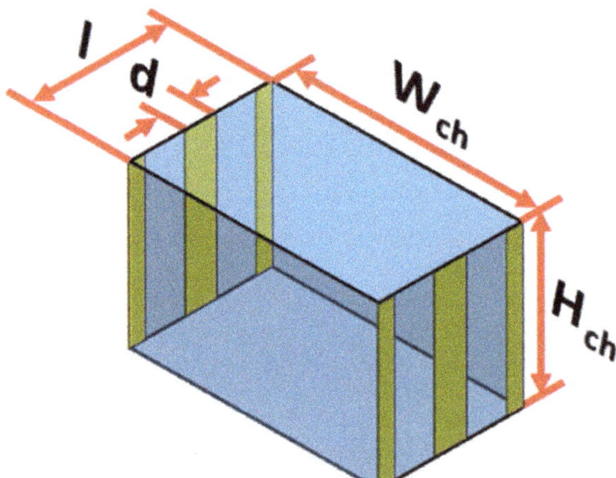

Figure 3. Schematic of the repeating unit of the proposed microfluidic device.

The performance of microfluidic devices employed for separation is quantified in terms of separation efficiency (SE), Equation (15), and separation purity (SP), Equation (16) [13]. SE and SP are quantified using the position of the microparticles at the exit of the microchannel. SE is the ratio of the number of microparticles of a particular size at the outlet to the number of microparticles of the same size at the inlet. SE represents the percentage of the total number of microparticles of a particular size separated using the device compared with the number of microparticles of the same size in the heterogeneous sample. SP is the ratio of the number of microparticles of a particular size at the outlet of the device to the total number of microparticles at the outlet.

$$SE(A) = \frac{\#\ of\ microparticles\ of\ size\ \prime A \prime\ at\ outlet}{\#\ of\ microparticles\ of\ size\ \prime A \prime\ at\ inlet} \tag{15}$$

$$SP(A) = \frac{\#\ of\ microparticles\ of\ size\ \prime A \prime\ at\ outlet}{\#\ of\ microparticles\ of\ all\ sizes\ at\ outlet} \tag{16}$$

All studies are done by uniformly releasing several microparticles from the inlet of the microchannel and subsequently tracking their trajectories to determine *SE* and *SP*; this approach has been previously adopted by researchers [13,25,26]. One of the assumptions of this model is that the microparticles do not experience Brownian motion. This assumption is acceptable as long as the microparticles are greater than 1 µm as established in literature [27–29]. Additionally, it is assumed that there is no particle-to-particle interaction inside the microchannel and this is a reasonable assumption as long as the sample handled in the microfluidic device is dilute and it is often the case when employing microfluidic devices [30].

3. Results and Discussions

Figure 4 shows the trajectory of the microparticles as the sample is being processed inside the microfluidic device. The sample introduced at the inlet of the microchannel consists of equal numbers of 2.5 µm (radius) and 5 µm (radius) polystyrene microparticles. Figure 4a,b show the path of 2.5 µm and 5 µm microparticles; two different figures are used so that the paths of each size of microparticles are clearly visible. For this study, microparticles of a particular size are uniformly released from the inlet of the microchannel; microparticles are released from 81 locations across the inlet of the microchannel. It can be noticed that all 2.5 µm microparticles are attracted to and captured on the electrodes while the 5 µm polystyrene microparticles travel through the microchannel unaffected. This depicts the ability of the microfluidic device to achieve separation based on size. The following parts of this section details the study carried out to understand the influence of operating and geometric parameters on SE and SP. The operating and geometric parameters considered include applied voltage, electrode dimensions, volumetric flow rate, and number of electrodes.

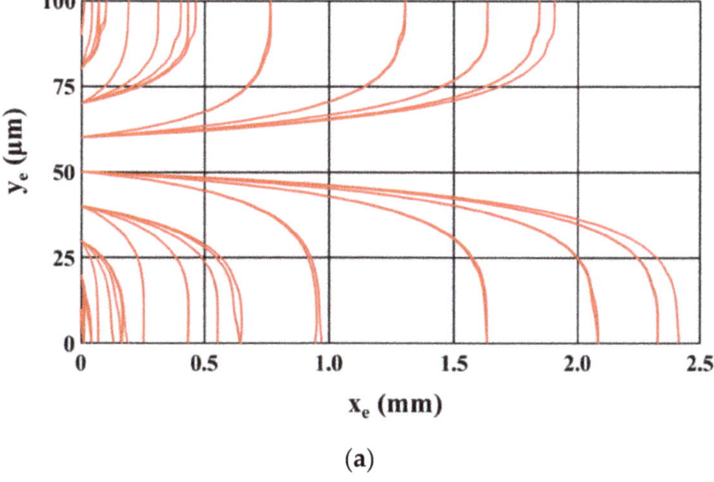

(**a**)

Figure 4. *Cont.*

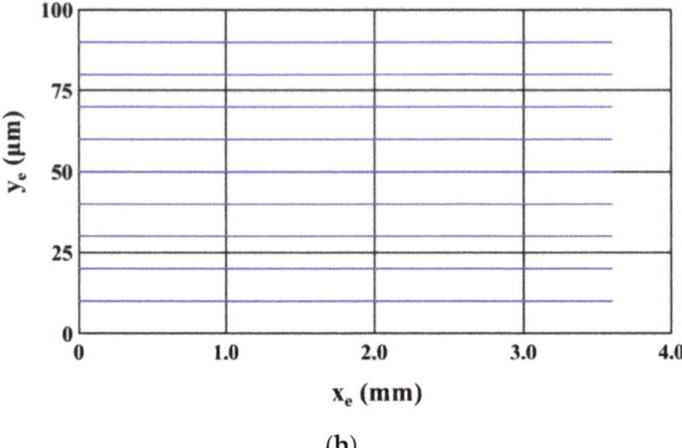

(b)

Figure 4. Top view of trajectory of microparticles (**a**) 2.5 μm and (**b**) 5 μm (V_1 = 20 V$_{pp}$, V_2 = 12.5 V$_{pp}$, Q_m = 200 μL/h, l = 120 μm, d = 30 μm, n = 30, N = $N_{cr,5}$ = 192.8 kHz).

Figure 5 shows the influence of electrode dimensions on SE and SP. It can be noticed that SE of 5 μm microparticle is independent of electrode dimensions for all applied electric voltages. While inside the microchannel, the 5 μm microparticles do not experience DEP as it needs to just remain suspended in the medium thereby making the SE of the same independent of the dimensions of the electrodes. On the other hand, SE of 2.5 μm microparticles depends on the electrode dimensions irrespective of the applied voltages (except for certain electrode dimensions operating at V_{pp1} = V_{pp2}). With the increase in electrode dimensions, the SE of 2.5 μm microparticles increases for all applied electric voltages. The residence time, of 2.5 μm microparticles, increases with the increase in electrode dimensions and this leads to a greater number of microparticles being captured on the electrodes thereby leading to the observed increase in SE, of 2.5 μm microparticles, for a specific applied electric voltage; residence time is the time spent by a microparticle inside the microchannel. For certain cases of equal applied electric voltages, all 2.5 μm microparticles released from the center of the inlet of the microchannel experience same pDEP from both sets of IDT electrodes due to which they remain uncaptured. This coupled with the fact that the number of 2.5 μm microparticle captured and subsequently released remain the same, the SP of the same remains independent of the electrode dimensions.

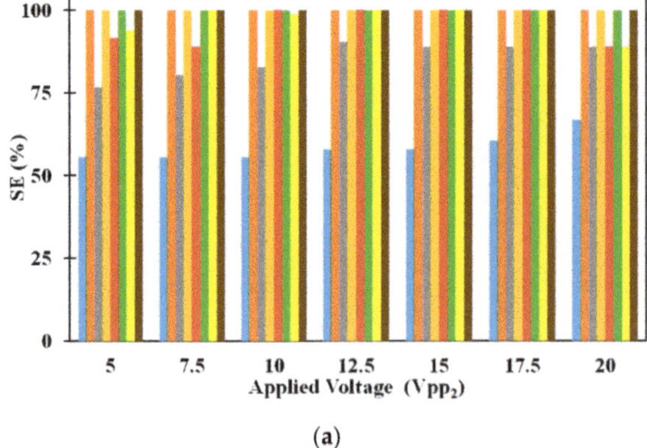

(a)

Figure 5. *Cont.*

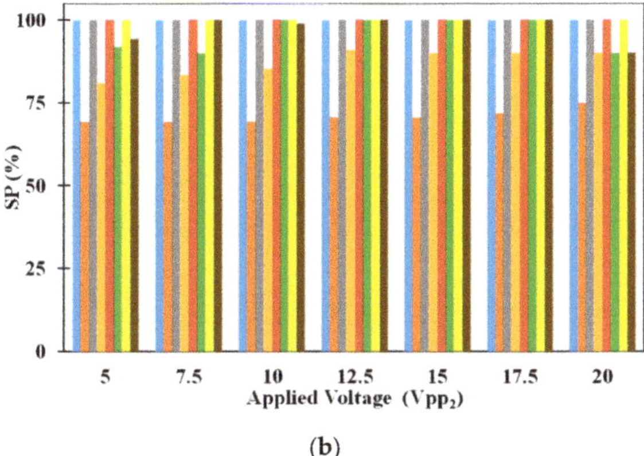

(b)

Figure 5. Influence of electrode dimensions on (**a**) SE and (**b**) SP for l = 40 μm and d = 10 μm (■: 2.5 μm, ■ 5 μm), l = 80 μm and d = 20 μm (■: 2.5 μm, ■ 5 μm), l = 120 μm and d = 30 μm (■: 2.5 μm, ■ 5 μm), and l = 160 μm and d = 40 μm (■: 2.5 μm, ■ 5 μm); n = 30, Q_m = 200 μL/h, W_{ch} = 100 μm, H_{ch} = 100 μm, V_{pp1} = 20 V_{pp}, $N = N_{cr,5}$ = 192.8 kHz.

The SP of both 2.5 μm and 5 μm microparticles are shown in Figure 5a as well. It can be noticed that changes in electrode dimensions do not influence the SP of 2.5 μm microparticles for all applied electric voltages. SP of 2.5 μm microparticles is calculated only using the number of 2.5 μm microparticles captured on the electrodes and as these are released for collection after processing of the sample, they appear at the exit without the presence of 5 μm microparticles thereby achieving SP of 100%. On the other hand, SP of 5 μm microparticles is influenced by electrode width and this is related to the number of 2.5 μm microparticles captured on the electrodes. As the number of the 2.5 μm microparticles captured on the electrodes increase due to increase in applied electric voltage or electrode width, the number of 2.5 μm polystyrene microparticles collected along with 5 μm microparticles decreases thereby leading to increase in the SP of the same.

Figure 6a provides the comparison between SE of 2.5 μm and 5 μm microparticles for different volumetric flow rates. It can be noticed that increase in volumetric flow rate decreases the SE of 2.5 μm when all other parameters are held constant. On the other hand, changes in volumetric flow rate, irrespective of the other parameters, does not influence SE of 5 μm microparticle. With the increase in volumetric flow rate, the residence time of all microparticles in the microchannel decreases. This reduction in residence time reduces the number of 2.5 μm microparticles captured on the electrodes which in turn negatively affects the SE of the same. Regarding the SE of 5 μm microparticle, it is below 100% for low volumetric flow rates; the increased residence time, associated with low volumetric flow rates, leads to the fall of several 5 μm microparticles to the bottom surface of the microchannel thereby preventing all of them from reaching the exit of the microchannel. At high volumetric flow rates, the residence time is low to prevent the fall of any 5 μm microparticles to the bottom surface of the microchannel thereby allowing all to remain suspended in the medium and subsequently reach the exit of the microchannel. Also, it can be noticed that increase in applied voltage increases the SE of 2.5 μm microparticles for a particular flow rate. Increase in applied voltage increases the pDEP force acting on the 2.5 μm microparticles and this leads to the observed increase in capture of the same with the increase in applied electric voltages. Additionally, it can be observed that the increase in applied electric voltages does not affect the SE of 5 μm microparticle since the separation of these microparticles is not dependent on them being captured on the electrodes. The SE of 2.5 μm microparticle is independent of the volumetric flow rate at applied electric voltages of 20 V_{pp}; this is because the microparticles in the vertical plane through the center the microchannel experience equal pDEP from both sets of electrodes thereby preventing them from being attracted to the electrodes and subsequently captured on the same.

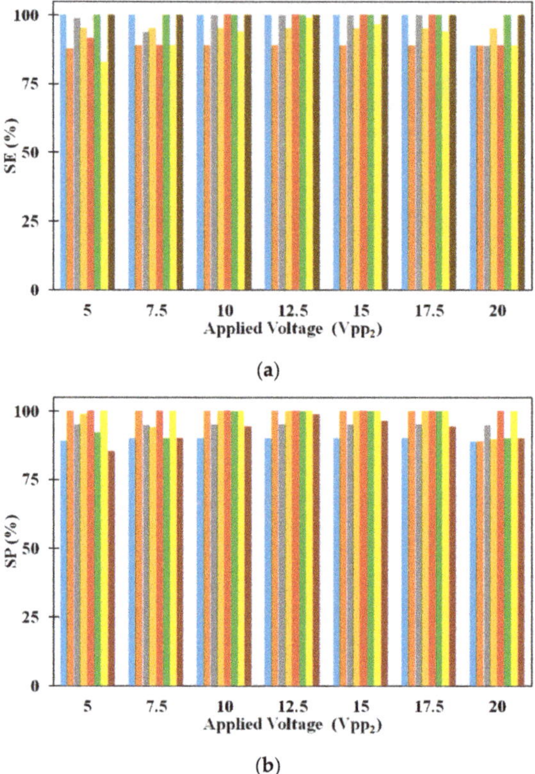

(a)

(b)

Figure 6. Influence of volumetric flow rate on (**a**) SE and (**b**) SP for $Q_m = 50$ μL/h (■: 2.5 μm, ■ 5 μm), $Q_m = 100$ μL/h (■: 2.5 μm, ■ 5 μm), $Q_m = 200$ μL/h (■: 2.5 μm, ■ 5 μm), and $Q_m = 300$ μL/h (■: 2.5 μm, ■ 5 μm); $n = 30$, $W_{ch} = 100$ μm, $H_{ch} = 100$ μm, $V_{pp1} = 20$ V_{pp}, $N = N_{cr,5} = 192.8$ kHz.

The effect of volumetric flow rate on SP of 2.5 μm and 5 μm microparticles are shown in Figure 6b. The SP of 2.5 μm microparticle is influenced by the volumetric flow rate at low and high volumetric flow rates, respectively. At low volumetric flow rates, several 5 μm microparticles drop to the bottom surface of the microchannel and these are collected along with the 2.5 μm microparticles captured on the electrodes thereby causing the SP of 2.5 μm microparticles to be dependent on volumetric flow rate. On the other hand, no 5 μm microparticles drop to the bottom surface of the microchannel at high volumetric flow rates and this maintains the SP of 2.5 μm microparticles at 100%. Regarding the SP of 5 μm microparticles, they are influenced by volumetric flow rate. With the increase in volumetric flow rate, for a specific applied electric voltage, the number of 2.5 μm microparticles captured on the electrodes decrease. The uncaptured 2.5 μm microparticles appear at the exit of the microchannel along with 5 μm microparticles thereby affecting the SP of 5 μm microparticles. Applied electric voltage also influences the SP of 5 μm microparticles; an increase in applied electric voltage increases the SP of 5 μm microparticles. This is because with the increase in applied electric voltage the number of 2.5 μm microparticles captured on the electrodes increase there by reducing the number of 2.5 μm microparticles appearing at the exit along with 5 μm microparticles and subsequently enhancing SP of 5 μm microparticles.

The influence of number of electrodes on the SE of 2.5 μm and 5 μm microparticles is provided in Figure 7a. It can be noticed that increase in the number of electrodes increases the SE of 2.5 μm microparticle irrespective of the applied electric voltages. With the increase in the number of electrodes, the length of the microchannel increases which increases the residence time of the microparticles thereby leading to increase in the SE of 2.5 μm microparticles. It can also be noticed that an increase in applied electric voltage, for a specific number of electrodes, increases the SE of 2.5 μm microparticle. With the increase in applied electric voltages, the pDEP forces attracting the microparticles to the electrodes increase and this is the reason for the observed increase in SE of 2.5 μm microparticle. SE of 2.5 μm microparticle is independent of the number of electrodes when the applied voltages are 20 V_{pp}. At 20 V_{pp}, the microparticles in the vertical plane passing through the center of the microchannel experience same pDEP force from both sets of electrodes due to which they remain uncaptured and subsequently the associated SE lower than 100%. At equal applied electric voltages of 20 V_{pp}, the pDEP force is strong to capture all but those microparticles in the vertical plane through the center of the microchannel and thus, the SE is independent of the number of electrodes. On the other hand, the SE of 5 μm microparticles is independent of the number of electrodes; to achieve separation of the heterogeneous mixture, the 5 μm microparticles just need to remain suspended in the microchannel and this is unaffected by the number of electrodes thereby maintaining the SE of the constant at 100%.

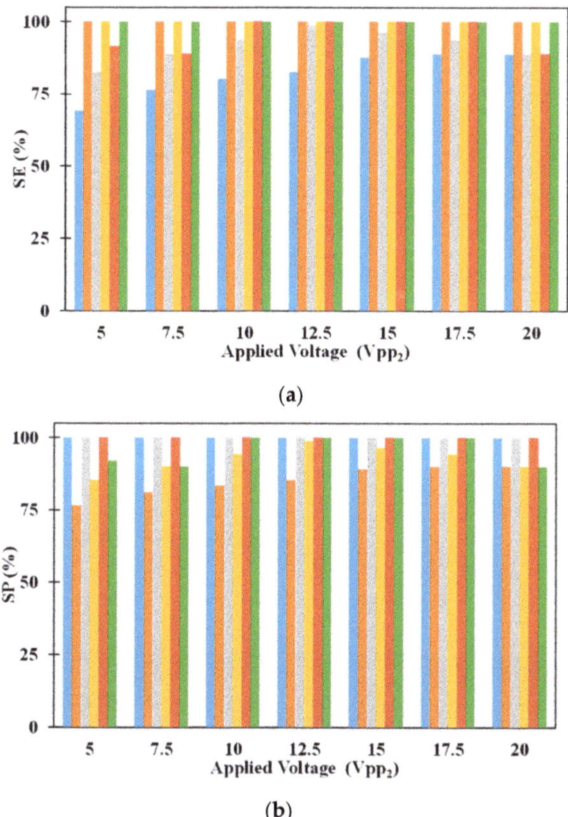

Figure 7. Influence of number of electrodes on (**a**) SE and (**b**) SP for $n = 10$ (■: 2.5 μm, ■ 5 μm), $n = 20$ (■: 2.5 μm, ■ 5 μm), and $n = 30$ (■: 2.5 μm, ■ 5 μm); $Q_m = 200$ μL/h, $W_{ch} = 100$ μm, $H_{ch} = 100$ μm, $V_{pp1} = 20\ V_{pp}$, $N = N_{cr,5} = 192.8$ kHz.

Figure 7b also shows the variation of SP, of, 2.5 µm and 5 µm microparticles, with number of electrodes. It is very clear from Figure 7b that SP of 2.5 µm microparticles are independent of the number of electrodes and applied electric voltages. This is the result of the SE of 5 µm microparticle being independent of the number of electrodes. As SE of 5 µm microparticle is 100% for all cases, no 5 µm microparticles will be collected with the 2.5 µm microparticles thereby making the same 100%. On the other hand, the SP of 5 µm microparticle varies with number of electrodes and applied electric voltages. In several instances, all 2.5 µm microparticles are not captured on the electrodes due to which many are collected, at the outlet, along with 5 µm microparticles and this makes the SP of 5 µm microparticles dependent on the number of number of electrodes and voltages. For equal applied electric voltages of 20 V_{pp}, the SP of 5 µm microparticle is constant irrespective of the number of electrodes. At equal electric voltages of 20 V_{pp}, none of the 2.5 µm microparticles in the vertical plane passing through the center of the microchannel are captured on the electrodes and subsequently they appear at the exit along with the 5 µm microparticles. As the number of 2.5 µm microparticles appearing at the exit of microchannel, along with 5 µm microparticle, is always same, the SP of 5 µm remains constant.

4. Device Limitations

For both equal and unequal applied electric voltages, there exists a vertical plane in the microchannel in which the DEP force is zero. This implies that the small microparticles in this vertical plane will not be attracted to the electrodes thereby influencing the SE of small microparticle and the SP of the big microparticle. This is the limitation of the microfluidic device; nevertheless, the device still exhibits good SE and SP as can be seen from the previous section.

5. Conclusions

This document presents a dielectrophoretic microfluidic device with multiple finite sized vertical electrodes, arranged in interdigitated transducer configuration, for separation of microparticles based of size. The electrodes on one side of the wall are controlled independently from the electrodes on the other side of the wall. In this device, microparticles of a specific size are captured on the electrodes while bigger microparticles pass through the microchannel unaffected to realize separation. The influence of parameters (operating and geometric) on the performance metrics, separation efficiency and separation purity, are quantified in this document. Separation efficiency of the small microparticles increase with the increase in applied electric voltage, width of electrodes, and number of electrodes while it reduces with the increase in the volumetric flow rate; the variation in separation efficiency can be attributed to the residence of the microparticles in the device. The separation efficiency of the 5 µm microparticle is always 100% as separation does not depend on the dielectrophoretic force. The separation purity of the 2.5 µm microparticles is independent of the applied electric voltage, electrode dimensions and the number of electrodes; however, volumetric flow rate influences separation purity. Separation purity of 5 µm microparticles are influenced by applied electric voltages, width of electrodes, volumetric flow rate, and number of electrodes and it is observed to increase with the increase in these parameters; this improvement in separation purity of 5 µm microparticles is due to the increase in separation efficiency of 2.5 µm microparticles with the increase in the parameters.

Author Contributions: Conceptualization, F.A. and B.M.; methodology, F.A, B.M. and A.H.-A; software, F.A. and B.M; formal analysis, F.A. and B.M.; investigation, F.A., B.M. and A.H.-A; funding acquisition, A.H.-A. All authors have read and agreed to the published version of the manuscript.

Funding: This research was funded in part by grants from UAEU (31N233 and 31N256).

Conflicts of Interest: Authors have no conflict of interest to declare.

Nomenclature

d: width of repeating unit (m or μm)
E: electric field
H_0: initial displacement along microchannel height (m or μm)
H: height (m or μm)
g: acceleration due to weight (m/s^2)
F: force (N)
K_s: surface conductance (S)
L: length of the electrode (m or μm)
N: number of microparticles for determining ΔW and ΔH (-)
n: number of electrodes (-)
Q: flow rate (m^3/s or μL/h)
$Re(f_{CM})$: Clausius–Mossotti factor (-)
r: radius (m or μm)
SE: separation efficiency
SP: separation purity
t: time (s)
u: velocity of medium (m/s)
V: voltage (V)
W_0: initial displacement along microchannel width (m or μm)
W: width (m or μm)
X: displacement vector (m or μm)
x: displacement in the x-direction (m)
y: displacement in the y-direction (m)
z: displacement in the z-direction (m)
Greek alphabet
ε: permittivity (-)
ε_0: permittivity of free space (F/m)
σ: conductivity (S/m)
ρ: density (kg/m^3)
ω: operating frequency (rad/s)
Subscripts
2.5: 2.5 μm microparticle
5: 5 μm microparticle
ch: microchannel
cr: cross-over
Drag: drag
DEP: dielectrophoresis
e: entity
RMS: root mean square
m: medium

References

1. Mathew, B.; Weiss, L. MEMS heat exchangers. In *Materials and Failures in MEMS and NEMS*; Tiwari, A., Raj, B., Eds.; Wiley: New York, NY, USA, 2015; pp. 63–120.
2. Convery, N.; Gadegaard, N. 30 years of microfluidics. *Micro Nano Eng.* **2019**, *2*, 76–91. [CrossRef]
3. Sajeesh, P.; Sen, A.K. Particle separation and sorting, in microfluidic devices: A review. *Microfluid. Nanofluid.* **2014**, *17*, 1–52. [CrossRef]
4. Khoshmanesh, K.; Nahavandi, S.; Baratchi, S.; Mitchell, A.; Kalantar-zadeh, K. Dielectrophoretic platforms for bio-microfluidic systems. *Biosens. Bioelectron.* **2011**, *26*, 1800–1814. [CrossRef] [PubMed]
5. Li, M.; Li, W.H.; Zhang, J.; Alici, G.; Wen, W. A review of microfabrication techniques and dielectrophoretic microdevices for particle manipulation and separation. *J. Phy. D Appl. Phys.* **2014**, *47*, 063001. [CrossRef]
6. Qian, C.; Huang, H.; Chen, L.; Li, X.; Ge, Z.; Chen, T.; Yang, Z.; Sun, L. Dielectrophoresis for bioparticle manipulation. *Int. J. Mol. Sci.* **2014**, *15*, 18281–18309. [CrossRef] [PubMed]

7. Alazzam, A.; Mathew, B.; Khashan, S. Microfluidic platforms for bio-applications. In *Advanced Mechatronics and MEMS Devices II*; Zhang, D., Wei, B., Eds.; Springer: Cham, Switzerland, 2017; pp. 253–282.

8. Zhang, H.; Chang, H.; Neuzil, P. DEP-on-a-chip: Dielectrophoresis applied to microfluidic platforms. *Micromachines* **2019**, *10*, 423.

9. Lee, S.H.; Yun, G.-Y.; Koh, Y.; Lee, S.-H.; Kim, Y.-K. Fabrication of a 3 dimensional dielectrophoresis electrode by a metal inkjet printing method. *Micro Nano Syst. Lett.* **2013**, *1*, 5. [CrossRef]

10. Rao, L.; Cai, B.; Yu, X.-L.; Guo, S.-S.; Liu, W.; Zhao, X.-Z. One-step fabrication of 3D silver paste electrodes into microfluidic devices for enhanced droplet-based cell sorting. *AIP Adv.* **2015**, *5*, 057134. [CrossRef]

11. Chiraco, M.S.; Bianco, M.; Amato, F.; Primiceri, E.; Ferrara, F.; Arima, V.; Maruccio, G. Fabrication of interconnected multilevel channels in a monolithic SU-8 structure using a LOR sacrificial layer. *Microelectron. Eng.* **2016**, *164*, 30–35. [CrossRef]

12. Mathew, B.; Alazzam, A.; Khashan, S.; Stiharu, I.; Dagher, S.; Furlani, E.P. Fabrication of microfluidic device with 3D embedded flow-invasive microelements. *Microelectron. Eng.* **2018**, *187–188*, 27–33. [CrossRef]

13. Alnaimat, F.; Ramesh, S.; Adams, S.; Parks, N.; Lewis, C.; Wallace, K.; Mathew, B. Model-based performance study of dielectrophoretic flow separator. *IEEE Sens. Lett.* **2019**, *3*, 1–4. [CrossRef]

14. Çetin, B.; Kang, Y.; Wu, Z.; Li, D. Continuous particle separation by size via AC-dielectrophoresis using a lab-on-a-chip device with 3-D electrodes. *Electrophoresis* **2009**, *30*, 766–772. [CrossRef] [PubMed]

15. Wang, L.; Lu, J.; Marchenko, S.A.; Monuki, E.S.; Flanagan, L.A.; Lee, A.P. Dual frequency dielectrophoresis with interdigitated sidewall electrodes for microfluidic flow-through separation of beads and cells. *Electrophoresis* **2009**, *30*, 782–791. [CrossRef] [PubMed]

16. Çetin, B.; Li, D. Continuous particle separation based on electrical properties using alternating current dielectrophoresis. *Electrophoresis* **2009**, *30*, 3124–3133. [CrossRef] [PubMed]

17. Kang, Y.; Çetin, B.; Wu, Z.; Li, D. Continuous particle separation with localized AC-dielectrophoresis using embedded electrodes and an insulating hurdle. *Electrochim. Acta* **2009**, *54*, 1715–1720. [CrossRef]

18. Faraghat, S.A.; Hoettges, K.F.; Steinbach, M.K.; van der Veen, D.R.; Brackenbury, W.J.; Henslee, E.A.; Labeed, F.H.; Hughes, M.P. High-throughput, low-loss, low-cost, and label-free cell separation using electrophysiology-activated cell enrichment. *PNAS* **2017**, *114*, 4591–4596. [CrossRef]

19. Mathew, B.; Alazzam, A.; Abutayeh, M.; Gawanmeh, A.; Khashan, A. Modeling the trajectory of microparticles subjected to dielectrophoresis in a microfluidic device for field flow fractionation. *Chem. Eng. Sci.* **2015**, *138*, 266–280. [CrossRef]

20. Mathew, B.; Alazzam, A.; Khashan, S.; El-Khasawneh, B. Path of microparticles in a microfluidic device employing dielectrophoresis for hyperlayer field flow fractionation. *Microsyst. Technol.* **2016**, *22*, 1721–1732.

21. Mathew, B.; Alazzam, A.; Abutayeh, M.; Stiharu, I. Model-based analysis of a dielectrophoretic microfluidic device for field-flow fractionation. *J. Sep. Sci.* **2016**, *39*, 3028–3036. [CrossRef]

22. Alazzam., A.; Hilal-Alnaqbi, A.; Alnaimat, F.; Ramesh, S.; Al-Shibli, M.; Mathew, B. Dielectrophoresis based microfluidic devices for field flow fractionation. *Med. Devices Sens.* **2018**, *1*, e10007. [CrossRef]

23. Dalili, A.; Samiei, E.; Hoorfar, M. A review of sorting, separation and isolation of cells and microbeads for biomedical applications: Microfluidic approaches. *Analyst* **2019**, *144*, 87–113. [CrossRef] [PubMed]

24. Burklund, A.; Zhang, J.X.J. Microfluidics-based organism isolation from whole blood: An emerging tool for bloodstream infection diagnosis. *Ann. Biomed. Eng.* **2019**, *47*, 1657–1674. [CrossRef] [PubMed]

25. Hemmatifar, A.; Saidi, M.S.; Sadeghi, A.; Sani, M. Continuous size-based focusing and bifurcation microparticle streams using a negative dielectrophoretic system. *Microfluid. Nanofluid.* **2013**, *14*, 265–276. [CrossRef]

26. Sahin, M.A.; Çetin, B.; Ozer, M.B. Investigation of effect of design and operating parameters on acoustophoretic particle separation via 3D device-level simulations, *Microfluid. Nanofluid.* **2020**, *24*, 8. [CrossRef]

27. Gascoyne, P.R.C.; Vykoukal, J. Particle separation by dielectrophoresis. *Electrophoresis* **2002**, *23*, 1973–1983. [CrossRef]

28. Castellanos, A.; Ramos, A.; Gonzalez, A.; Green, N.G.; Morgan, H. Electrohydrodynamics and dielectrophoresis in microsystems: Scaling and laws. *J. Phys. D Appl. Phys.* **2003**, *36*, 2584–2596. [CrossRef]

29. Lei, U.; Lo, Y.J. Review of the theory of generalized dielectrophoresis. *IET Nanobiotechnol.* **2011**, *5*, 86–106. [CrossRef]

30. Loth, E. Numerical approaches for motion of dispersed particles, droplets and bubbles. *Progress Energy Combust. Sci.* **2000**, *26*, 161–223. [CrossRef]

 *micromachines*

Article

Evaluation of Performance and Tunability of a Co-Flow Inertial Microfluidic Device

Amanda Bogseth [1], **Jian Zhou** [1,2] **and Ian Papautsky** [1,2,*]

1   Department of Bioengineering, University of Illinois at Chicago, Chicago, IL 60607, USA; abogse2@uic.edu (A.B.); jzhou88@uic.edu (J.Z.)
2   University of Illinois Cancer Center, Chicago, IL 60612, USA
*   Correspondence: papauts@uic.edu; Tel.: +1-312-413-3800

Received: 28 January 2020; Accepted: 5 March 2020; Published: 10 March 2020

Abstract: Microfluidics has gained a lot of attention for biological sample separation and purification methods over recent years. From many active and passive microfluidic techniques, inertial microfluidics offers a simple and efficient method to demonstrate various biological applications. One prevalent limitation of this method is its lack of tunability for different applications once the microfluidic devices are fabricated. In this work, we develop and characterize a co-flow inertial microfluidic device that is tunable in multiple ways for adaptation to different application requirements. In particular, flow rate, flow rate ratio and output resistance ratio are systematically evaluated for flexibility of the cutoff size of the device and modification of the separation performance post-fabrication. Typically, a mixture of single size particles is used to determine cutoff sizes for the outlets, yet this fails to provide accurate prediction for efficiency and purity for a more complex biological sample. Thus, we use particles with continuous size distribution (2–32 μm) for separation demonstration under conditions of various flow rates, flow rate ratios and resistance ratios. We also use A549 cancer cell line with continuous size distribution (12–27 μm) as an added demonstration. Our results indicate inertial microfluidic devices possess the tunability that offers multiple ways to improve device performance for adaptation to different applications even after the devices are prototyped.

Keywords: microfluidics; particle separation

---

1. Introduction

Microfluidic systems have emerged as viable alternatives to the conventional benchtop methods for separation and purification of cells [1–3]. Such systems are generally classified as either active, those relying on electric, magnetic or acoustic forces or passive, those relying on hydrodynamic forces. While active systems offer more accurate and selective manipulation of cells, their throughput is inherently low. Thus, passive systems, such as deterministic lateral displacement (DLD) [4,5], pinched flow fractionation (PFF) [6,7], hydrodynamic filtration [8,9] and inertial focusing [10–12], have gained popularity. In particular, inertial focusing is especially attractive due to simpler device designs since separation occurs by hydrodynamic forces only, while DLD and PFF require specific geometric designs [3]. Inertial focusing also has relatively high throughputs as compared with hydrodynamic filtration [9,13]. A number of recent reviews on inertial microfluidics have highlighted the underlying physical principles and have described the promising applications from enrichment of particles to medical diagnoses [3,14–16].

Development of microfluidic devices generally begins with validation and optimization using microparticles. This is especially true in inertial microfluidics, where microparticles have been and still are used to demonstrate new devices concepts, understand device performance and elucidate device physics [3,17–29]. These experiments often determine the separation efficiency and the resulting purity, allowing for optimization of flow conditions without wasting any of the biological sample.

For biological applications, the size of the particles within the sample must be known so a device can be made to fractionate the sample into groups with specific cutoff sizes. Additionally, depending on the application, the samples may have a wide size range requiring multiple devices with various channel lengths. To avoid designing multiple channels, an inertial microfluidic device can be 'tuned' to sort the sample with a specific cutoff size using output channel resistance. Wang and Papautsky [30] demonstrated that output resistance and flow rate can be used to dynamically change the cutoff size of the outlets in a vortex separator. Tu et al. [10] demonstrated that output resistance changes the quality of concentrating a cell sample, as an alternative to centrifugation. These findings offer a simpler way to process diverse samples within the same device.

In this paper, we develop and characterize a co-flow inertial microfluidic device which shows tunability of separation by multiple parameters including flow rate, flow rate ratio and resistance ratio. Low-cost commercial microbeads of a continuous diameter range (2–32 µm) were separated into three outlets in the device (Figure 1a). The results show the possibility of separation refinement and flexibility of cutoff size. The results also suggest constraints for the refinement of a continuous range of particles because of the minimal size differences between particles near the cutoff size of an outlet. As shown previously [9,29], changing the resistance ratio of the outputs had the largest effect on flexibility of cutoff size, while flow rate and flow rate ratio had a smaller effect. Using these data, particles of sizes 7 µm, 15 µm and 26 µm were used to mimic biological sample and test the cutoff size requirements. An optimized separation is illustrated in Figure 1b, with the smallest particles exiting through outlet 1, the largest particles traveling to outlet 3, and those in between entering outlet 2. These experiments led to higher efficiency than the continuous range of particles because the large diameter difference between each particle led to specific equilibrium positions away from streamlines. These three-sized particle analyses validated the data demonstrated by the continuous range of particles, suggesting that the lower-cost alternative provides a better approximation of the device performance given that biological components, like cells, come in a large distribution of sizes.

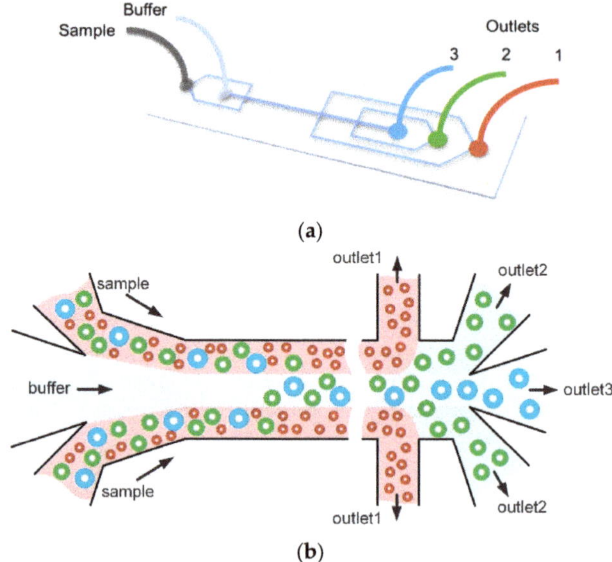

Figure 1. Illustrations of device setup and particle behaviors in device. (**a**) Schematic of device. The buffer enters in the first inlet and the sample behind, resulting in the sample nearest the walls and the buffer in the center. (**b**) Schematic of particle migration in an ideal situation with the smallest particles exiting in outlet 1, the largest in outlet 3 and those in between in outlet 2.

2. Experimental Methods

2.1. Microfabrication

Microfluidic devices were fabricated in polydimethylsiloxane (PDMS) using the standard soft lithography process with dry photoresist masters, as we detailed previously [31]. Briefly, 3″ silicon wafers were dehydrated for 15 min on a 225 °C hotplate, laminated with a 50 μm thick ADEX film (DJ Microlaminates Inc., Sudbury, MA, USA) and baked for 5 min on a 65 °C hotplate. Next, the wafers were exposed to UV light (I-line 365 nm, Optical Associates Inc., San Jose, CA, USA) for 33 s at 10 mW/cm^2 through a mask plate in hard contact. The wafers were developed in cyclohexanone (98%, Acros Organics, Pittsburg, PA, USA), washed with isopropyl alcohol (IPA) and deionized (DI) water, air dried and baked for 90 min on a 170 °C hotplate. A mixture of a 10:1 ratio of PDMS (Sylgard 184, Dow Corning, Midland, MI, USA) and curing agent was cast on the master, degassed for 90 min in a vacuum oven and cured for 120 min at 80 °C. Devices were cut out using a scalpel and inlet and outlet ports were cored using a biopsy punch with a diameter of 1.5 mm. Finally, devices were bonded to standard microscope glass slides using an oxygen plasma treatment at 10 W for 20 s (PE-50, Plasma Etch Inc., Carson City, NV, USA), baked for 60 min at 80 °C and allowed to cool to room temperature before use.

2.2. Sample Preparation

For experiments involving non-fluorescent polymethyl methacrylate (PMMA) microparticles, a saline buffer was first prepared by mixing 2 g of NaCl (Thermo Fisher Scientific Inc., Waltham, MA, USA) with 10 mL of DI water. The PMMA microparticles (Cospheric LLC, Santa Barbara, CA, USA) with continuous size range 2–32 μm in diameter were then mixed with the prepared saline buffer at a concentration of ~3 million particles/mL (0.134 g in 50 mL of saline buffer). Buffer solution was used in preparing sample solution in order to match the particle density of 1.2 g/cm^3 to achieve neutral buoyancy and minimize particle sedimentation in the syringe during experiments.

For experiments involving fluorescent polystyrene microparticles, a saline buffer was first prepared by mixing 1.6 g of NaCl (Fisher Scientific Inc., Waltham, MA, USA) with 10 mL of DI water to match the particle's density of 1.06 g/cm^3. The polystyrene particles of diameter 7.32 μm and 15.45 μm (Bangs Laboratories Inc., Fishers, IN, USA), 18.67 μm and 26.3 μm (Polysciences Inc., Warrington, PA, USA) were mixed with the prepared saline buffer at a concentration of ~5 million particles/mL. Tween 80 was added at 0.1% *v/v* (Fisher Scientific, Waltham, MA, USA) to all particle solutions to minimize aggregation and avoid clogging.

2.3. Flow Experiments

Syringes carrying sample and buffer solution was attached to the device using tubing of diameter 1.5 mm. Using syringe pumps, buffer solution was used to prime each channel before use. Particle solution was then added into the channel at an appropriate flow rate (150–750 μL·min^{-1}) for ~2 min to allow stabilization of the flow before samples were collected. An inverted microscope (Olympus IX83 with Andor Zyla 5.5 camera, Andor Technology Ltd., Belfast, UK) was used to image samples in Bright Field and Fluorescence. A high-speed camera (FASTCAM Mini AX 200, Photron USA Inc., San Diego, CA, USA) at 1.05 μs exposure rate was used to capture bright-field images of the particles inside the microchannel. Images were compiled and analyzed using ImageJ®. Particle sizes of obtained samples were measured using software CellSens (Olympus Corp., Tokyo, Japan) for n = 300 particles.

2.4. Cell Culture

Human non-small cell lung cancer cell line A549 was cultured in 25 cm^2 flasks in completed Roswell Park Memorial Institute (RPMI) medium (Corning, Corning, NY, USA) containing 10% fetal bovine serum (FBS) (Gemini Bio, West Sacramento, CA, USA) and 1% antibiotic antimycotic solution (Sigma-Aldrich, St. Louis, MO, USA) in an incubator at 37 °C with 5% CO_2. Cells were passaged once 70% confluency was reached to provide stable growing conditions. In preparation for experiments, the cells were extracted from flasks using 2 mL of 0.25% trypsin-ethylenediaminetetraacetic acid (EDTA) (Gibco, Waltham, MA, USA), incubated for 6 min and centrifuged for 5 min at 300 g. The cell pellet was re-suspended in 0.1% phosphate-buffered saline (PBS) at 100,000 cells/mL to be separated in the microfluidic device. After separation, cell sizes were measured using Cellsens software (Olympus Corp., Tokyo, Japan) and the viability of the cells were determined using Trypan Blue stain (ThermoFischer Scientific Inc., Waltham, MA, USA). After separation, cells were fixed using 80% ethanol for 30 min at −20 °C and stained using Hoechst 33342 (ThermoFisher Scientific Inc., Waltham, MA, USA) at a concentration of 1:300 for cell cycle analysis using flow cytometry conducted on the Gallios Flow Cytometer machine (Beckman Coulter, Brea, CA, USA). Gated data from the flow cytometer was analyzed using the Kaluza software (Beckman Coulter, Brea, CA, USA).

3. Results and Discussion

3.1. Inertial Focusing and Hydrodynamic Separation

As with other inertial microfluidic devices, the microfluidic chip we use in this work relies on inertial migration to achieve size-based separation of particles [13,32]. Thus, particles flowing downstream migrate across streamlines and order deterministically at equilibrium positions near channel walls. This ordering is caused by the balance of lift forces arising from the curvature of the velocity profile (the shear-induced lift F_s) and the interaction between microparticles and the channel wall (the wall-induced lift F_w). Under influence of these forces, microparticles rapidly equilibrate along each sidewall into bands where these two dominant lift forces balance each other. Once this initial equilibrium is reached, microparticle motion near channel sidewalls is dominated by the rotation-induced lift force F_Ω, which drives them towards the center of channel sidewalls. In our co-flow microfluidic system, this means that particles can migrate out of the sample flow and into the central buffer flow. The size-selective aspect of the device arises from the inertial lift forces but more specifically from the rotation induced lift force ($F_\Omega \propto a^3$) in our low aspect ratio channel [32]. Consequently, the larger particles migrate across the streamlines faster than the smaller particles.

Representative results illustrating device operation are presented in Figure 2. The randomly distributed particles in a sample of continuous size range 2–32 μm, are initially confined to the sidewalls. Under the influence of the shear-induced lift forces, as discussed above, the larger particles rapidly migrate to the center, while the smaller particles remain within the sample flow near sidewalls (Figure 2a). A symmetrical outlet system was used to remove the smallest particles, with diameter smaller than the set cutoff size ($a < a_{c1}$), leaving the larger ones in the buffer flow. Further downstream, the largest particles which are focused near the channel centerline and have a larger diameter than the second cutoff size ($a > a_{c2}$), exit through outlet 3. The mid-sized particles with diameters between the two cutoff sizes ($a_{c1} < a < a_{c2}$) exit through outlet 2 (Figure 2b). Histograms in Figure 2c confirm the cutoff sizes as $a_{c1} = 12$ μm and $a_{c2} = 18$ μm. The initial sample shows the visible size differences in the particles, with majority being a larger size (Figure 2d). More importantly, these results illustrate that particles with a size difference of ~2 μm can be separated with 70%–80% efficiency. With larger size difference of ~3 μm, separation efficiency is even higher, >90%.

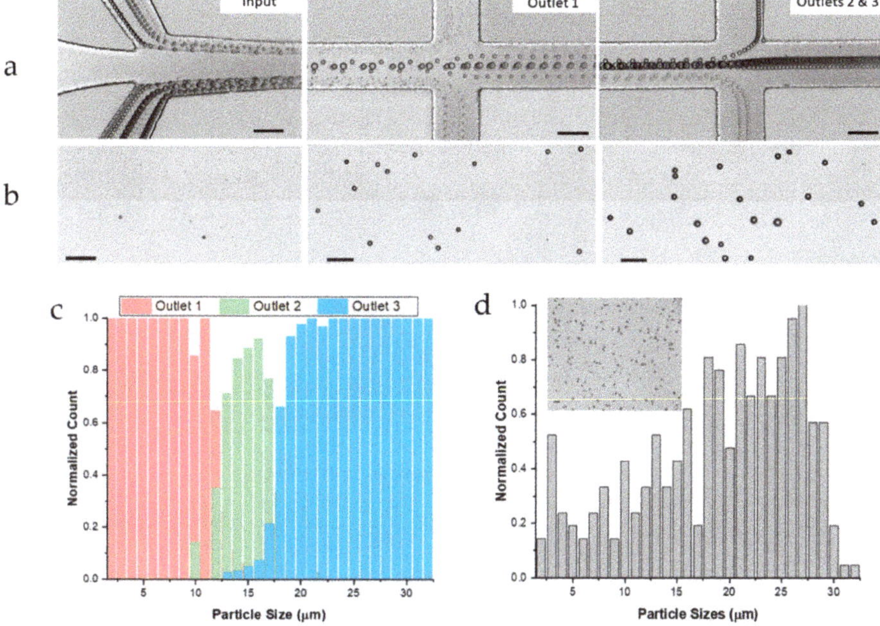

Figure 2. Separation of a continuous particle sample sized 2–32 μm in diameter at 300 μL·min⁻¹, 1:4 flow ratio and 0.59 resistance ratio. Scale bars indicate 100 μm. (**a**) Bright field images at input and each outlet (1.05 μs exposure, stack *n* = 800). (**b**) Outlet samples. (**c**) The percentage of particles in each outlet for each size. (**d**) Initial particle sample sizes and bright field image.

3.2. Analysis of Flow Rate

Flow rate affects throughput and viability of cells post-separation [10,33]. Thus, flow rate optimization is a crucial step when developing a microfluidic device. Flow rates in the range 150–750 μL·min⁻¹ (*Re* 40–200) were used to analyze the effect of flow speed on the flexibility of cutoff size and separation quality. For a high throughput system, flow rates lower than 150 μL·min⁻¹ would not deliver sufficient throughput. Flow rates higher than 750 μL·min⁻¹ would lead to deformation of PDMS causing disruptions in the microchannel. It has also been previously reported that efficiencies of separation at high flow rates are inadequate [34]. The separation quality was evaluated on the bases of the refinement of the 30 μm range sample fractionated within the outlets. Under the optimal 300 μL·min⁻¹ flow rate, the device was able to decrease the distributions of the initial 30 μm size range to 20 μm, 17 μm and 21 μm size ranges at outlets 1, 2 and 3 respectively (Figure 3a).

In general, increase of the total flow rate leads to the slight elevation of both cutoff sizes (Figure 3b). The cutoff size of outlet 1 changed only by ±1 μm. This negligible change is likely due to the inability of inertial forces to influence smaller particles within the limited downstream length of the microchannel. The cutoff size of outlet 2 increased from 19 μm to 23 μm as the flow rate increased, showing that the flow rate had an impact on larger particle's focusing positions. This is mainly a time-limited case where increasing the flow rate resulted in less time for the particles to focus causing the larger particles to enter into outlet 2. This increased the average particle sizes and cutoff size of outlet 2. These results suggest that a microfluidic device's cutoff sizes can only be changed with restriction to the size range of the particles.

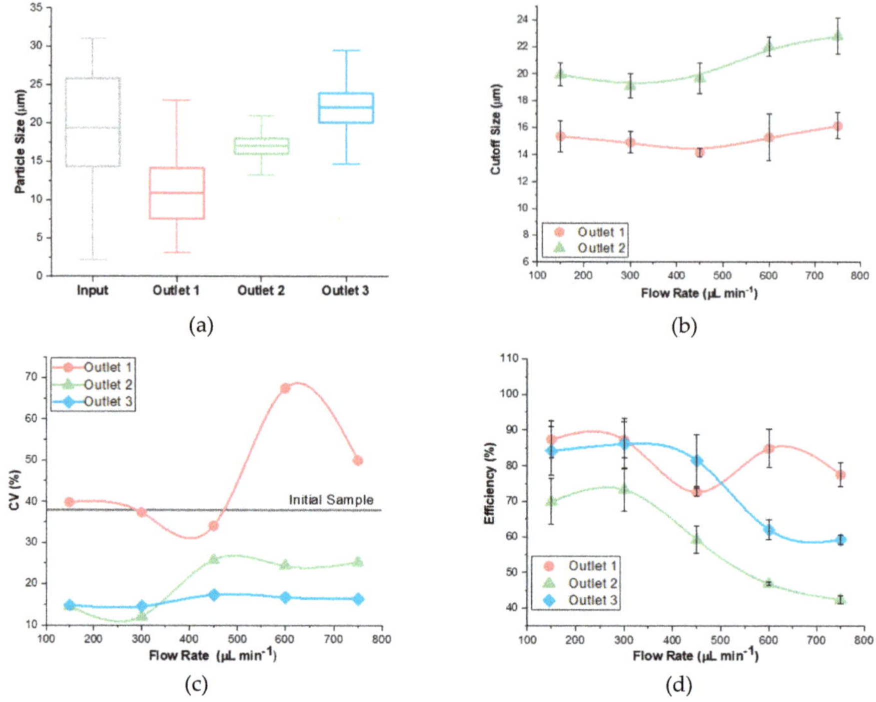

Figure 3. Sample flow rate impacts separation performance. (**a**) Fractionation of input sample into 3 outlets (300 μL·min^{-1}, 1:2 ratio, 0.59 resistance ratio). The whiskers on the box plots show the full range of particle sizes. (**b**) Cutoff size stays constant in outlet 1 but increases in outlet 2 with flow rate. (**c**) Coefficient of variance is low at a low flow rate but increases with flow rate. (**d**) Efficiency of separation is high at a low flow rate but decreases rapidly beyond 300 μL·min^{-1}.

The coefficient of variation (CV) of particles at each outlet was reduced at the optimal 300 μL min^{-1} flow rate (Figure 3c). Outlet 1 always had a higher CV than outlets 2 and 3 because the cutoff size was ~15 μm, which was the middle of the initial particle distribution. This reduced the ability to decrease the refinement of smaller particles because this device would always include half of the sample in outlet 1. Furthermore, the CV of outlet 1 was the same or higher than the initial sample for four of the data points. This is because the initial sample had a large standard deviation and a large average, while outlet 1 had a large standard deviation and a small average. Consequently, the CV does not give the correct impression when observing the refinement of the particles. The CV of the initial sample was smaller than outlet 1 but the distributions of the initial sample and outlet 1 were different size ranges. Therefore, the box plot is a better representation of refinement showing the outlet ranges and establishing the quality of device performance. However, the CV is important because it is able to compare the distributions of all three outlets relative to each other.

The separation efficiency of the device decreased with increased flow rate in all three outlets (Figure 3d). This efficiency is defined as the percent fraction of target particles in their target outlets. At higher flow rates, particles have less time to achieve full focusing, effectively increasing the cutoff size (Figure 3b). Lower efficiency at higher flow rate is likely due to the increased sensitivity of performance on particle's initial lateral positions. Since the lateral migration velocity scales with the square of downstream velocity ($U_L \propto U_f^2$) [32], slightly smaller particles initially located near the center might achieve focusing and exit into outlet 3, leading to the contamination of the outlet and thus decreased efficiency. Outlet 1 had the highest efficiency because it collected the smallest particles which were unable to cross the streamlines. The efficiency of outlet 2 was the lowest because the particle sizes overlapped with outlets 1 and 3. As a result, a lower flow rate resulted in better separation in terms of a high efficiency for all three outlets.

3.3. Analysis of Flow Rate Ratio

Flow rate ratio can impact the separation performance of a co-flow inertial system by controlling initial lateral positions of the suspended particles [20]. In this co-flow system, we defined flow rate ratio as the sample flow rate over the buffer flow rate. According to the size-dependent lateral migration and similar migration time in a given channel, a smaller ratio would be preferred as it gives similar initial lateral positions of particles regardless of size difference since all of them would be confined near sidewalls in narrow sample streams [20]. Six ratios were chosen to analyze the impact of flow rate ratio on separation quality and cutoff size flexibility (Figure 4a). The results confirm that a lower flow rate ratio produced better refinement of particles (Figure 4b). The comparison of all outlets can be found in Appendix A (Figure A1). As compared to the initial 30 μm range, outlets 1, 2 and 3 decreased to a 15 μm, 11 μm and 17 μm range, respectively. These decreased ranges showed better refinement at a 1:4 ratio compared to the 1:2 ratio above. While the 1:6 ratio also resulted in similar refinement and efficiency, the processing throughput is significantly lower. Therefore, the 1:4 ratio was used as the optimal condition for our co-flow channel.

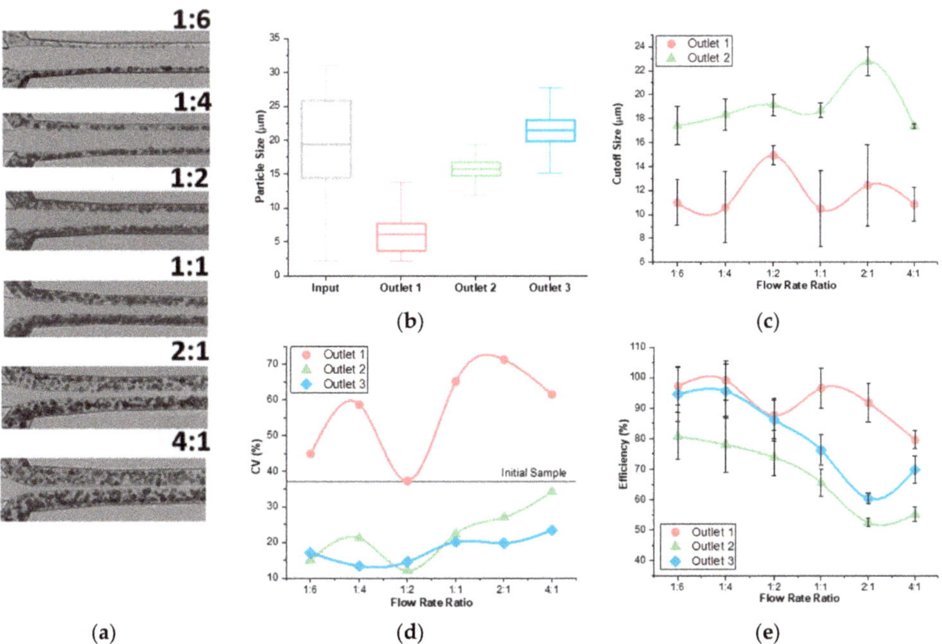

Figure 4. Influence of flow rate ratio on device performance. (**a**) Bright field images of flow show the boundary lines gets closer to the center and the width of the initial particle positions increases as the flow rate ratio increases. (**b**) The distribution of particles decreases from initial distribution in outlets for a flow rate of 300 μL·min^{-1}, at a 1:4 ratio and a 0.59 resistance ratio. (**c**) Cutoff size in outlets 1 and 2 change negligibly when flow rate ratio increases. (**d**) Coefficient of variance increases with increasing flow rate ratio. (**e**) Efficiency of separation decreases when flow rate ratio increases.

Our results further show the flow rate ratio slightly affects the cutoff sizes (Figure 4c). The trends that the cutoff sizes displayed were relevant to the conditions in the device. From the 1:4 to the 1:2

ratio, the cut off size increased for both outlets. The 1:2 ratio had a larger initial width for the sample, with random particle positions. As discussed above, the difference in particle initial lateral position warrants varying lateral migration distance even for particles with identical sizes. This caused some larger particles to go into outlet 1, increasing the cutoff size of the 1:2 ratio. Furthermore, as the flow rate ratio increased from the 2:1 to 4:1 ratio the cutoff size decreased. In a 4:1 ratio the sample initially filled up most of the channel (Figure 4a). With random initial positions of the particles, smaller particles were near the channel center more than when there was a smaller flow rate ratio. This caused smaller particles to enter into outlets 2 and 3 which decreased the cutoff size. Overall, our results suggest the influence of flow rate ratios on cutoff sizes of the device is small.

The CV indicated improved results with a lower flow rate ratio (Figure 4d). The trend shows an increase in CV as the flow rate ratio increased. Due to increased distributions at higher flow rate ratios, the efficiency showed a decreasing trend as flow rate ratio increased (Figure 4e). At a high flow rate ratio, the widely-distributed sample at the inlet yielded situations where particles near the channel wall did not have sufficient downstream length to migrate to their focusing positions, causing them to enter outlets prematurely. Additionally, some smaller particles would focus in the center and enter into outlets further down the channel. This caused a low efficiency because both small and large particles were entering into non-target outlets. Thus, these data suggest that a lower flow rate ratio results in better separation in terms of high efficiency of all three outlets.

3.4. Analysis of Resistance Ratio

In addition to flow conditions, fluidic resistance of channels can be used to tune a microfluidic device to optimize separation. Fluidic resistance can be manipulated in multiple ways, including changing the external features of tubing and syringe pumps or internal features of aspect ratios or adding vortices in the channel [10,30,35,36]. In this work, we define the resistance ratio, $\sigma = r/R$, as the ratio of downstream lengths between outlets 1 and 2 (Figure 5a) [35]. We tested five different resistance ratios to assess the impact on cutoff size flexibility and separation quality. The results show that resistance ratio has a larger impact on cutoff size as compared to flow rate and flow rate ratio (Figure 5b). As σ increases, streamlines separating flows between side outlets and the main channel move closer to the sidewalls, resulting in a progressively smaller fraction of flow exiting through the side outlets. Since the particles equilibrate with the larger ones closer to the center and the smaller ones closer to the sidewalls, the shift in the separation streamline causes smaller diameter particles to enter outlet 1, resulting in a decreased cutoff diameter as a function of resistance ratio. The same occurs at outlet 2, with cutoff size also decreasing but by a smaller amount. In fact, the two outlets are coupled, with their cutoff sizes changing in tandem. These results illustrate that changing the external conditions offers more flexibility in tuning particle separation in the device.

The concept of modifying the external channel resistance to tune separation quality is not entirely new, as it was previously demonstrated by us [29] and others [9]. It may be tempting to draw comparisons with our previous work on tunable vortex separators, where resistance ratio was also used for adjustments of the cutoff size. However, while the approach was the same, the results were different. This is because the vortex separators rely on inertial focusing, which leads the larger particles to focus closer to the sidewall and smaller particles closer to the channel centerline [29]. In this case, increasing resistance ratio between side and main channels yields an increase in cutoff size. Herein, the co-flow microfluidic system causes the larger particles to focus closer to the center and thus yields a decrease in the cutoff size as a function of resistance ratio.

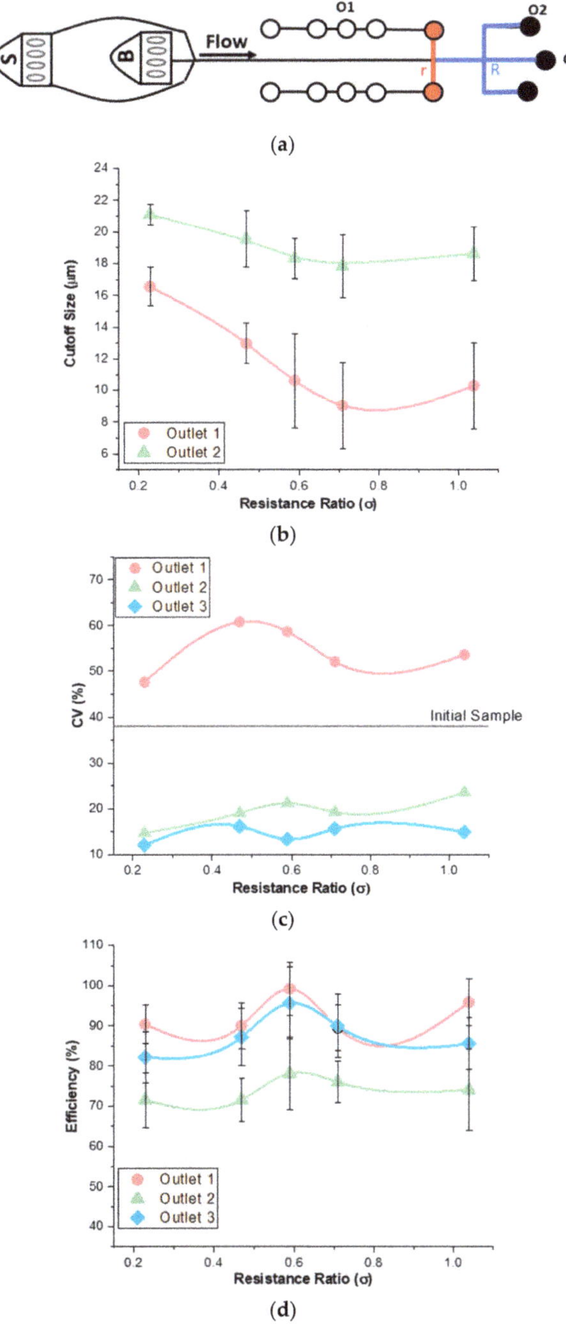

(a)

(b)

(c)

(d)

Figure 5. Influence of resistance ratio on device performance. (**a**) Schematic of resistance modification in outlet 1. (**b**) Cutoff size rapidly decreases in outlets 1 and 2 when resistance decreases. (**c**) Coefficient of variance (CV) decreases in outlet 1 and increases in outlets 2 and 3 as resistance ratio increases. (**d**) Efficiency of separation is highest when resistance is at $\sigma = 0.59$.

The CV decreased as σ increased in outlet 1 and increased as σ increased for outlets 2 and 3 (Figure 5c). The trend of the cutoff size explains this. As σ increased, the cutoff size decreased which allowed a smaller range of particles to enter into outlets 1 and 2. Therefore, the particle sizes resulted in smaller standard deviations and averages. This resulted in lower ratios and a smaller CV. For outlets 2 and 3, the CV increased early on and then saturated at ±4%. Looking at outlet 2, the cutoff size decreased rapidly at first and then leveled off, so these curves match in describing the refinement of particles. These CV results did change significantly for outlet 1, suggesting that changing the resistance of the outlet did affect the refinement of separation.

The efficiency data suggests that a mid-σ is the best for the particles to focus efficiently. Insignificant changes occurred for efficiency trend except for the peak in the middle at σ = 0.59 (Figure 5d). In these experiments, the 1:4 flow rate ratio used gave each particle the same amount of length to focus. If the efficiency was entirely based on the particle's lateral distance, then it would be the same for each σ. However, there was another mechanism that changed the efficiency which was the boundary line that allowed particles into the outlets. This boundary line only changes with alterations of the channel length. When σ was small, the boundary line moved closer to the channel center. This caused larger particles to enter into outlet 1. When σ was high, the boundary line moved closer to the channel wall. This caused only smaller particles to enter into outlet 1. When the resistance was in the middle of the channel, we hypothesize that the boundary line was at a point where the large particles have all focused and the small particles have not started focusing yet which caused the high efficiency. This is further evidenced in the cutoff size data explained above.

3.5. Separation of a Complex Particle Mixture

Following the determination of cutoff size flexibility, we evaluated the cutoff size of the device by flowing single sized particles individually through the channel to determine if they would separate as predicted. Using σ = 0.59 with a cutoff size a_{c1} = 12 µm and a_{c2} = 18 µm, we expected that the 7.32 µm particles would exit through outlet 1, the 15.45 µm particles would bimodally split into outlets 2 and 3 and the 18.67 µm and 26.3 µm particles would go into outlet 3. It is important to note the 15.45 µm particles had an initial size range of 15–22 µm and an average of 16.5 µm. Therefore, these particles would not exit in one outlet because their size range was in the middle of the cutoff size for outlet 2. The 7.32 µm particles were in the range 6–11 µm with an average of 9 µm. The 18.67 µm particles had a range of 16–21 µm and an average of 19.5 µm. The 26.3 µm particles had a range of 26–30 µm with an average of 27.5 µm. The blockage ratios for each particle were 14%, 30%, 37% and 52% for 7.32 µm, 15.45 µm, 18.67 µm and 26.3 µm, respectively. The experiments resulted with the 7.32 µm, 18.67 µm and 26.3 µm particles having an efficiency >95% (not shown) within their target outlets (Figure 6). The 15.45 µm particles split bimodally into outlets 2 and 3 with the cutoff size calculated as 18.32 µm, which agreed with the cutoff size calculated in Figure 4c (Figure 6b).

We next challenged the device with a mixture of 7.32 µm, 15.45 µm and 26.3 µm diameter beads with the resistance ratio of σ = 0.59. Mixtures of two or three particle sizes are typically used to define cutoff sizes of a microfluidic device to prepare for biological sample [18,20,21,37]. Using a mixture also allowed for considerations of particle-on-particle interactions because of higher concentrations in the channel. For σ = 0.59, the results showed very high efficiency (>85%) for outlets 1 and 3 but as discussed above, the 15.45 µm beads yielded low purities for outlets 2 and 3 and consequently decreased the efficiency for outlet 2 (Figure 7a).

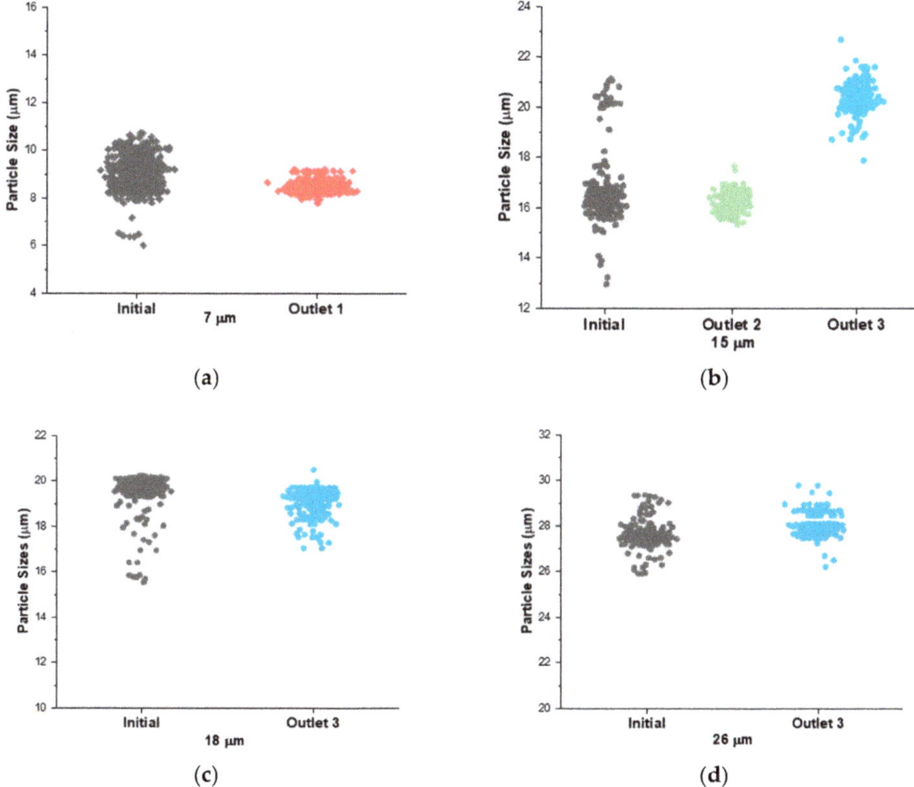

Figure 6. Single sized particles individually run through the device at 300 μL·min^{-1}, 1:4 ratio, 0.59 resistance ratio. (**a**) All 7.32 μm particles went into outlet 1. (**b**) The 15.45 μm particles split between outlet 2 and outlet 3 with minimal overlap. (**c**) All 18.7 μm particles went into outlet 3. (**d**) All 26.3 μm particles went into outlet 3.

3.6. Application to Cell Cycle Synchronization

The ability to study cells at a specific cell cycle phase is important for elucidating cellular mechanisms. Cell activity leading up to cell division can be characterized in four phases: gap phase G1, DNA synthesis phase S, gap phase G2 and mitosis phase M. Investigating the cell checkpoint at the G1 phase, which leads to cell proliferation or cell death, can allow for a better understanding of how a cell becomes cancerous. It has previously been shown that cell size is correlated to cell phase [38]. Thus, we used the microfluidic chip developed in this work to demonstrate enrichment of the G1 cell cycle phase of the A549 non-small cell lung cancer cell line. The A549 cells are approximately 12–27 μm in diameter and can be modeled with a continuous distribution of microparticles, as we have done earlier in this work. The chip was operated at 600 μL·min^{-1}, 1:4 ratio and σ = 0.21. Figure 8a shows representative images of the microchannel with small cells exiting through outlet 1, larger cells exiting through outlet 2 and the largest cells exiting through outlet 3. The analysis of the three outlets using flow cytometry in Figure 8b clearly shows an enrichment of G1 phase from a 2.36 ratio in the control sample to 6.30 and 3.37 ratios for outlets 1 and 2, respectively. Here, the enrichment ratio is defined as percent fraction of gated G1/G2 cells. Outlet 3 collected both G1 and G2 phases, with an enrichment ratio of 1.68. The viability of cells post-separation was > 90%, which is comparable to post-separation viability reported by others [11,38], suggesting the microfluidic device can be used for separation of cells. In addition, the higher flow rate increased throughput, which is an added advantage of this method compared to the standard drug-induced methods for cell cycle synchronization [38,39].

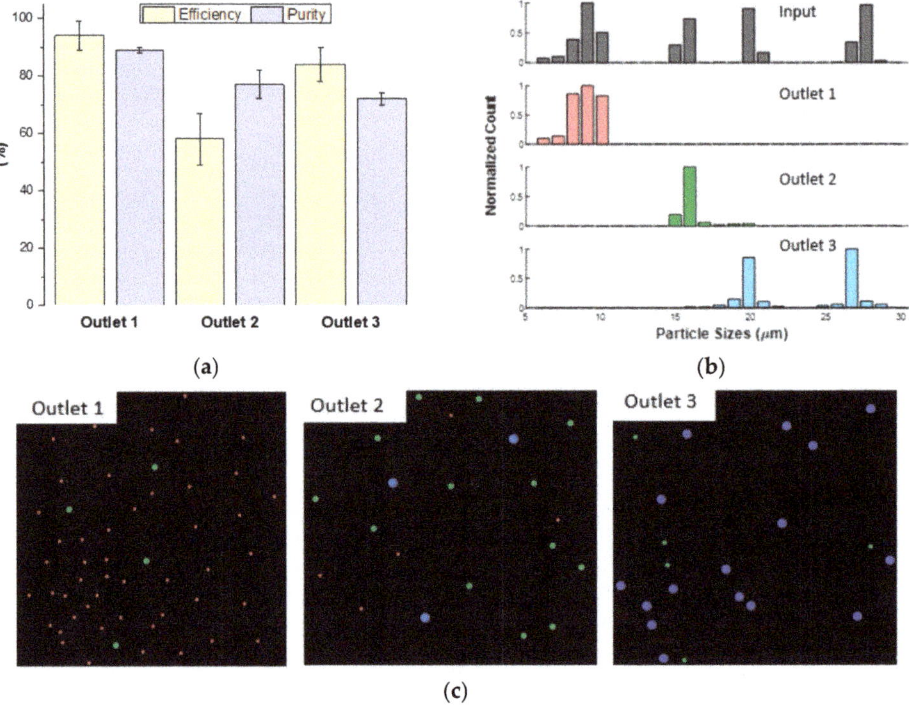

Figure 7. Mixed particle experiment of 7 μm (red), 15 μm (green) and 26 μm (blue) at 300 μL min^{-1}, 1:4 ratio, 0.59 resistance ratio to show device optimization. (**a**) Purity and efficiency of outlets. (**b**) Initial particle sizes and particle sizes in outlets after experiments. (**c**) 10X images show the purity based on the amount of colors in each outlet.

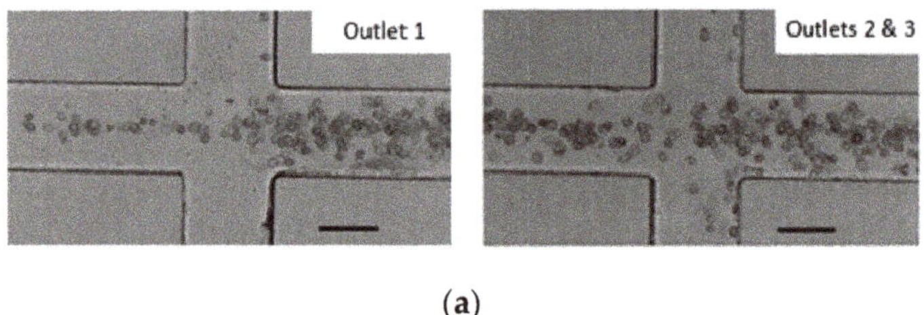

(**a**)

Figure 8. *Cont.*

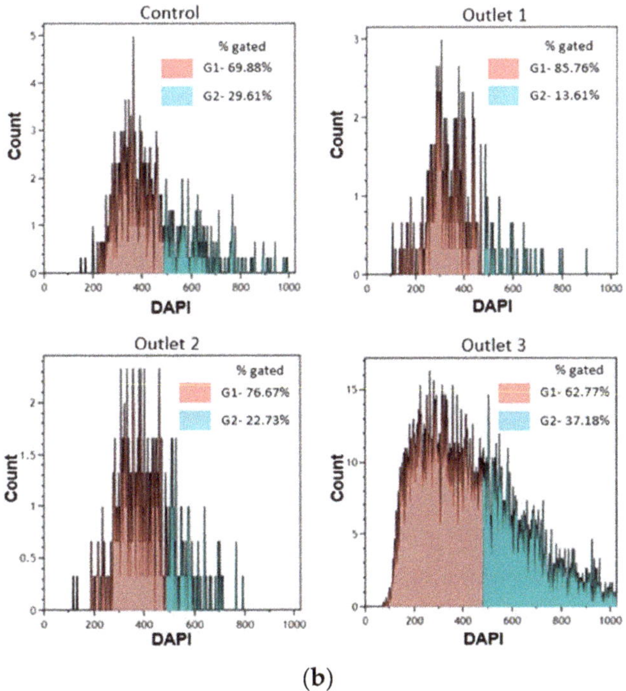

(b)

Figure 8. Representative results of the cell cycle synchronization. Flow conditions were 600 µL·min^{-1}, 1:4 ratio and σ = 0.21. (**a**) Bright field images show cells exiting through outlets 1 and 2. Scale bar is 100 µm. (**b**) 4′,6-diamidino-2-phenylindole (DAPI) area analysis of cells shows increase of G1 phase in outlets 1 and 2 and an increase in G2 phase in outlet 3 as compared to the control.

4. Conclusions

In conclusion, we have comprehensively shown the impact of flow rate, flow rate ratio and output resistance on the tunability of cutoff size and the separation quality of a continuous range of microparticles in a co-flow device. Both flow rate and flow rate ratio offer convenient ways to fine tune the cutoff size of a given co-flow channel for separation. The former is more effective in fine tuning larger sized particles while better efficiency is achieved at lower flow rate. The latter is important for tuning separation quality as smaller ratio giving more uniform initial particle positions. The resistance ratio provides the most significant tunability in terms of cutoff size. Consequently, this also has a large effect on the refinement of separation due to the large flexible range of cutoff sizes. Overall, our results show that a lower flow rate, flow rate ratio and a mid-level resistance ratio provides the best separation efficiency and quality of all three outlets. Testing the device with a typical mixture of single sized particles validated the cutoff size showing that the continuous particles can be used to predict performance of the device.

Author Contributions: A.B., J.Z. and I.P conceived and designed the experiments; A.B. performed the experiments; A.B and J.Z. analyzed the data; A.B., J.Z. and I.P. wrote the paper; I.P. supervised the work. All authors have read and agreed to the published version of the manuscript.

Funding: We gratefully acknowledge partial support from the NSF Center for Advanced Design and Manufacturing of Integrated Microfluidics (CADMIM), NSF awards IIP-1841473 and IIP-1738617.

Conflicts of Interest: Authors declare no conflict of interest.

Appendix A

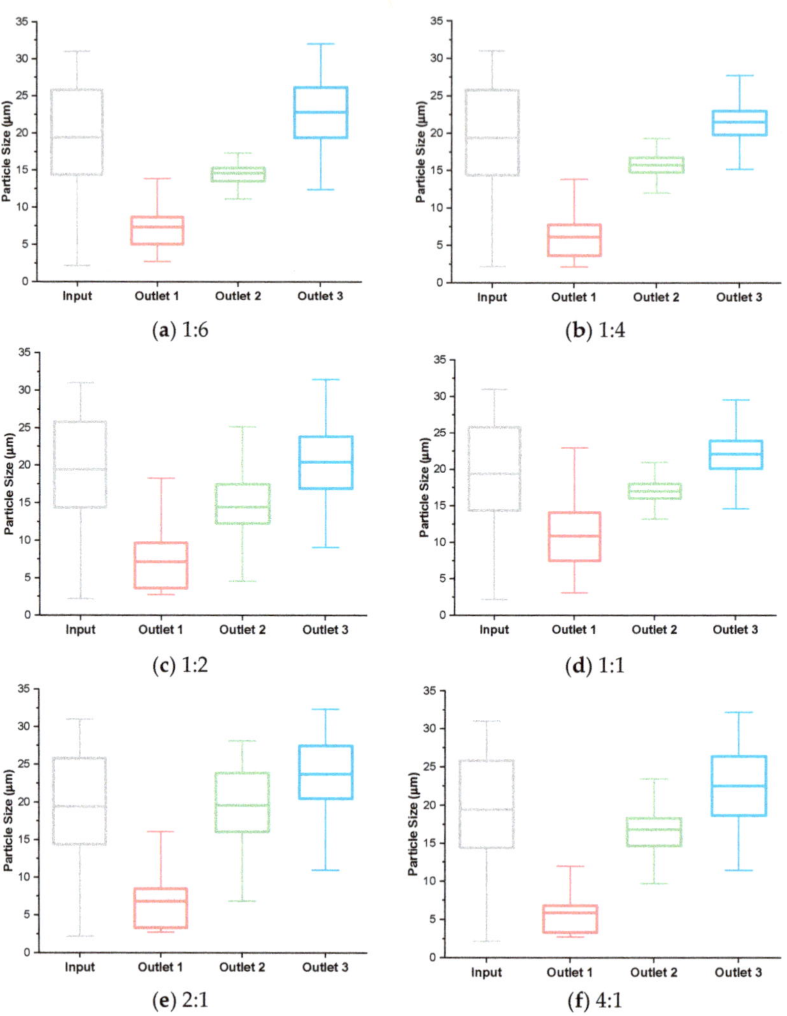

Figure A1. Distribution of particles within the outlets for various flow rate ratios. (**a**) 1:6 ratio had a larger distribution in outlet 3 compared to the 1:4 ratio. (**b**) 1:4 ratio had the most optimal refinement of the input distribution in all three outlets. (**c**) 1:2 ratio had a larger distribution in outlet 1 compared to the 1:4 ratio. (**d**) 1:1 ratio had a larger distribution in outlet 1 compared to the 1:4 ratio. (**e**) All outlets had larger distributions compared to the 1:4 ratio. (**f**) All outlets had larger distributions compared to the 1:4 ratio.

References

1. Shields, C.W., IV; Reyes, C.D.; López, G.P. Microfluidic cell sorting: A review of the advances in the separation of cells from debulking to rare cell isolation. *Lab Chip* **2015**, *15*, 1230–1249. [CrossRef] [PubMed]
2. Nguyen, N.-T.; Hejazian, M.; Ooi, C.; Kashaninejad, N. Recent advances and future perspectives on microfluidic liquid handling. *Micromachines* **2017**, *8*, 186. [CrossRef]

3. Zhou, J.; Mukherjee, P.; Gao, H.; Luan, Q.; Papautsky, I. Label-Free microfluidic sorting of microparticles. *APL Bioeng.* **2019**, *3*, 041504. [CrossRef] [PubMed]

4. Huang, L.R. Continuous particle separation through deterministic lateral displacement. *Science* **2004**, *304*, 987–990. [CrossRef] [PubMed]

5. McGrath, J.; Jimenez, M.; Bridle, H. Deterministic lateral displacement for particle separation: A review. *Lab Chip* **2014**, *14*, 4139–4158. [CrossRef] [PubMed]

6. Yamada, M.; Nakashima, M.; Seki, M. Pinched flow fractionation: Continuous size separation of particles utilizing a laminar flow profile in a pinched microchannel. *Anal. Chem.* **2004**, *76*, 5465–5471. [CrossRef]

7. Bhagat, A.A.S.; Hou, H.W.; Li, L.D.; Lim, C.T.; Han, J. Pinched flow coupled shear-modulated inertial microfluidics for high-throughput rare blood cell separation. *Lab Chip* **2011**, *11*, 1870. [CrossRef]

8. Yamada, M.; Seki, M. Hydrodynamic filtration for on-chip particle concentration and classification utilizing microfluidics. *Lab Chip* **2005**, *5*, 1233. [CrossRef]

9. Pamme, N. Continuous flow separations in microfluidic devices. *Lab Chip* **2007**, *7*, 1644. [CrossRef]

10. Tu, C.; Zhou, J.; Liang, Y.; Huang, B.; Fang, Y.; Liang, X.; Ye, X. A flexible cell concentrator using inertial focusing. *Biomed. Microdevices* **2017**, *19*, 83. [CrossRef]

11. Warkiani, M.E.; Guan, G.; Luan, K.B.; Lee, W.C.; Bhagat, A.A.S.; Kant Chaudhuri, P.; Tan, D.S.-W.; Lim, W.T.; Lee, S.C.; Chen, P.C.Y.; et al. Slanted spiral microfluidics for the ultra-fast, label-free isolation of circulating tumor cells. *Lab Chip* **2014**, *14*, 128–137. [CrossRef] [PubMed]

12. Nivedita, N.; Papautsky, I. Continuous separation of blood cells in spiral microfluidic devices. *Biomicrofluidics* **2013**, *7*, 054101. [CrossRef] [PubMed]

13. Di Carlo, D. Inertial microfluidics. *Lab Chip* **2009**, *9*, 3038. [CrossRef] [PubMed]

14. Amini, H.; Lee, W.; Di Carlo, D. Inertial microfluidic physics. *Lab Chip* **2014**, *14*, 2739. [CrossRef] [PubMed]

15. Martel, J.M.; Toner, M. Inertial focusing in microfluidics. *Annu. Rev. Biomed. Eng.* **2014**, *16*, 371–396. [CrossRef] [PubMed]

16. Zhang, J.; Yan, S.; Yuan, D.; Alici, G.; Nguyen, N.-T.; Ebrahimi Warkiani, M.; Li, W. Fundamentals and applications of inertial microfluidics: A review. *Lab Chip* **2016**, *16*, 10–34. [CrossRef]

17. Warkiani, M.E.; Khoo, B.L.; Wu, L.; Tay, A.K.P.; Bhagat, A.A.S.; Han, J.; Lim, C.T. Ultra-Fast, label-free isolation of circulating tumor cells from blood using spiral microfluidics. *Nat. Protoc.* **2016**, *11*, 134–148. [CrossRef]

18. Khojah, R.; Stoutamore, R.; Di Carlo, D. Size-Tunable microvortex capture of rare cells. *Lab Chip* **2017**, *17*, 2542–2549. [CrossRef]

19. Wu, Z.; Willing, B.; Bjerketorp, J.; Jansson, J.K.; Hjort, K. Soft inertial microfluidics for high throughput separation of bacteria from human blood cells. *Lab Chip* **2009**, *9*, 1193. [CrossRef]

20. Zhou, J.; Kulasinghe, A.; Bogseth, A.; O'Byrne, K.; Punyadeera, C.; Papautsky, I. Isolation of circulating tumor cells in non-small-cell-lung-cancer patients using a multi-flow microfluidic channel. *Microsyst. Nanoeng.* **2019**, *5*, 8. [CrossRef]

21. Sarkar, A.; Hou, H.W.; Mahan, A.E.; Han, J.; Alter, G. Multiplexed affinity-based separation of proteins and cells using inertial microfluidics. *Sci. Rep.* **2016**, *6*, 23589. [CrossRef] [PubMed]

22. Lee, M.G.; Shin, J.H.; Bae, C.Y.; Choi, S.; Park, J.-K. Label-Free cancer cell separation from human whole blood using inertial microfluidics at low shear stress. *Anal. Chem.* **2013**, *85*, 6213–6218. [CrossRef] [PubMed]

23. Wang, X.; Zandi, M.; Ho, C.-C.; Kaval, N.; Papautsky, I. Single stream inertial focusing in a straight microchannel. *Lab Chip* **2015**, *15*, 1812–1821. [CrossRef] [PubMed]

24. Mukherjee, P.; Wang, X.; Zhou, J.; Papautsky, I. Single stream inertial focusing in low aspect-ratio triangular microchannels. *Lab Chip* **2019**, *19*, 147–157. [CrossRef]

25. Guan, G.; Wu, L.; Bhagat, A.A.; Li, Z.; Chen, P.C.Y.; Chao, S.; Ong, C.J.; Han, J. Spiral microchannel with rectangular and trapezoidal cross-sections for size based particle separation. *Sci. Rep.* **2013**, *3*, 1475. [CrossRef]

26. Bhagat, A.A.S.; Kuntaegowdanahalli, S.S.; Papautsky, I. Continuous particle separation in spiral microchannels using dean flows and differential migration. *Lab Chip* **2008**, *8*, 1906. [CrossRef]

27. Fan, L.-L.; Yan, Q.; Zhe, J.; Zhao, L. Single particle train ordering in microchannel based on inertial and vortex effects. *J. Micromech. Microeng.* **2018**, *28*, 065011. [CrossRef]

28. Sonmez, U.; Jaber, S.; Trabzon, L. Super-Enhanced particle focusing in a novel microchannel geometry using inertial microfluidics. *J. Micromech. Microeng.* **2017**, *27*, 065003. [CrossRef]

29. Mach, A.J.; Di Carlo, D. Continuous scalable blood filtration device using inertial microfluidics. *Biotechnol. Bioeng.* **2010**, *107*, 302–311. [CrossRef]

30. Wang, X.; Papautsky, I. Size-Based microfluidic multimodal microparticle sorter. *Lab Chip* **2015**, *15*, 1350–1359. [CrossRef]

31. Mukherjee, P.; Nebuloni, F.; Gao, H.; Zhou, J.; Papautsky, I. Rapid prototyping of soft lithography masters for microfluidic devices using dry film photoresist in a non-cleanroom setting. *Micromachines* **2019**, *10*, 192. [CrossRef] [PubMed]

32. Zhou, J.; Papautsky, I. Fundamentals of inertial focusing in microchannels. *Lab Chip* **2013**, *13*, 1121. [CrossRef] [PubMed]

33. Nivedita, N.; Garg, N.; Lee, A.P.; Papautsky, I. A high throughput microfluidic platform for size-selective enrichment of cell populations in tissue and blood samples. *Analyst* **2017**, *142*, 2558–2569. [CrossRef] [PubMed]

34. Kim, J.; Lee, J.; Wu, C.; Nam, S.; Di Carlo, D.; Lee, W. Inertial focusing in non-rectangular cross-section microchannels and manipulation of accessible focusing positions. *Lab Chip* **2016**, *16*, 992–1001. [CrossRef] [PubMed]

35. Reece, A.E.; Oakey, J. Long-Range forces affecting equilibrium inertial focusing behavior in straight high aspect ratio microfluidic channels. *Phys. Fluids* **2016**, *28*, 043303. [CrossRef]

36. Wang, X.; Yang, X.; Papautsky, I. An integrated inertial microfluidic vortex sorter for tunable sorting and purification of cells. *Technology* **2016**, *4*, 88–97. [CrossRef]

37. Yang, D.; Leong, S.; Lei, A.; Sohn, L.L. High-Throughput Microfluidic Device for Rare Cell Isolation. *Proc SPIE Int Soc Opt Eng* **2015**, *9518*, 95180E.

38. Lee, W.C.; Bhagat, A.A.S.; Huang, S.; Van Vliet, K.J.; Han, J.; Lim, C.T. High-Throughput cell cycle synchronization using inertial forces in spiral microchannels. *Lab Chip* **2011**, *11*, 1359. [CrossRef]

39. Davis, P.K.; Ho, A.; Dowdy, S.F. Biological methods for cell-cycle synchronization of mammalian cells. *BioTechniques* **2001**, *30*, 1322–1331. [CrossRef]

 *micromachines*

Article

Nucleotide Identification in DNA Using Dielectrophoresis Spectroscopy

Fleming Dackson Gudagunti, Logeeshan Velmanickam, Dharmakeerthi Nawarathna and Ivan T. Lima Jr. *

Department of Electrical and Computer Engineering, North Dakota State University, Fargo, ND 58102, USA; Fleming.Gudagunti@ndsu.edu (F.D.G.); Logeeshan.Velmanicka@ndus.edu (L.V.); Dharmakeerthi.Nawara@ndsu.edu (D.N.)
* Correspondence: Ivan.Lima@ndsu.edu; Tel.: +1-701-231-6728

Received: 14 November 2019; Accepted: 26 December 2019; Published: 28 December 2019

Abstract: We show that negative dielectrophoresis (DEP) spectroscopy is an effective transduction mechanism of a biosensor for the detection of single nucleotide polymorphism (SNP) in a short DNA strand. We observed a frequency dependence of the negative DEP force applied by interdigitated electrodes to polystyrene microspheres (PM) with respect to changes in both the last and the second-to-last nucleotides of a single-strand DNA bound to the PM. The drift velocity of PM functionalized to single-strand DNA, which is proportional to the DEP force, was measured at the frequency range from 0.5 MHz to 2 MHz. The drift velocity was calculated using a custom-made automated software using real time image processing technique. This technology for SNP genotyping has the potential to be used in the diagnosis and the identification of genetic variants associated with diseases.

Keywords: single nucleotide polymorphism; dielectrophoresis; spectroscopy; bioelectronics

---

1. Introduction

Genetic markers are used to follow the inheritance patterns of chromosomal regions from generation to generation and are used in identifying the genetic variants associated with human diseases [1]. The most common genetic variation is single nucleotide polymorphism (SNP), which is due to the differences of a single base substitution. In every DNA sequence, SNP represents a difference in a single nucleotide [2]. SNP may replace a single nucleotide with another nucleotide [3]. Previous research showed that SNP predicts an individual's risk of developing certain diseases such as cardiovascular disease, type 2 diabetes mellitus [4], autoimmune disease, Alzheimer's disease [5], cancer, and an individual's response to certain drugs [6–10]. These small differences can be used to track an individual's susceptibility to environmental factors such as toxins [11]. Since SNPs are stable over generations, they are excellent genetic markers. It is important to explore the role of SNPs in the genetic analysis of diseases, as they would enable the identification of complex diseases and genetic disorders [12]. Even though SNPs have been shown to be an important factor in genetic variation [13,14], detecting SNPs is still expensive and time-consuming with existing techniques.

SNP genotyping can be performed with DNA sequencing methods [15]. Mapping and assembly with quality (MAQ) maps shotgun reads can build the assemblies by using value-based scores to derive genotype calls of the consensus sequence of a diploid genome. This technique is based on the Bayesian statistical model that includes error probability and mapping qualities from the quality scores of the sequence. Although minimum order quality (MOQ) [16] is efficient and highly sensitive, the high probability of sequencing errors in this method makes it less reliable than other methods.

The Short Oligonucleotide Analysis Program (SOAP) package is a resequencing tool that compares raw sequencing reads with the reference genome to calculate the probability of each possible genotype [17]. This method incorporates data quality, alignment, and recurring experimental errors, making this method complex with large acquisition time.

TaqMan PCR is a fast and reliable tool for genotyping. There are few applied biosystem instruments such as real time PCR available for the processing of TaqMan SNP genotyping assays. The endpoint read can be performed on applied biosystems such as real time PCR. The results from this method are highly accurate and reproducible, although the biosystem involved is expensive and time consuming. The amplification of the alleles is done using two pairs of primers. This process involves overlapping of the primer pairs so that it matches the pairs but not to the alternative allele for the SNP [15–18]. The conventional techniques for SNP detection rely on gel electrophoresis for the fragment analysis [19–21]. Several techniques include oligonucleotide ligation [22–24], extension of primers [25–27], allele-specific DNA hybridization [28], or electrochemical typing [29,30].

To address the need for newer technologies for SNP detection that could potentially simplify at least one of the steps in the SNP detection, we developed a rapid SNP genotyping method based on dielectrophoresis (DEP) spectroscopy that does not rely on fluorescence. DEP spectroscopy has the potential to be a cost-effective transduction mechanism for SNP detection in short single-stranded DNA (ssDNA), since it replaces the fluorescence tagging that is used in the last step of most of the existing techniques. DEP has been widely used for manipulation, separation, and characterization of cells, DNA, viruses, and colloid particles (at both micro and nano scales) in various microfluidic platforms, since it has numerous unique advantages [31].

In a previous study, we showed that DEP spectroscopy was an effective method for the detection of the pancreatic cancer biomarker CA 19-9 in serum [32,33]. In this study, we show that this method, which does not rely on fluorescence labeling, can also be used in SNP genotyping in a short DNA strand, and it also has the potential to be used in sequencing.

2. Theory

Dielectrophoresis (DEP) [34] is a physical phenomenon in which the coulomb force acts on electrically polarized dielectric particles in a non-uniform electric field. Since the electric field is non-uniform, the dielectric particles can experience translational force towards the region with higher electric field if they are more polarizable than the medium. This phenomenon is denoted positive DEP. If the medium is more polarizable than the dielectric particle, the dielectric particles suffer a net force towards the region with lower electric field. This phenomenon is denoted negative DEP [34–37]. Since the DEP force is strongly dependent on the characteristics of the particles and the molecules bound to their surfaces, DEP has been used to isolate, concentrate, or separate different types of target particles [38–43]. However, the physics behind the frequency-dependence of the DEP force on the molecular structures is still not well understood [40,44,45].

3. Materials and Methods

3.1. Sample Preparation

The sample preparation consisted of the following steps:

Step 1: Streptavidin attachment to biotinylated polystyrene microspheres (PM). Biotinylated PM with 750 nm diameter were purchased from Spherotech Inc (Lake Forest, IL, USA). The first step in the preparation of samples consisted of binding the biotinylated PM with the antigen Streptavidin purchased from Vector Labs Inc. Biotin acted as a conjugate to the protein Streptavidin, and they formed a strong bond with very high affinity. This process was done first by a 3 µL Streptavidin solution into a 10 µL biotinylated PM solution at 70 °F in a centrifuge tube to have 100% binding [46], according to the manufacturer's recommendation. This procedure recommended by the manufacturer might not produce 100% binding. However, inaccuracies in the protocol recommended by the manufacturer should not have affected the fact that the DEP force depends on both the frequency and the nucleotides near the end of a DNA sequence, since the same protocol was used in all the experiments with different

DNA sequences, and four batches were prepared for each sequence to validate these results. The total volume was set to 400 μL by adding 0.01× phosphate-buffered saline (PBS) solution with conductivity 0.01 S/m. Then, the sample was uniformly mixed using a vortex machine and left on a shaker for 20 min for the Streptavidin–biotin binding process. After 20 min, the tube was centrifuged at 5000 rpm for 14 min to remove the unbound Streptavidin molecules and the buffer.

Step 2: Biotinylated DNA attachment to the biotinylated PM + Streptavidin. Each Streptavidin molecule can bind up to 4 biotin molecules. One binding site of each Streptavidin molecule was used to bind that Streptavidin molecule with the PM. The remaining three Streptavidin binding sites bound with three biotinylated DNA. First, 3 μL of biotinylated DNA purchased from The Midland Certified Reagent Company was added into 397 μL of 0.01× Tris-EDTA to have 100% binding of the biotinylated DNA with all the Streptavidin molecule binding sites, according to the manufacturer's requirement. Then, this sample was added into the solution with Streptavidin–biotin PM and uniformly mixed using a vortex machine. After that, the sample was kept on a shaker for 20 min at 70 °F for the biotinylated DNA to bind with the Streptavidin molecules of the PM. After 20 min, the tube was centrifuged at 5000 rpm for 14 min to remove the unbound DNA antibody molecules and the buffer.

Step 3: Preparation of the solution for the experiments. First, 200 μL of 0.01× TE buffer was added to the centrifuge tube and mixed uniformly. Then, 10 μL of this sample solution was pipetted on to the microelectrodes and used for each experiment.

The ssDNA sequences with difference in last nucleotide were:

5′-(biotin) TGTTGTGCG*A*-3′
5′-(biotin) TGTTGTGCG*T*-3′
5′-(biotin) TGTTGTGCG*G*-3′
5′-(biotin) TGTTGTGCG*C*-3′

The ssDNA sequences with difference in the second-to-last nucleotide were:

5′-(biotin) TGTTGTGC*A*C-3′
5′-(biotin) TGTTGTGC*T*C-3′
5′-(biotin) TGTTGTGC*C*C-3′

3.2. DEP Spectroscopy: Hardware and Software

To measure the DEP spectrum to characterize the nucleotide in ssDNA, we used the same experimental setup that we used to measure the concentration of CA 19-9 in spiked serum [33]. We developed a software package using Microsoft foundation classes in visual C++ in the Windows operating system. The software application controlled a USB video class (UVC) standard compliant ProScope Digital Microscope 5 MP Camera (ProScope, Wilsonville, OR, USA) and a Tektronix AFG series function generator (Tektronix Beaverton, Beaverton, OR, USA). With the application controlling the function generator, the frequency was swept to measure the DEP force applied to functionalized PM as a function of the frequency. The experimental set up is shown in Figure 1. The microscope camera extracted the pixel information of the live video using a real-time image processing algorithm. The region of interest to observe the DEP effect was set by placing rectangular boxes (as shown in Figure 1c) in which the pixel and the color information were extracted, and with this data, the drift velocity due to DEP force as a function of frequency of the electric field was calculated using the algorithm.

The direction of movement for the dielectric particle due to DEP depended upon the relative polarizability of the particle and the medium as well as on the presence of a large gradient of the electric field intensity produced by electrodes [47]. In the experiments, we used the interdigitated electrodes shown in Figure 1b. The electrode was fabricated on a commercially available glass wafer with standard fabrication procedures including photolithography, metal sputtering, and lift-off procedures using 1000 Å thick gold film in the microfabrication facilities at North Dakota State University. Using COMSOL, we verified the maximum electric field of the electrode was 1.8×10^4 V/m and the gradient of the electric field intensity was as high as 3×10^{12} V^2/m^3.

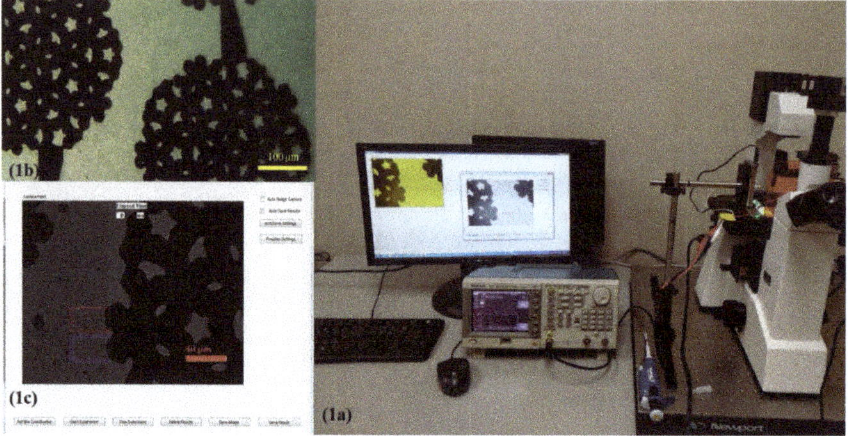

Figure 1. Single nucleotide polymorphism (SNP) sensor prototype: (**a**) experimental setup, (**b**) interdigitated electrode used in the experiments. The electrode is visible as the darker region in the picture. The yellow scale bar indicates 100 μm. Multi-colored hollow rectangles depict the regions of interest for measurements and analysis. (**c**) Interface of Microsoft Windows application for dielectrophoresis (DEP) spectroscopy. The red scale bar indicates 50 μm.

3.3. Spectrum Measurement

The drift velocity was measured from the point of application of positive DEP to the negative DEP at the applied frequency. Positive DEP was used to attract the PM to the convex edge of the electrodes. That was done through the application of a low frequency such as 10 kHz to the electrodes. Negative DEP was used to repel the PM from the convex edge of the electrodes to enable drift velocity measurements, which were produced with frequencies on the order of hundreds or thousands of kHz. This process enabled an indirect measurement of the negative DEP force, since the DEP force was proportional to the drift velocity due to the viscosity of the solution. The custom-made software was programmed to sweep a set of alternating frequency generating positive and negative DEP. The experiments were conducted with frequencies from 0.5 MHz to 2 MHz in linear steps with 10 V peak-to-peak.

A preliminary set of experiments were conducted to determine the choice of time intervals for the application of negative and positive DEP. The software automatically switched the frequency to a higher value after 1000 ms, producing negative DEP effect for 40 ms. The cycle is repeated with the application of positive DEP and the following frequency that produced negative DEP until the stop frequency is reached. The system acquired the frames at 0 ms and at 40 ms after the application of a frequency that produced negative DEP to measure the average drift velocity of the PM. The function generator used in the experiments was a Tektronix AFG series that was controlled by the computer via USB port. The negative DEP spectrum was measured indirectly through the average drift velocity of the PM as a function of the frequency, since the drift velocity is proportional to the negative DEP force.

4. Results

We validated our method to detect SNPs using DEP spectroscopy experiments with our custom-made image processing software for observation and data acquisition. The principle of operation of our method is shown in Figure 2. The initial frequency was set to 10 kHz to produce positive DEP (< 50 kHz), and the sweeping frequency from 0.5 MHz to 2 MHz produced negative DEP (> 250 kHz). Figure 2a clearly shows the application of positive DEP, which concentrates PMs on the edge of the convex regions of the electrodes up to 0 ms, when negative DEP was applied. Figure 2b shows the position of the PM band while it was being repelled away from the convex regions of the electrode due to negative DEP after 40 ms. We developed a system that can calculate the drift velocity of the PM layer to indirectly measure the negative DEP force as a function of the applied frequency.

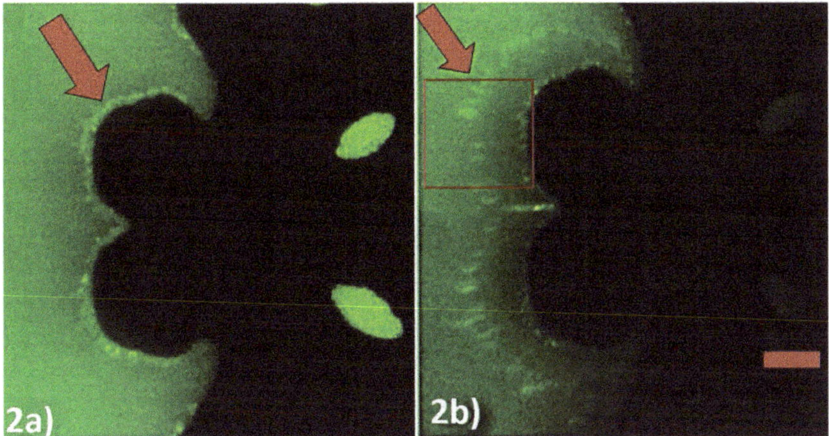

Figure 2. Experimental demonstration of negative DEP effect through time-lapse images captured through DEP spectroscopy application. The electric field was changed from 10 kHz to 0.5 MHz with 10 V peak-to-peak at $t = 0$ ms. (**a**) $t = 0$ ms and (**b**) $t = 40$ ms. The ssDNA sequence used for this was 5′-(biotin) TGTTGTGCGA-3′. The interdigitated electrode is visible as the darker region in the picture. The bright layer visible on the edge of the electrode is formed by the accumulation of PMs. The scale bar indicates 25 µm. The rectangular box depicts the region of interest processed to extract the position of the PM band as a function of the time.

The region of interest, which is depicted as the rectangular box in the Figure 1c, was processed to extract the position of the PM band as a function of the time while a frequency that produced negative DEP was applied. Figure 3 represents the light intensity of the images captured shortly after the application of negative DEP and 40 ms later. The average location of the PM was the center of mass x_{CM} of the light intensity curves. The change in the center of mass of the curves resulted from the movement of the PM band due to the application of a frequency that produced negative DEP. The center of mass was an effective way to measure the location of the PM band to determine the velocity of the PM band. The formula used to calculate the center of mass of the light intensity curve is given by:

$$x_{CM} = \frac{\sum_{i=1}^{N} I_i x_i}{\sum_{i=1}^{N} I_i}$$

where I_i is the value of the light intensity at the distance x_i from the convex edge of the electrode. Only the region whose intensity exceeded two-thirds of the peak intensity was included in the calculation of the center of mass.

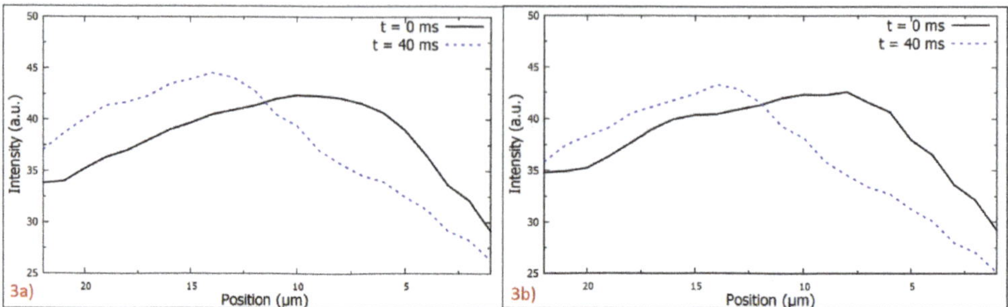

Figure 3. Light intensity as a function of the pixel position at $t = 0$ ms, when negative DEP was applied, and at $t = 40$ ms (**a**) for the ssDNA sequence 5′-(biotin) TGTTGTGCGG-3′, which corresponded to the results shown in Figure 2, and (**b**) for the ssDNA sequence 5′-(biotin) TGTTGTGCGA-3′ in 10 µL at the frequency 0.5 MHz. The convex edge of the electrode is located on the right side of the polystyrene microspheres (PM) layer, as shown in Figure 2.

A relationship was observed between the drift velocity of the PM functionalized with ssDNA, which was proportional to the negative DEP force, as a function of frequency and the type of nucleotide in the last and in the second-to-last nucleotide. The ssDNA sequence considered in these experiments was 5′-(biotin) TGTTGTGCGA-3′, and its variations in the last and in the second-to-last nucleotide are listed in Section 3. A clear dependence was observed between the negative DEP spectrum and the nucleotide sequence. The resulting DEP spectra are shown in Figures 4 and 5 for changes in the last nucleotide and Figures 6 and 7 for changes in the second-to-last nucleotide. Each DEP spectrum curve shown in these figures required only 68 s to be obtained during the experiments. Figure 5 shows a higher resolution spectrum for change in last nucleotide, and Figure 7 shows the high-resolution spectrum for change in the second-to-last nucleotide over the same frequency range shown in Figure 5.

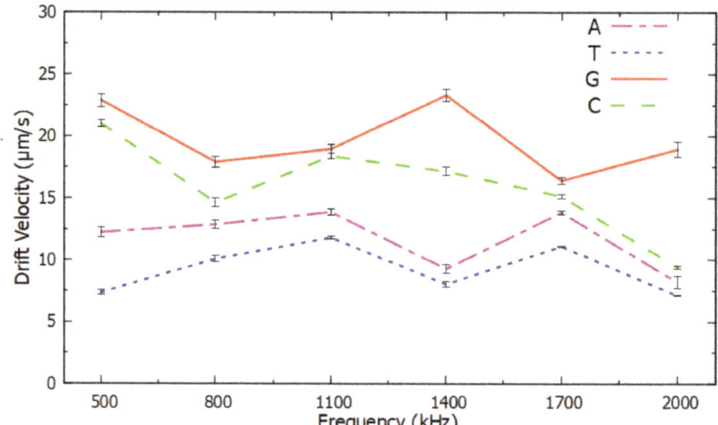

Figure 4. Negative DEP spectra for different nucleotides at the end of ssDNA sequences averaged over six measurements per frequency. All the other nucleotides were the same in the ssDNA sequences. The error bars show the confidence interval of the average value.

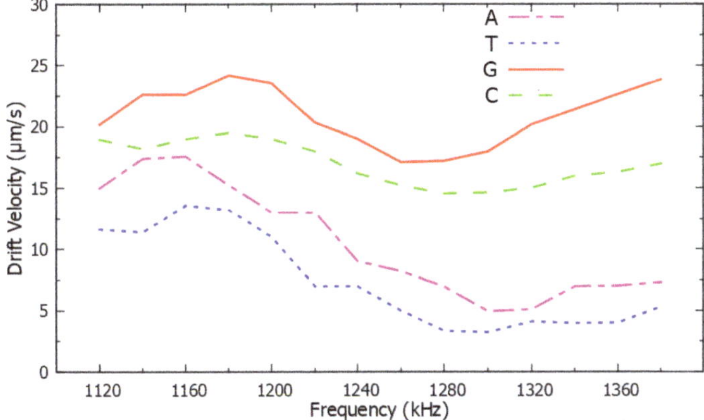

Figure 5. High resolution negative DEP spectra from 1120 kHz to 1380 kHz for the same ssDNA sequences used in Figure 4.

The difference in the negative DEP spectra shown in Figures 4–7 resulted only from changes in a single nucleotide of the ssDNA sequence. No other parameters and conditions were changed throughout the experiments. The difference in the nucleotides near the end of the ssDNA bound to the PM caused a sufficiently different polarizability of the PM to affect the real part of the Clausius–Mossotti factor that determines the DEP force [35,36]. The narrow error bars shown with plus and minus one standard deviation of the average of six measurements in Figures 4 and 6 are indications of the high accuracy and the repeatability of the measurements. Since there was no overlap in the confidence intervals of different curves in Figures 4 and 6, these results are sufficient to determine the type of the last and the second-to-last nucleotide of these ssDNA sequences. Therefore, these results demonstrate the potential use of this method as a novel transduction mechanism for SNP detection in the last and in the second-to-last nucleotides in ssDNA sequences. The DEP spectra shown in Figures 4–7 can be used as the calibration curve for the SNP detection in the sequences investigated. The results with guanine in the second-to-last position of the ssDNA sequence were not shown in Figures 6 and 7 because they were undermined by the hairpin effect that led to clustering of the PM, which is a limitation of this method.

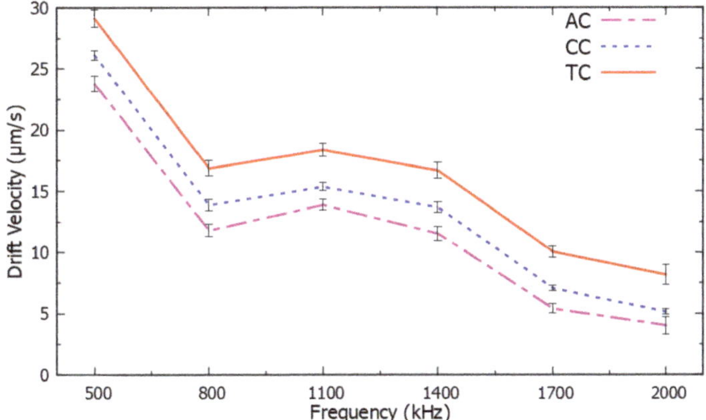

Figure 6. Negative DEP spectra for different nucleotides adjacent to the last nucleotide of ssDNA sequences averaged over six measurements per frequency. All the other nucleotides were the same in the ssDNA sequences. The error bars show the confidence interval of the average value.

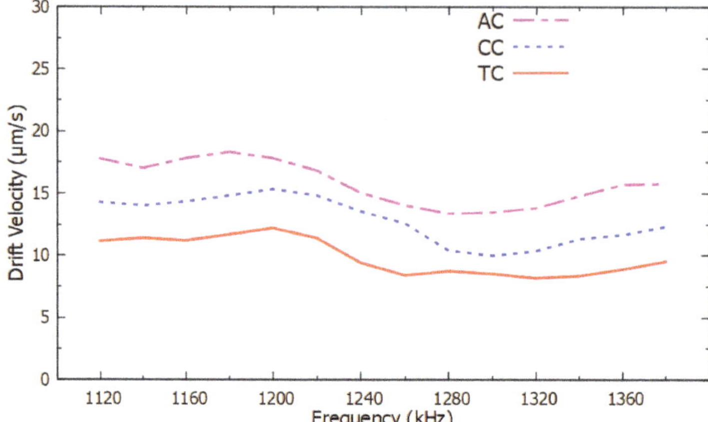

Figure 7. High resolution negative DEP spectra from 1120 kHz to 1380 kHz for the same ssDNA sequences used in Figure 6.

5. Conclusions

We demonstrated that the negative DEP spectroscopy method is an effective transduction method to accurately detect SNPs in short DNA strands. The negative DEP spectrum was measured using a custom-made real time image processing technique to detect the drift velocity of PM bound to ssDNA in microstructured electrodes, since the drift velocity is proportional to the DEP force. The frequency-dependent velocity of repulsion due to negative DEP on a set of PMs bound to ssDNA that we measured using image processing had a strong dependence on the last and on the second-to-last nucleotides in the ssDNA sequences that we investigated. This technique does not require the use of fluorescent labels, which eliminates the washing step of unbound fluorescent molecules in the preparation process and consequently does not require a careful calibration of the light source intensity and the photodetector sensitivity. The shape of the negative DEP spectrum curve used in our method for SNP genotyping is a measurement that does not depend on the light intensity, the number of PM, or the sensitivity of the microscope camera. The DEP force on dielectric microspheres depends on the polarizability of the microspheres, the polarizability of the ssDNA bound to the microspheres, and the polarizability of the molecules in the solution. The polarization mechanism of DNA is still not fully understood [36,48]. We are currently improving this method by designing a microstructured electrode array that is optimized for negative DEP spectroscopy. We plan to investigate the practical application of this method to detect SNPs by studying a large number of ssDNA with different nucleotide sequences and different lengths to assess the effectiveness of this method in SNP genotyping and DNA sequencing for clinical applications.

6. Patent

I. T. Lima Jr., D. Nawarathna, F. D. Gudagunti, and L. Velmanickam, "Method for Detecting and Quantifying Biological Molecules using Dielectrophoresis," Patent Application No. US 2019/0234,902, August 1, 2019.

Author Contributions: Conceptualization, F.D.G., D.N. and I.T.L.J.; Formal analysis, F.D.G. and L.V.; Funding acquisition, D.N. and I.T.L.J.; Investigation, F.D.G. and L.V.; Methodology, F.D.G., L.V., D.N. and I.T.L.J.; Project administration I.T.L.J.; Writing–original draft, F.D.G.; Writing–review & editing, F.D.G., L.V., D.N. and I.T.L.J. All authors have read and agreed to the published version of the manuscript.

Funding: The authors acknowledge financial support from ND EPSCoR (#46200) and North Dakota State University.

Conflicts of Interest: The authors declare no conflict of interest.

References

1. Chen, C.; Keith, J.; Chen, C.C.; Schwender, H.; Keith, J.; Nunkesser, R.; Mengersen, K.; Macrossan, P. Methods for identifying SNP interactions: A review on variations of logic regression, random forest and bayesian logistic. *IEEE/ACM Trans. Comput. Biol. Bioinform.* **2011**, *8*, 1580–1591. [CrossRef] [PubMed]

2. Bush, W.S.; Moore, J.H. Chapter 11: Genome-wide association studies. *PLoS Comput. Biol.* **2012**, *8*, 1–11. [CrossRef] [PubMed]

3. Shen, Y.; Wan, Z.; Coarfa, C.; Drabek, R.; Chen, L.; Ostrowski, E.A.; Liu, Y.; Weinstock, G.M.; Wheeler, D.A.; Gibbs, R.A.; et al. A SNP discovery method to assess variant allele probability from next-generation resequencing data. *Genome Res.* **2010**, *20*, 273–280. [CrossRef] [PubMed]

4. Yasukochi, Y.; Sakuma, J.; Takeuchi, I.; Kato, K.; Oguri, M.; Fujimaki, T.; Horibe, H.; Yamada, Y. Two novel susceptibility loci for type 2 diabetes mellitus identified by longitudinal exome-wide association studies in a Japanese population. *Genomics* **2019**, *111*, 34–42. [CrossRef]

5. Roy, J.; Mallick, B. Altered gene expression in late-onset Alzheimer's disease due to SNPs within 3'UTR microRNA response elements. *Genomics* **2017**, *109*, 177–185. [CrossRef]

6. Sato, M.; Soma, M.; Nakayama, T.; Kanmatsuse, K. Dopamine D1 receptor gene polymorphism is associated with essential hypertension. *Hypertension* **2000**, *36*, 183–186. [CrossRef]

7. Koptides, M.; Mean, R.; Stavrou, C.; Pierides, A.; Demetriou, K.; Nakayama, T.; Hildebrandt, F.; Fuchshuber, A.; Deltas, C.C. Novel NPR1 polymorphic variants and its exclusion as a candidate gene for medullary cystic kidney disease (ADMCKD) type 1. *Mol. Cell. Probes* **2001**, *15*, 357–361. [CrossRef]

8. Nakayama, T.; Asai, S.; Sato, N.; Soma, M. Genotype and haplotype association study of the STRK1 region on 5q12 among Japanese: A case-control study. *Stroke* **2006**, *37*, 69–76. [CrossRef]

9. Brænne, I.; Zeng, L.; Willenborg, C.; Tragante, V.; Kessler, T.; Consortium, C.; Consortium, C.; Willer, C.J.; Laakso, M.; Wallentin, L.; et al. Genomic correlates of glatiramer acetate adverse cardiovascular effects lead to a novel locus mediating coronary risk. *PLoS ONE* **2017**, *12*, e0182999. [CrossRef]

10. Rusu, V. Hoch type 2 diabetes variants disrupt function of SLC16A11 through two distinct mechanisms. *Cell* **2017**, *170*, 199–212. [CrossRef]

11. Maher, B. Personal genomes: The case of the missing heritability. *Nature* **2008**, *456*, 18–21. [CrossRef] [PubMed]

12. Toma, T.T.; Williams, Z.; Dawson, J.; Adjeroh, D. What Can One Chromosome Tell us about human Biogeographical Ancestry? In Proceedings of the 2017 IEEE International Conference on Bioinformatics and Biomedicine (BIBM), Kansas City, MO, USA, 13–16 November 2017; pp. 188–193.

13. He, J.; Zelikovsky, A. Informative SNP selection methods based on SNP prediction. *IEEE Trans. Nanobiosci.* **2007**, *6*, 60–67. [CrossRef] [PubMed]

14. LaFramboise, T. Single nucleotide polymorphism arrays: A decade of biological, computational and technological advances. *Nucleic Acids Res.* **2009**, *37*, 4181–4193. [CrossRef] [PubMed]

15. Kwok, P.Y.; Chen, X. Detection of single nucleotide polymorphisms. *Curr. Issues Mol. Biol.* **2003**, *5*, 43–60.

16. Talukder, A.K.; Gandham, S.; Prahalad, H.A.; Bhattacharyya, N.P. Cloud-MAQ: The Cloud-enabled Scalable Whole Genome Reference Assembly Application. In Proceedings of the 2010 Seventh International Conference on Wireless and Optical Communications Networks-(WOCN), Colombo, Sri Lanka, 25 May 2010; pp. 0–4.

17. Li, R.; Li, Y.; Fang, X.; Yang, H.; Wang, J.J.; Kristiansen, K.; Wang, J.J. SNP detection for massively parallel whole-genome resequencing-ppt. *Genome Res.* **2009**, *19*, 1124–1132. [CrossRef]

18. Cheng, Y.H.; Kuo, C.N.; Lai, C.M. Effective natural PCR-RFLP primer design for SNP genotyping using teaching-learning-based optimization with elite strategy. *IEEE Trans. Nanobioscience* **2016**, *15*, 657–665. [CrossRef]

19. Hall, J.G.; Eis, P.S.; Law, S.M.; Reynaldo, L.P.; Prudent, J.R.; Marshall, D.J.; Allawi, H.T.; Mast, A.L.; Dahlberg, J.E.; Kwiatkowski, R.W.; et al. Sensitive detection of DNA polymorphisms by the serial invasive signal amplification reaction. *Proc. Natl. Acad. Sci. USA* **2000**, *97*, 8272–8277. [CrossRef]

20. Schmalzing, D.; Belenky, A.; Novotny, M.A.; Koutny, L.; Salas-Solano, O.; El-Difrawy, S.; Adourian, A.; Matsudaira, P.; Ehrlich, D. Microchip electrophoresis: A method for high-speed SNP detection. *Nucleic Acids Res.* **2000**, *28*, E43. [CrossRef]

21. Ross, P.; Hall, L.; Smirnov, I.; Haff, L. High level multiplex genotyping by MALDI-TOF mass spectrometry. *Nat. Biotechnol.* **1998**, *16*, 1347. [CrossRef]

22. Landegren, U.; Kaiser, R.; Sanders, J.; Hood, L. A ligase-mediated gene detection technique. *Science* **1988**, *241*, 1077–1080. [CrossRef]

23. Wu, D.Y.; Wallace, R.B. The ligation amplification reaction (LAR)—Amplification of specific DNA sequences using sequential rounds of template-dependent ligation. *Genomics* **1989**, *4*, 560–569. [CrossRef]

24. Tobe, V.O.; Taylor, S.L.; Nickerson, D.A. Single-well genotyping of diallelic sequence variations by a two-color ELISA-based oligonucleotide ligation assay. *Nucleic Acids Res.* **1996**, *24*, 3728–3732. [CrossRef] [PubMed]

25. Huh, Y.S.; Lowe, A.J.; Strickland, A.D.; Batt, C.A.; Erickson, D. Surface-enhanced Raman scattering based ligase detection reaction. *J. Am. Chem. Soc.* **2009**, *131*, 2208–2213. [CrossRef] [PubMed]

26. Xu, Y.; Karalkar, N.B.; Kool, E.T. Nonenzymatic autoligation in direct three-color detection of RNA and DNA point mutations. *Nat. Biotechnol.* **2001**, *19*, 148–152. [CrossRef] [PubMed]

27. Sando, S.; Abe, H.; Kool, E.T. Quenched auto-ligating dnas: Multicolor identification of nucleic acids at single nucleotide resolution. *J. Am. Chem. Soc.* **2004**, *126*, 1081–1087. [CrossRef]

28. Wallace, R.B.; Shaffer, J.; Murphy, R.F.; Bonner, J.; Hirose, T.; Itakura, K. Hybridization of synthetic oligodeoxyribonucleotides to phi chi 174 DNA: The effect of single base pair mismatch. *Nucleic Acids Res.* **1979**, *6*, 3543–3557. [CrossRef]

29. Liu, G.; Wan, Y.; Gau, V.; Zhang, J.; Wang, L.; Song, S.; Fan, C. An enzyme-based E-DNA sensor for sequence-specific detection of femtomolar DNA targets. *J. Am. Chem. Soc.* **2008**, *130*, 6820–6825. [CrossRef]

30. Huang, Y.; Zhang, Y.-L.; Xu, X.; Jiang, J.-H.; Shen, G.-L.; Yu, R.-Q. Highly specific and sensitive electrochemical genotyping via gap ligation reaction and surface hybridization detection. *J. Am. Chem. Soc.* **2009**, *131*, 2478–2480. [CrossRef]

31. Swami, N.; Chou, C.F.; Ramamurthy, V.; Chaurey, V. Enhancing DNA hybridization kinetics through constriction-based dielectrophoresis. *Lab Chip* **2009**, *9*, 3212–3220. [CrossRef]

32. Kirmani, S.A.M.; Gudagunti, F.D.; Velmanickam, L.; Nawarathna, D.; Lima, I.T. Negative dielectrophoresis spectroscopy for rare analyte quantification in biological samples. *J. Biomed. Opt.* **2017**, *22*, 037006. [CrossRef]

33. Gudagunti, F.D.; Velmanickam, L.; Nawarathna, D.; Lima, I.T. Label-free biosensing method for the detection of a pancreatic cancer biomarker based on dielectrophoresis spectroscopy. *Chemosensors* **2018**, *6*, 33. [CrossRef]

34. Yahya, N.N.; Aziz, N.A.; Buyong, M.R.; Majlis, B.Y. Size-based particles separation utilizing dielectrophoresis technique. In Proceedings of the 2017 IEEE Regional Symposium on Micro and Nanoelectronics (RSM), Penang, Malaysia, 23–25 August 2017; pp. 10–13.

35. Pethig, R. Dielectrophoresis: Status of the theory, technology, and applications. *Biomicrofluidics* **2010**, *4*, 1–35. [CrossRef] [PubMed]

36. Viefhues, M.; Eichhorn, R. DNA dielectrophoresis: Theory and applications a review. *Electrophoresis* **2017**, *38*, 1483–1506. [CrossRef] [PubMed]

37. Khoshmanesh, K.; Nahavandi, S.; Baratchi, S.; Mitchell, A.; Kalantar-zadeh, K. Dielectrophoretic platforms for bio-microfluidic systems. *Biosens. Bioelectron.* **2011**, *26*, 1800–1814. [CrossRef]

38. Negr, F.; Cavallini, A. Effect of dielectrophoretic forces on nanoparticles. *IEEE Trans. Dielectr. Electr. Insul.* **2017**, *24*, 1708–1717. [CrossRef]

39. Altinagac, E.; Ozcan, S.S.; Genc, Y.; Kizil, H.; Trabzon, L. Biological particle manipulation: An example of Jurkat enrichment. In Proceedings of the 2015 IEEE 10th International Conference NanoMicro Engineered and Molecular System NEMS 2015, Hong Kong, 7 November 2015; Volume 10, pp. 25–27.

40. Camacho-alanis, F.; Ros, A. HHS public access. *Bioanalysis* **2015**, *7*, 353–371. [CrossRef]

41. Yafouz, B.; Kadri, N.A.; Ibrahim, F. Dielectrophoretic manipulation and separation of microparticles using microarray dot electrodes. *Sensors* **2014**, *14*, 6356–6369. [CrossRef]

42. Cui, L.; Holmes, D.; Morgan, H. The dielectrophoretic levitation and separation of latex beads in microchips. *Electrophoresis* **2001**, *22*, 3893–3901. [CrossRef]

43. Green, N.G.; Morgan, H. Dielectrophoresis of submicrometer latex spheres. 1. experimental results. *J. Phys. Chem. B* **1999**, *103*, 41–50. [CrossRef]

44. Weng, P.Y.; Chen, I.A.; Yeh, C.K.; Chen, P.Y.; Juang, J.Y. Size-dependent dielectrophoretic crossover frequency of spherical particles. *Biomicrofluidics* **2016**, *10*, 011909. [CrossRef]

45. Chu, C.-H.; Sarangadharan, I.; Regmi, A.; Chen, Y.-W.; Hsu, C.-P.; Chang, W.-H.; Lee, G.-Y.; Chyi, J.-I.; Chen, C.-C.; Shiesh, S.-C.; et al. Beyond the Debye length in high ionic strength solution: Direct protein detection with field-effect transistors (FETs) in human serum. *Sci. Rep.* **2017**, *7*, 5256. [CrossRef] [PubMed]

46. Green, N.M. Avidin. 3. The nature of the biotin-binding site. *Biochem. J.* **1963**, *89*, 599–609. [CrossRef] [PubMed]
47. Rohani, A.; Moore, J.H.; Kashatus, J.A.; Sesaki, H.; Kashatus, D.F.; Swami, N.S. Label-free quantification of intracellular mitochondrial dynamics using dielectrophoresis. *Anal. Chem.* **2017**, *89*, 5757–5764. [CrossRef] [PubMed]
48. Camacho-Alanis, F.; Ros, A. Protein dielectrophoresis and the link to dielectric properties. *Bioanalysis* **2015**, *7*, 353–371. [CrossRef] [PubMed]

 *micromachines*

Article

Numerical Study of Paramagnetic Elliptical Microparticles in Curved Channels and Uniform Magnetic Fields

Christopher Sobecki, Jie Zhang and Cheng Wang *

Department of Mechanical and Aerospace Engineering, Missouri University of Science and Technology, 400 W. 13th St., Rolla, MO 65409, USA; cas3n3@mst.edu (C.S.); jzn39@mst.edu (J.Z.)
* Correspondence: wancheng@mst.edu

Received: 30 October 2019; Accepted: 25 December 2019; Published: 28 December 2019

Abstract: We numerically investigated the dynamics of a paramagnetic elliptical particle immersed in a low Reynolds number Poiseuille flow in a curved channel and under a uniform magnetic field by direct numerical simulation. A finite element method, based on an arbitrary Lagrangian-Eulerian approach, analyzed how the channel geometry, the strength and direction of the magnetic field, and the particle shape affected the rotation and radial migration of the particle. The net radial migration of the particle was analyzed after executing a π rotation and at the exit of the curved channel with and without a magnetic field. In the absence of a magnetic field, the rotation is symmetric, but the particle-wall distance remains the same. When a magnetic field is applied, the rotation of symmetry is broken, and the particle-wall distance increases as the magnetic field strength increases. The causation of the radial migration is due to the magnetic angular velocity caused by the magnetic torque that constantly changes directions during particle transportation. This research provides a method of magnetically manipulating non-spherical particles on lab-on-a-chip devices for industrial and biological applications.

Keywords: microparticles; paramagnetic; magnetic field; direct numerical simulation; curved channel; low Reynolds number

---

1. Introduction

Applying magnetic fields to separate magnetic micro- and nanoparticles by shape immersed in a fluid is a long searched for achievement in biomedical and industrial applications, such as cell separation [1,2], drug deliverance [3,4], mining ores [5], and waste management [6]. The method of magnetic separation in these industries stems from the use of magnetophoresis, i.e., magnetic forces. The production of magnetic forces are due to the shape, size, and magnetic susceptibility of the particle and the spatial non-uniform magnetic field [7].

In previous studies, passive methods to separate particles by shape and size have been studied through inertial effects in curved channels by the primary flow (influenced by the Reynolds number) and secondary flow (influenced by the Deans number). Some of the various curved channel designs include single curved [8,9], serpentine [10–12], spiral [13–18], and wave channels [19,20]. The separation of particles by shape and size are caused by vortices due to the cross-sectional shape of the channel including trapezoidal [13,15,16,18], or rectangular [8–12,14,17,19,20], and caused by the magnitude of the Deans number, Reynolds number, and the hydraulic diameter. Additional strategies include a more active approach on particle separation or particle movement in curved channels, such as dielectrophorsis [21,22] and magnetophoresis [23].

Recent experimental [24,25], theoretical [26], and numerical [27–31] studies, however, have validated a non-traditional strategy to manipulate the dynamics of non-spherical magnetic microparticles by coupling uniform magnetic fields and shear flows in straight channels. By applying a uniform magnetic field, the magnetic force is zero, but there exists a magnetic torque. The lateral migration of non-spherical particles relies on the coupling of the magnetic field, flow field, and particle-wall hydrodynamic interactions. With the understanding of the physics involved, Zhou et al. successfully separated ellipsoidal and spherical paramagnetic microparticles [24,25]. In the absence of a magnetic field, the ellipsoidal particle oscillates towards and away from the wall but results in a zero-net migration due to its symmetric rotation. In the presence of a magnetic field applied at an arbitrary direction, the symmetry of rotation is broken and the ellipsoidal particle has a net lateral migration towards or away from the wall. The migratory behavior is based on an oscillatory motion (weak magnetic regime) or a non-oscillatory motion as the ellipsoidal particle is pinned at a stable steady angle (strong magnetic regime). For a paramagnetic spherical particle, the magnetic torque is zero and its rotation cannot be manipulated by the uniform magnetic field [26]. Therefore, a spherical particle behaves the same with or without a magnetic field and its net migration remains zero. Additionally, the Reynolds number is less than one, thus making the lateral migration of a non-spherical particle highly dependent on its rotational behavior. As a result, when an ellipsoidal and a spherical particle are immersed in a low Reynolds number fluid flow in a straight channel and under a uniform magnetic field, it is feasible to separate particles by shape rather than relying on magnetic forces (where the strength rapidly decreases further away from the magnetic source), and inertial focusing (relies on the cross-sectional vortices and equilibrium positions for different shaped and sized particles).

Although recent experimental studies provide a useful strategy to separate particles by shape, it is still difficult to manage a well-controlled experiment, especially for curved channels. If we want to conduct a particle-focused experiment, the difficulty of studying a single particle in a curved channel is based on the many turns of a serpentine channel, gradual increasing radii of spiral channels, and large radii of a single curved channel. Additionally, it is difficult to observe the particle position in the cross-section of the channel. In other words, the difficulty of a curved channel experiment is based on optics.

On the other hand, numerical simulations are used as powerful tools to observe the transportation, rotation, and radial migration behavior of a particle. The particle dynamics are dependent on many parameters including the size and shape of the particle, the height, width, and shape of the channel cross-sections, low fluid velocities, radii of the curved channel, and the strength and direction of the magnetic field. For example, Harding et al. conducted a numerical and experimental study on a spherical particle transporting in a low Reynolds number flow in a curved channel by investigating the size of the particle and the cross-sectional shape, radii, width, and height of the channel [32,33]. In the experiments given, the larger particles migrated towards the inner channel wall and the results of the numerical study, conducted by Harding and Betozzi, concurred with the experimental findings [32]. In their article, however, their Reynolds number is large (greater than 50), whereas our Reynolds number in this paper is less than one and our particle is non-spherical.

Even though there are no published articles for non-spherical particles in curved channels and under uniform magnetic fields, it is an important subject for science and engineering regarding theoretical, experimental, and numerical analyses. The popular and revolutionary work of G.B. Jeffery investigated the simple shear flow acting on the particle in the absence of a channel wall [34] and the application of a uniform magnetic field, carried out by Zhou et al., has become recognizable for the separation of particles. Today, there have been many advancements towards computer simulations to study non-spherical particles in shear flows, Poiseuille flows, and under a uniform magnetic field. Some of the recent and successful numerical simulations, to study particles in Poiseuille and Couette flows and under a uniform magnetic field, were computed by Zhang et al. and Cao et al. by applying the arbitrary Lagrangian-Eulerian (ALE) algorithm for the finite element methods (FEM) and by using the direct numerical simulation (DNS) [27,30,31]. Further simulations from this method were successfully compared with the experimental and theoretical results of a neutrally buoyant particle under a uniform magnetic field [25,26,30].

In this paper, we focus on a two-dimensional study for an elliptical particle in low Reynolds number, Poiseuille flows in a curved channel, and under a uniform magnetic field. Therefore, the elliptical particle becomes dependent on the wall lift and the shear lift forces (hydrodynamic force), and the magnetic and hydrodynamic torques (to study its orientation). In the hydrodynamic section, we analyze the dynamics of the particle in two ways: one for a π periodic rotation and the other discussing the transportation throughout the upper half of the curved channel. Due to the hydrodynamic torque and force, our numerical simulations demonstrate the rotation and migration behavior of the elliptical particle. For its rotational behavior, we observe whether or not the particle executes a symmetric rotation, and we study its angular velocity. By analyzing a single periodic rotation, we examine the particle's net radial migration and conclude the radial migration at the exit of the curved channel. We also analyze how different parameters affect the particle dynamics including: the particle aspect ratio, radii of the channel walls, and initial positions. Finally, we evaluate one of these parameters to show how the uniform magnetic field strength and direction affect the particle dynamics along with the aspect ratio of the particle, initial particle-wall distances, and the channel geometry. By applying a uniform magnetic field, we can examine the symmetrical property of the particle's angular velocity, the oscillatory motion of the particle, and its average radial velocity. We apply a DNS by using a FEM, based on an ALE approach to analyze the coupling of the magnetic and the flow field, and to solve the dynamics and the transportation of the particle and the flow field (affected by the particle dynamics) in a curved channel. The hydrodynamic force and the magnetic and hydrodynamic toques are computed by a COMSOL FEM solver (5.2a, COMSOL Inc., Burlington, MA, USA) to find the orientation and the radial displacement motions by Newton's second law of physics and Euler's laws of motion.

2. Materials and Methods

2.1. Simulation Method and Mathematical Models

We place a neutrally buoyant prolate elliptical particle in a Poiseuille flow of an incompressible Newtonian fluid, with density ρ_f and dynamic viscosity η_f, as observed in Figure 1. The computational domain consists of a particle domain Γ and a fluid domain Ω enclosed by the boundary ABCD. The computational domain is bounded in a curved channel with an average radius $R_{avg} = \frac{R_{in}+R_{out}}{2}$ where the inner and outer radii of the walls are R_{in} and R_{out}, respectively, and W is the width of the channel. The particle has an aspect ratio $AR = a/b$ where a and b are the major and minor semi-axes lengths of the particle, respectively. A uniform magnetic field strength, $\mathbf{H}_0$, is applied at a direction α. The particle-wall separation distance, r_p, is defined as the difference between the resultant length of the particle center of mass (x_p, y_p) from the origin, O, and R_{in}. The particle position in the curved channel, θ_p, is defined as the arc-tangent of the particle center of mass from the origin. In this study, $\theta_p = 0°$ indicates that the particle's center of mass is at $-R_{out} < x_p < -R_{in}$ and $y_p = 0$ μm (the channel entrance), $\theta_p = 90°$ indicates that $R_{in} < y_p < R_{out}$ and $x_p = 0$ μm (halfway point), and $\theta_p = 180°$ indicates that $R_{in} < x_p < R_{out}$ and $y_p = 0$ μm (the channel exit). The variable ϕ'_p, is defined as the lab frame angle of the particle between its major axis and the positive y'-axis. The y'-axis is perpendicular to the channel curve and is dependent on the particle θ_p position and the global frame of the particle orientation ϕ_p (the angle between its major axis and the positive y-axis):

$$\phi'_p = \phi_p - \theta_p + 90°. \tag{1}$$

The directions $\phi'_p = 0°$ and $\phi'_p = 90°$ indicate that the semi-major axis of the particle is perpendicular and parallel to the channel wall, respectively. In this study, we define the change in radial position as $\Delta r_{p\pi} = r_{p\pi} - r_{p0}$ as the difference between the particle-wall distance after one periodic rotation (ϕ'_p rotates from 0° to 180°). The variable $r_{p\pi}$ is the radial distance at the end of its periodic rotation $\phi'_p = 180°$ and r_{p0} is the initial radial position at the beginning of the periodic rotation

$\phi'_p = 0°$. We have another change in radial position as $\Delta r_{p\theta} = r_{p\theta} - r_{p0}$ where $r_{p\theta}$ is the radial position at the exit of the curved channel $\theta_p = 180°$ and r_{p0} is the radial position at the entrance of the curved channel $\theta_p = 0°$. Therefore, in our simulations the initial radial position r_{p0} happens at the orientation $\phi'_p = 0°$ and at the position in the channel $\theta_p = 0°$. The particle either has a net radial migration towards the channel center, $\Delta r_{p\pi} > 0$ µm ($\Delta r_{p\theta} > 0$ µm), towards the channel wall, $\Delta r_{p\pi} < 0$ µm ($\Delta r_{p\theta} < 0$ µm), or neither, $\Delta r_{p\pi} = 0$ µm ($\Delta r_{p\theta} = 0$ µm).

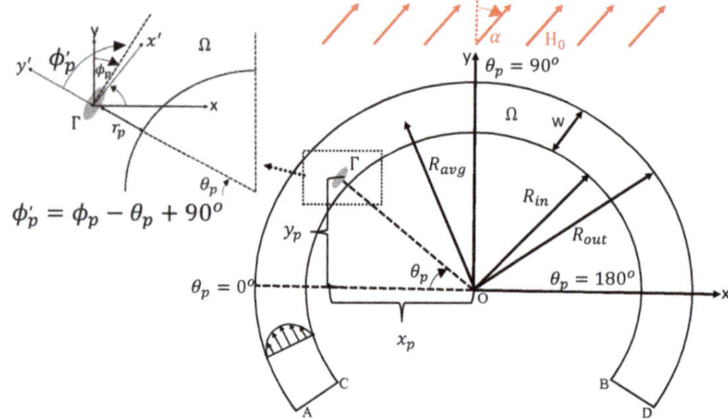

Figure 1. Schematic view of the numerical model of an elliptical particle that is neutrally buoyant in a Poiseuille flow and under a uniform magnetic field with strength H_0 and applied at direction α. The fluid and particle domains are Ω and Γ, respectively. The orientation of the particle is denoted as $\phi'_p = \phi_p - \theta_p + 90°$ such that the laboratory frame is the $x' - y'$ axes, and y' is perpendicular to the channel wall, where ϕ_p is the global direction of the particle ($x - y$ axes), and θ_p is the particle position inside the channel with respect to the origin of both curves, O. The particle-wall separation distance is denoted by r_p, and the channel geometry of the curve is denoted by the average radius $R_{avg} = \frac{R_{in}+R_{out}}{2}$, where R_{in} and R_{out} are the inner and outer radii, respectively, and W is the channel width.

The flow field, $\mathbf{u}$, is governed by the continuity and the Navier–Stokes equations for an incompressible Newtonian flow with a no-slip condition on the channel walls AD and CB. The inlet of the channel is at AC and becomes a fully developed laminar flow, and BD is the outlet where the normal pressure is zero:

$$\nabla \cdot \mathbf{u} = 0, \tag{2}$$

$$\rho_f[\frac{\partial \mathbf{u}}{\partial t} + (\mathbf{u} \cdot \nabla)\mathbf{u}] = -\nabla p + \nabla \cdot v_f(\nabla \mathbf{u} + (\nabla \mathbf{u})^T), \tag{3}$$

where p is the pressure and t is time.

Since there is also a no-slip condition on the surface of the elliptical particle, the fluid velocity on the the particle surface is a function of the particle angular and translation velocities:

$$\mathbf{u} = \mathbf{U}_p + \omega_p \times (\mathbf{x}_s - \mathbf{x}_p), \tag{4}$$

where $\mathbf{U}_p$ and ω_p are the translational and rotational velocities of the particle, respectively, and $\mathbf{x}_s$ and $\mathbf{x}_p$ are the vector positions of the particle surface and center of mass.

The hydrodynamic force and torque acting on the particle are also given:

$$\mathbf{F}_h = \int (\tau_h \cdot \mathbf{n})dS, \tag{5}$$

$$\mathbf{T}_h = \int (\tau_h \times (\mathbf{x}_s - \mathbf{x}_p) \cdot \mathbf{n})dS, \tag{6}$$

where $\tau_h = \eta_f(\nabla \mathbf{u} + (\nabla \mathbf{u})^T)$ is the hydrodynamic stress tensor. The governing equations of the magnetic field are given from Maxwell's equations:

$$\nabla \times \mathbf{H} = 0, \tag{7}$$

$$\nabla \cdot \mathbf{B} = 0, \tag{8}$$

where $\mathbf{H}$ and $\mathbf{B}$ are the magnetic field strength and flux density, respectively. It should be mentioned that the fluid is non-magnetic and the channel walls, AD and CB, are magnetically insulated. When a uniform magnetic field is applied onto a particle, the magnetic torque is non-zero but the magnetic force is negligible. When calculating the magnetic torque, we assume that the paramagnetic particle is homogeneous, isotropic, and linearly magnetizable:

$$\mathbf{T}_m = \mu_0 V_p \chi_p \mathbf{H}^- \times \mathbf{H}_0, \tag{9}$$

where μ_0 is the magnetic permeability of free space, χ_p is the paramagnetic particle magnetic susceptibility, V_p is the particle volume, and the magnetic field strengths are inside, $\mathbf{H}^-$, and outside, $\mathbf{H}_0$, of the particle.

For the hydrodynamic force and the hydrodynamic and magnetic torques acting on the particle, we have a set of equations:

$$m_p \frac{d\mathbf{U}_p}{dt} = \mathbf{F}_h, \tag{10}$$

$$\mathbf{I}_p \frac{d\omega_p}{dt} = \mathbf{T}_h + \mathbf{T}_m, \tag{11}$$

where m_p and $\mathbf{I}_p$ are the mass and the moment of inertia of the particle. The time-dependent position of the particle center of mass $C_p(t) = (x_p, y_p)$ and orientation ϕ_p are calculated by:

$$\mathbf{C}_p(t) = \mathbf{C}_p(0) + \int_0^t \mathbf{U}_p(s)ds, \tag{12}$$

$$\phi_p(t) = \phi_p(0) + \int_0^t \omega_p(s)ds, \tag{13}$$

where $\mathbf{C}_p(0)$ and $\phi_p(0)$ are the initial position and orientation of the particle.

The rotation and radial migration of the paramagnetic particle are affected by the hydrodynamic force and the coupling of hydrodynamic and magnetic torques. Likewise, every channel, radial, and orientation positions of the particle will cause a change in the magnetic and hydrodynamic torques and the fluid flow. Given the geometry of the channel and the calculations that include the radial position and orientation of the particle, we use direct numerical simulation (DNS) from the finite element method (FEM), and an arbitrary Lagrangian-Eulerian (ALE) method for the coupling on the particle, fluid flow, and the uniform magnetic field [27,30]. Our simulations were solved by numerical modeling using a commercial FEM solver COMSOL Multiphysics. Similar to previous articles, we use a stationary solver for parametric sweep analysis to simulate the magnetic field inside and outside of the particle, and calculate the magnetic torque on the particle [27,30]. We then apply a time-dependent solver for a particle-fluid interaction model, and we imported a variable that represents the magnetic torque. We use a piecewise function to activate the uniform magnetic field and the magnetic torque on the particle at a time, t, when $\theta_p \approx 0°$. To estimate an accurate calculation on the particle surface by the torques and the hydrodynamic force, we use a fine quadratic triangular mesh around the particle and a finer quadratic triangular mesh at the tip of the particle.

The accuracy and convergence of our simulation is based on the number of elements in the computational domain Ω, particle surface Γ, and the time step Δt. As part of our numerical simulation setup in Figure 1, we use 18,571 domain elements for Ω, and 152 boundary elements for Γ. The number of elements establishes time efficient calculations, while also ensuring accurate results of the particle transportation and rotation. For the fluid mechanics acting on the particle surface, we use the time-step function to have the Poiseuille flow to reach its average velocity and become fully developed before the particle approaches $\theta_p = 0°$ with a time step $\Delta t = 1 \times 10^{-5}$ s. When the particle position is $\theta_p \approx 0°$, we set $\phi'_{p0} \approx 0°$ since the semi-major axis of the particle is almost perpendicular to the channel wall. Additionally, we start our new initial radial position, r_{p0}, and our initial time at $t = 0$ s. Comparisons between the computational meshes for time step $\Delta t = 1 \times 10^{-5}$ s and comparisons between time steps for 18,571 domain elements and 152 boundary elements can be seen in Appendix B.

2.2. Material Properties

In the following computed simulations, the fluid property is a water-based material with a density of 1000 kg/m^3 and the dynamic viscosity is 1.002×10^{-3} Pa·s, the width of the channel is kept at 50 μm, and the inlet flow velocity is $U_{avg} = 2.5$ mm/s. The Reynolds number is defined as $R_e = \frac{\rho_f U_{avg} W}{\eta_f}$, and the value in the simulation setup is $R_e = 0.125$, thus placing the elliptical particle in a laminar flow where the fluid and particle inertia are small. Due to the low Reynolds number, we assume that any secondary flow is neglected for a two-dimensional study on a particle. The particle is a magnetic-doped polystyrene particle with a magnetic susceptibility of $\chi_p = 0.26$, and the fluid is non-magnetic. For some parts of our analysis, we will analyze elliptical particles with various aspect ratios, but they will have the same volume as a 7 μm-diameter circular particle.

3. Results and Discussion

3.1. Particle Dynamics in a Uniform Magnetic Field

In this section, we investigate the transportation and rotation of an elliptical particle in the presence of a uniform magnetic field near the inner wall whereas an elliptical particle near the outer wall is discussed in Appendix A. We keep $AR = 4$, $r_{p0} = 12$ μm, $R_{avg} = 175$ μm, and the magnetic field is applied at either $\alpha = 0°$ or $\alpha = 90°$. In our numerical simulation, we activate the magnetic field by using a piecewise function at a time when $\theta_p \approx 0°$. At this point, we establish our new initials (ϕ'_{p0} and r_{p0}) and set $t = 0$ s. We have compared some of our results to an elliptical particle in the absence of a magnetic field observed in Appendix C.

3.1.1. Magnetic Field at $\alpha = 0°$

In this section, we apply different magnetic field strengths at $\alpha = 0°$ to study the particle rotation and the particle-wall distance as the particle is transporting in a curved channel. Figure 2a,b, each exhibiting the particle's rotation and radial migration, respectively, as a function of θ_p, shows the comparison between them and $H_0 = 0$ A/m and $H_0 = 3000$ A/m. For both magnetic field strengths, we observe where in the channel the particle completes the first half of its rotation. At $H_0 = 0$ A/m, the first half rotation of a particle ($\phi'_p = 90°$) occurs in the first half of the curved channel ($\theta_p = 74°$). On the other hand, when $\alpha = 0°$ and $H_0 = 3000$ A/m, the particle approaches the first half of its rotation in the second half of the curve ($\theta_p = 122°$), indicating that the magnetic field affects the orientation of the particle during its transportation. We reintroduce the dimensionless variable τ as a function of the particle aspect ratio, radii and shape of the curve geometry, initial particle-wall distance, direction of the particle, and the direction and strength of the magnetic field. We see that in Figure 2d, a particle in the absence of a magnetic field executes symmetric rotations ($\tau = 0.50$), while a particle in a magnetic field experiences an asymmetric rotation ($\tau = 0.76$). We use the rotation of the particle to analyze the net radial migration of the particle. Due to the asymmetric rotation, the particle radially migrates toward the channel center after one periodic rotation ($\Delta r_{p\pi} > 0$ μm), seen in Figure 2e, and at the channel exit ($\Delta r_{p\theta} > 0$ μm), seen in Figure 2b, for the magnetic field strength $H_0 = 3000$ A/m applied at $\alpha = 0°$.

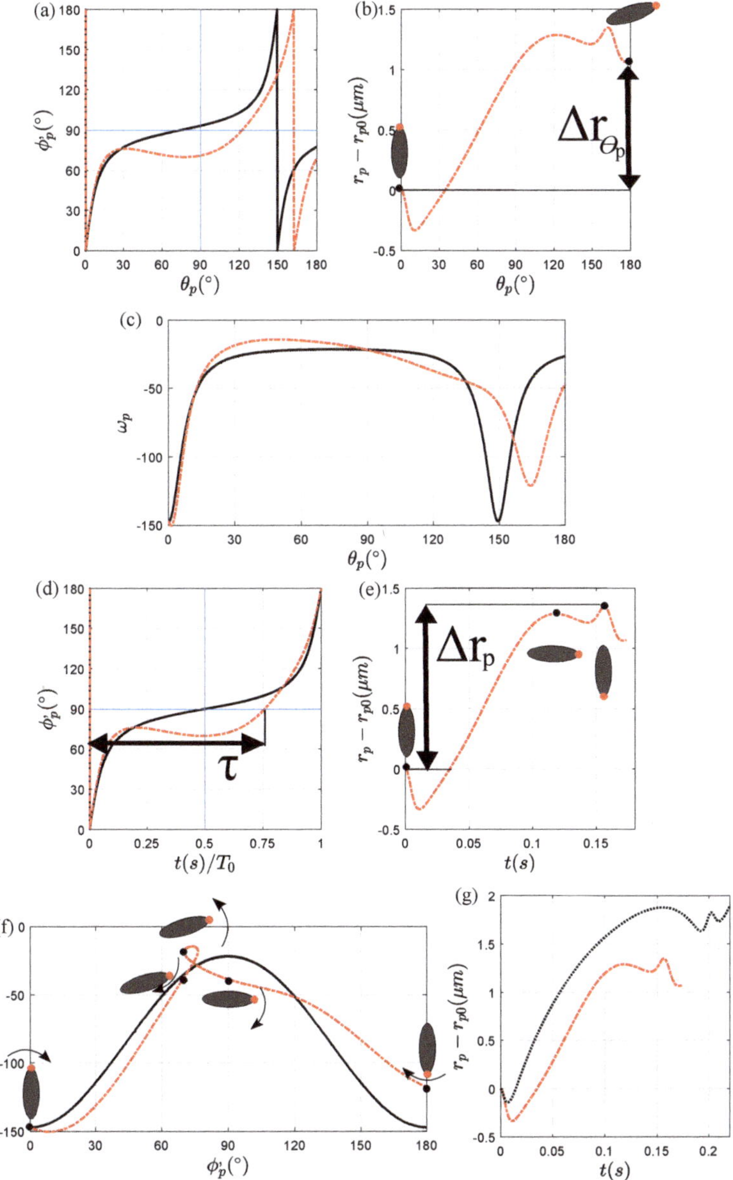

Figure 2. Particle dynamic and transportation comparison between $H_0 = 0$ A/m (solid black line), and $H_0 = 3000$ A/m applied at $\alpha = 0°$ (dot-dash red line) for (**a**) the particle's rotation, (**b**) the radial migration, and (**c**) the angular velocity in terms of θ_p; (**d**) the particle's rotation, (**e**) the radial migration, (**f**) the angular velocity in terms of time t, and (**g**) comparison between the particle-wall distances in a straight channel (dot black line) and in a curved channel (dot-dash red line).

The observation of an elliptical particle's dynamics under a magnetic field applied at $\alpha = 0°$ can thus be explained. In the absence of a magnetic field, the particle rotates due to the hydrodynamic torque. The particle, however, has a net zero migration after one periodic rotation, but migrates towards the channel wall at $\theta_p = 180°$, as shown in Appendix C. In Figure 2a, we see that the applied magnetic field strength and direction rotates the particle backwards towards $\phi'_p = 0°$ due to the direction and magnitude of the shear rate at the particle position inside the channel and its distance from the wall. In the range $31° \leq \theta_p \leq 78°$, the particle exposed to $H_0 = 3000$ A/m rotates toward $\phi'_p = 0°$ because the magnetic torque is stronger than the hydrodynamic torque (i.e., the particle is in a strong field regime) and the orientation of the particle at that θ_p range approaches towards its stable steady angle [25–27,30]. The rotation of the particle over time is thus caused by the hydrodynamic and magnetic angular velocities, shown in Figure 2c,f. The magnetic angular velocity, ω_m, is in the opposite direction of the hydrodynamic angular velocity, ω_h, in the first half of the particle orientation $0° < \phi'_p < 90°$ and is in the same direction for $90° < \phi'_p < 180°$. As a result, the particle will spend a longer time in the first half of its rotation compared to its second half (i.e., the particle rotation becomes more asymmetric). Thus, at a certain position θ_p, the particle orientation allows the particle to radially migrate away from the wall faster due to the rotational dynamics affected by the magnetic field and the wall lift force caused by the hydrodynamic interaction. Otherwise, the particle continues its periodic rotation in other parts of the channel since the magnetic field is either considered weak compared to the hydrodynamic angular velocity or both angular velocities are in the same direction.

The result of the particle net radial migration after one periodic rotation and at the exit of the curved channel can be seen in Figure 2b,e. Compared to a particle in the absence of a magnetic field, the particle exposed to a magnetic field strength $H_0 = 3000$ A/m migrates toward the channel center, $\Delta r_{p\pi} > 0$ µm. In this case, the oscillatory migration is positive for both ranges $0° < \phi'_p < 90°$ and $90° < \phi'_p < 180°$. For the particle transportation throughout the channel curve under a uniform magnetic field, the net radial migration results in $\Delta r_{p\theta} > 0$ µm compared to a particle in the absence of a magnetic field in Appendix C, Figure A5.

To further understand the particle angular velocity and its net radial migration, we must also understand its time-dependent periodic rotation and its rotation during the transportation throughout the curved channel. Figure 2d,e shows the particle rotation and radial migration with respect to dimensionless and dimensional time, respectively. In Figure 2d, when $H_0 = 0$ A/m, $\tau = 0.50$ ($T_0 = 0.15$ s) and is almost symmetric, whereas the magnetic field strength $H_0 = 3000$ A/m allows $\tau = 0.76$ ($T_0 = 0.16$ s), thus making the particle rotation more asymmetric. In Figure 2e, we see that increasing τ will also increase the net radial migration towards the channel center due to the magnetic field and the magnitude of the shear rate at the particle-wall distance. Since the shear rate decreases closer to the channel center, and the oscillatory motion is caused by the coupling between the magnetic field and the shear rate, a greater asymmetric rotation causes the particle to radially migrate further. Therefore, by adding the magnetic field at $\alpha = 0°$, the rotation affected by T_0, and τ results in an increase in the particle-wall distance.

The magnetic field's influence on the particle's rotation and radial migration is based on the magnetic and hydrodynamic torques, and consequentially, their angular velocities. We see in Figure 2f that the total particle angular velocity $\omega_p = \omega_h + \omega_m$ is a function of the particle direction ϕ'_p for one periodic rotation in the curved channel. When the magnetic field is applied at $\alpha = 0°$, there exists a 'loop' for the orientation $70° < \phi'_p < 76°$ in the range $31° \leq \theta_p \leq 78°$. In this case, the particle rotates backward because the magnetic torque is changing directions, as well as the hydrodynamic torque. By considering the particle channel position θ_p, its radial position r_p, and its orientation ϕ'_p, the magnetic angular velocity is greater than the hydrodynamic angular velocity. In Figure 2c, we compare the angular velocities between a particle with and without a magnetic field during its transportation. While a particle is executing is periodic rotation, the angular velocity profile is symmetric in the absence of a magnetic field. After the periodic rotation, the particle will continue to execute symmetric rotations due to the shear rate at the particle-wall distance. In the presence of a magnetic field, however, the angular velocity is largely asymmetric and the angular velocity at $\theta_p = 162.2°$ (when $\phi'_p = 180°$) has increased. Thus, the disproportionate angular velocity and the asymmetric periodic rotation contributes to the particle net radial migration.

We finally compare the particle-wall distances between a particle in a straight channel and a particle in a curved channel, as seen in Figure 2g. In both simulations, all conditions are the same except one channel is curved, while the other channel is straight. As we can see, we expect different results between a particle in a straight channel and a particle in a curved channel because θ_p is considered constant for a straight channel ($\theta_p = 90°$), whereas θ_p constantly changes throughout the curved channel. Therefore, since the magnetic torque constantly changes directions in a curved channel, the particle's rotational dynamics will be affected and, consequentially, so will its particle-wall distance. We notice that after one periodic rotate, both particles will migrate away from the wall, but a particle in a straight channel will migrate faster from the wall than a particle in a curved channel.

Now that we have established how the particle rotation and migration is affected by applying a magnetic field, we analyze how increasing the magnetic field strength affects the orientation and radial migration of an elliptical particle. Figure 3a,d shows the orientation of the particle, ϕ'_p, with respect to θ_p and time t, respectively. In Figure 3a, as the magnetic field strength increases from $H_0 = 1000$ A/m to $H_0 = 4000$ A/m, the rotation of the particle becomes more asymmetric, and the magnetic torque and angular velocity becomes more dominant for a wider range of θ_p. For a magnetic field strength $H_0 = 2000$ A/m, a backward orientation occurs in the range $52° \leq \theta_p \leq 61°$, $31° \leq \theta_p \leq 78°$ for $H_0 = 3000$ A/m, and $24° \leq \theta_p \leq 88°$ for $H_0 = 4000$ A/m. The particle is therefore exposed to a strong field regime for a larger portion of the curved channel. In Figure 3d, the periodic rotation time, T_0, increases. For an increasing magnetic field strength, the periodic rotation times are $T_0 = 0.147$ s for $H_0 = 0$ A/m, $T_0 = 0.154$ s for $H_0 = 1000$ A/m, $T_0 = 0.157$ s for $H_0 = 2000$ A/m, $T_0 = 0.156$ s for $H_0 = 3000$ A/m, $T_0 = 0.160$ s for $H_0 = 4000$ A/m. Consequentially, τ increases in (f) where the orientation becomes more asymmetrical. As a result, for an increasing magnetic field strength, both $\Delta r_{p\pi}$ and $\Delta r_{p\theta}$ increase as seen in Figure 3b,d, respectively. Similar to previous studies, we study the particle net radial migration by calculating the average radial migration velocities, $U_{r_{p\pi}} = \Delta r_{p\pi}/t$ and $U_{r_{p\theta}} = \Delta r_{p\theta}/t$, where t is replaced with T_0 or the time that it takes for the particle to transport through the channel [27,30]. The average radial velocities for different magnetic fields strengths are shown in Figure 3c for $U_{r_{p\pi}}$ (triangle symbol) and $U_{r_{p\theta}}$ (square symbol). We see that both radial velocities increase as the magnetic field strength increases, suggesting that the particle migrates faster from the inner wall due to the angular oscillatory motion in lower shear rates.

The rotation and migration behavior of an elliptical particle exposed to a magnetic field applied at $\alpha = 0°$ can be explained as follows. While a particle is traveling in the curved channel, the shear rate is constantly changing directions at every θ_p. Therefore, if we apply a magnetic field at $\alpha = 0°$, the magnetic torque acting on the particle also changes its direction and the particle has a different rotation and migration experience during its transportation. For example, when the position of the particle is at $\theta_p = 90°$, it experiences a magnetic torque where the magnetic field is applied at $\alpha = 0°$ because the magnetic torque is perpendicular to the shear flow. On the other hand, a particle at $\theta_p = 0°$ experiences a magnetic torque similar to the magnetic field applied at $\alpha = 90°$ because the magnetic torque is parallel to the shear flow. Therefore, in some ranges of θ_p, the particle experiences a backwards orientation because the particle is considered to be inside a strong magnetic field regime [25–27,30]. In all other regions, however, the particle is either in a weak field regime or the magnetic and hydrodynamic angular velocities are in the same direction. Thus, the effect on the particle orientation depends on the direction of the hydrodynamic and magnetic torques and the strength and direction of the magnetic field. It should also be mentioned that the particle orientation is influenced by its radial position; its orientation can be easily influenced if the radial position of the particle is in a low shear rate (closer to the channel center) versus a high shear rate (closer to the channel wall). After a full periodic rotation, some magnetic field strengths allows the particle to oscillate away from the wall due the channel geometry, the orientation of the particle, and the shear flow direction. Thus, by controlling the orientation of the particle, we also control its net radial migration and $\Delta r_{p\theta} > 0$ μm.

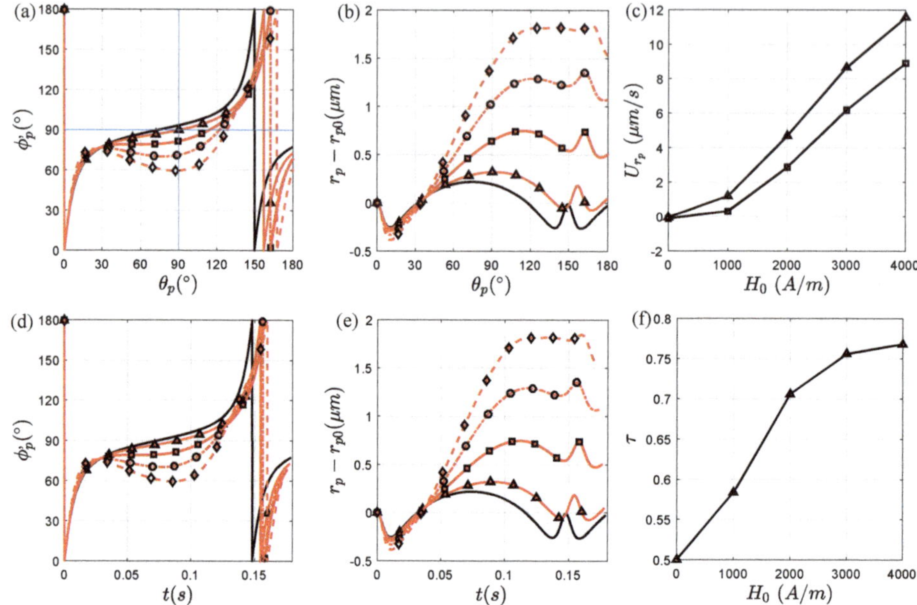

Figure 3. Effect of the magnetic field strength when it is applied at $\alpha = 0°$. The particle ($AR = 4$) is affected by the magnetic field when its center of mass is approximately at $\theta_p = 0°$ in (**a**,**d**) when its rotation is with respect to θ_p and time, respectively, for, $H_0 = 0$ A/m (solid black line), $H_0 = 1000$ A/m (triangle symbol), $H_0 = 2000$ A/m (square symbol), $H_0 = 3000$ A/m (circular symbol), and $H_0 = 4000$ A/m (diamond symbol). The radial particle-wall distance change of the particle with respect to (**b**) θ_p, and (**e**) time. (**c**) The radial velocities of the particle U_{rp} (triangle symbol) and $U_{r\theta}$ (square symbol) as functions of H_0. (**f**) The dimensionless parameter τ as a function of H_0. We see that as τ increases, the net migration of the particle increases.

3.1.2. Magnetic Field at $\alpha = 90°$

In this section, we apply a magnetic field at $\alpha = 90°$ to study the particle dynamics in a curved channel. Figure 4a,b, showing the particle orientation and radial migration, respectively, as a function of θ_p, compares them with the magnetic field strengths $H_0 = 0$ A/m and $H_0 = 3000$ A/m. The first half rotation of the particle, $\phi'_p = 90°$, occurs at $\theta_p = 74°$ for $H_0 = 0$ A/m and at $\theta_p = 42°$ for $H_0 = 3000$ A/m. One periodic rotation of the particle results in a positive net radial migration, as shown in Figure 4e. Similar to the magnetic field at $\alpha = 0°$, Figure 4d shows that, in the absence of a magnetic field, a particle executes a symmetric rotation ($\tau = 0.50$), while a particle exposed to a magnetic field strength of $H_0 = 3000$ A/m experiences a slight asymmetric rotation ($\tau = 0.53$). The reason for the particle periodic rotation and radial migration can be explained in a way similar to our analysis of $\alpha = 0°$. We see that in Figure 4f, for a range of θ_p, the magnetic angular velocity, ω_m, and the hydrodynamic angular velocity, ω_h, are in an opposite direction in the first half of the particle orientation and performs in the same direction in the second half of its orientation. We also notice that unlike $\alpha = 0°$, there does not exist a 'loop' during its periodic rotation even though the orientation is asymmetric with respect to $\phi_p = 90°$. For this range of θ_p, the particle will be considered to be in a weak field regime, but the particle spends more time in the first half of its rotation and therefore results in a positive radial migration.

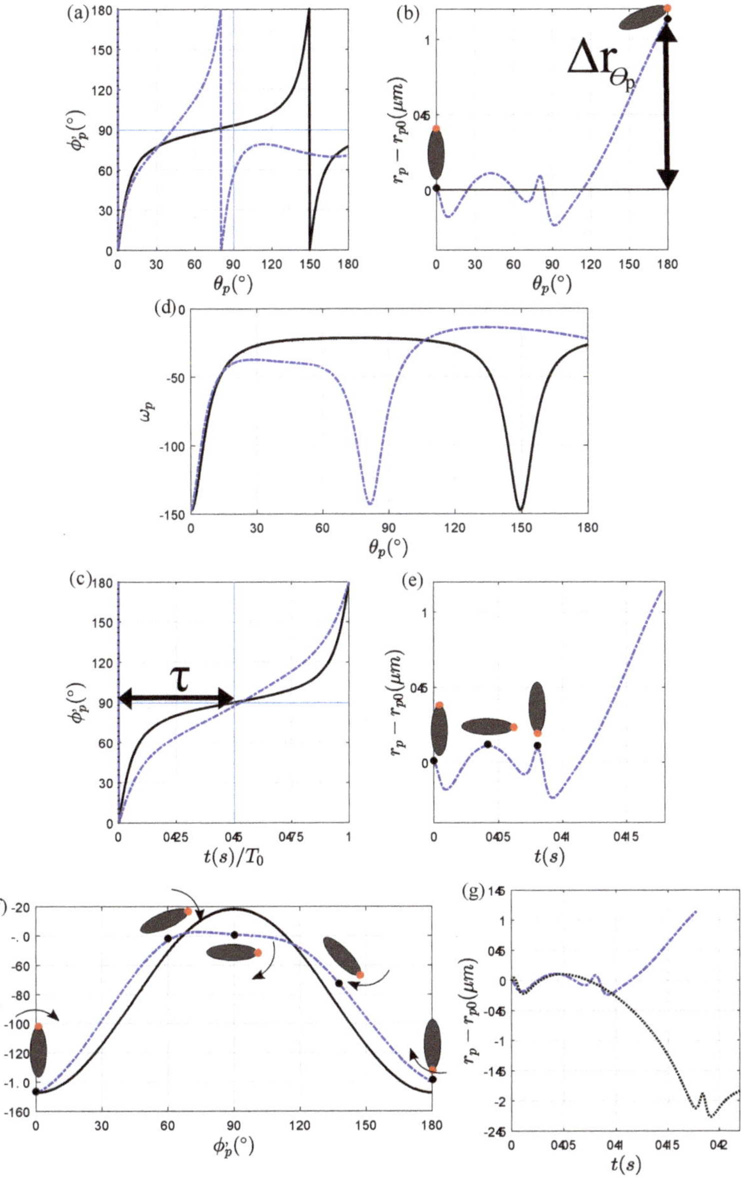

Figure 4. Particle dynamic and transportation comparison between $H_0 = 0$ A/m (solid black line), and $H_0 = 3000$ A/m applied at $\alpha = 90°$ (dot-dash blue line) for (**a**) the particle's rotation, (**b**) the radial migration, and (**c**) the angular velocity in terms of θ_p; (**d**) the particle's rotation, (**e**) the radial migration, (**f**) the angular velocity in terms of time t, and (**g**) comparison between the particle-wall distances in a straight channel (dot black line) and in a curved channel (dot-dash blue line).

The large net radial migration towards the channel center as the particle exits the curved channel, $\Delta r_{p\theta} > 0$ µm, can be explained by the particle transportation. Even though $\Delta r_{p\pi} = 0$ µm and $\Delta r_{p\theta} < 0$ µm in the absence of the magnetic field, we observe that the particle in a magnetic field strength of $H_0 = 3000$ A/m and at a direction of 90° behaves similarly as a particle exposed to a magnetic field strength of $H_0 = 3000$ A/m applied at $\alpha = 0°$. In Figure 4a, we see that the magnetic field strength of $H_0 = 3000$ A/m applied at $\alpha = 90°$ does not result in a backwards orientation in the same region as $\alpha = 0°$ due to the direction and magnitude of the shear rate in the range $31° < \theta_p < 78°$, r_p, and the strength and direction of the magnetic field. The particle, however, rotates backwards in the range $113° < \theta_p < 168°$ in Figure 4a,c. Compared to $\alpha = 0°$, the region where the particle ends up in a backwards orientation increases by $81° < \theta_p < 91°$. In Figure 4c, we compare the angular velocities between a particle with and without a magnetic field as it is transporting inside the curved channel. The angular velocity profile is symmetric in the absence of a magnetic field, whereas a magnetic field causes the angular velocity to become asymmetric in a magnetic field strength of $H_0 = 3000$ A/m applied at $\alpha = 90°$. For a certain position θ_p, the particle orientation ϕ_p' allows a positive net radial migration away from the wall faster due to the rotational dynamics affected by the magnetic field and the wall lift force. The particle is able to execute a periodic rotation in most parts of the channel given that either the magnetic torque is considered to be weak compared to the hydrodynamic torque or the magnetic and hydrodynamic torque are in the same direction. The result of the net radial migration throughout the curved channel can be seen in Figure 4b. Compared to a particle in the absence of a magnetic field, the particle exposed to a magnetic field strength of $H_0 = 3000$ A/m will result in a positive radial migration away from the inner channel wall, $\Delta r_{p\theta} > 0$ μm.

Similar to what we did in Section 3.1.1, we compare the particle-wall distance between a particle in a straight channel and a particle in a curved channel as seen in Figure 4g. We again see that since the magnetic torque constantly changes directions in a curved channel, the particle's rotational dynamics will be affected and consequentially, so will its particle-wall distance. We see that after one periodic rotate, a particle in a straight channel will migrate towards the wall, whereas a particle in a curved channel will radially migrate away from the wall. Therefore, we can conclude that the particle-wall distance will be greatly affected not only by different magnetic field strengths and directions, but also in a different channel geometry. Additionally, for magnetic field directions $\alpha = 0°$ and $\alpha = 90°$, and for the magnetic field strength $H_0 = 3000$ A/m, a particle in a curved channel will complete its periodic rotation faster than in a straight channel, resulting in a short particle-wall distance change in a curved channel than in a straight channel.

Now that we have analyzed how a magnetic field applied at $\alpha = 90°$ affects the particle orientation and migration, we evaluate how increasing the magnetic field strength affects the orientation and net radial migration of the particle. Figure 5a,d shows the orientation of the particle, ϕ_p', with respect to θ_p and time t, respectively. In Figure 5a, as the magnetic field strength increases from $H_0 = 1000$ A/m to $H_0 = 4000$ A/m the particle orientation becomes more asymmetric and the magnetic torque and angular velocity become more dominant for a wider range of θ_p near the channel exit. The particle is therefore exposed to a strong field regime for a large portion of the second half of the curved channel. In Figure 5d, the periodic rotation time T_0 decreases and, consequentially, τ increases in Figure 5f. The particle-wall distance also increases after one periodic rotation and upon exiting the channel curve in Figure 5b,e as a result of increasing H_0. In Figure 5c, the average radial velocities, $U_{r_{p\pi}}$ (triangle symbol) and $U_{r_{p\theta}}$ (square symbol), increase as the magnetic field strength increases, resulting in a fast radial migration away from the wall.

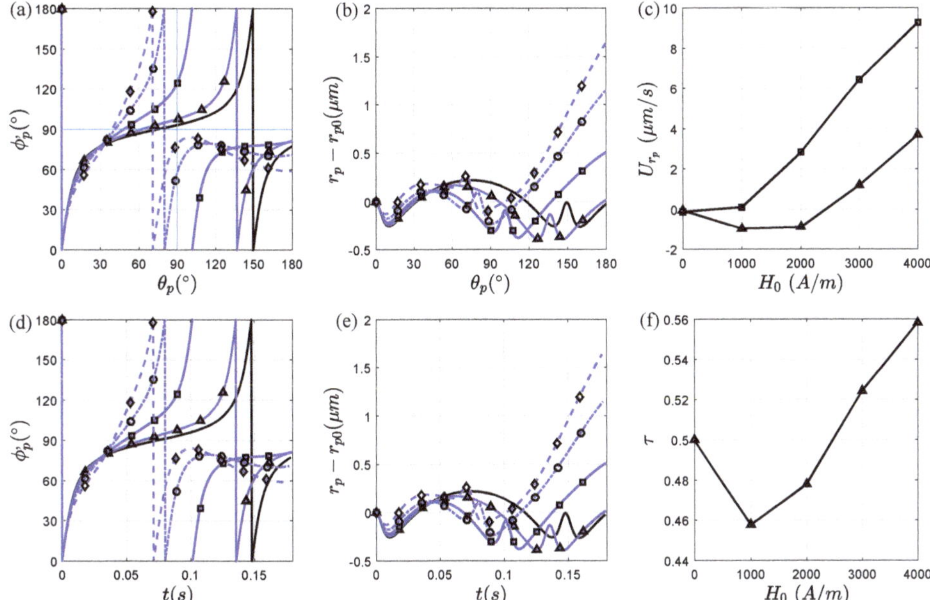

Figure 5. Effect of the magnetic field strength when it is applied at $\alpha = 90°$. The particle ($AR = 4$) is affected by the magnetic field when its center of mass is approximately at $\theta_p = 0°$ in (**a,d**) when its rotation is with respect to θ_p and time, respectively, for, $H_0 = 0$ A/m (solid black line), $H_0 = 1000$ A/m (triangle symbol), $H_0 = 2000$ A/m (square symbol), $H_0 = 3000$ A/m (circular symbol), $H_0 = 4000$ A/m (diamond symbol). The radial particle-wall distance change of the particle with respect to (**b**) θ_p, and (**e**) time. (**c**) The radial velocities of the particle U_{rp} (triangle symbol) and $U_{r\theta}$ (square symbol) as functions of H_0. (**f**) The dimensionless parameter τ as a function of H_0. We see that as τ increases, the net migration of the particle increases.

3.2. Effect Due to Changes in Parameters

By changing the aspect ratio, average radius, and/or initial position, the elliptical particle experiences a different orientation and net radial migration, as shown in Appendix C. For these parameters, we can explain what can occur in the presence of a magnetic field based on our observations for $AR = 4$, $R_{avg} = 175$ μm, and $r_{p0} = 12$ μm in Sections 3.1.1 and 3.1.2.

In the absence of a magnetic field, for larger aspect ratios, T_0 increases and the particle results in a zero net migration for all aspect ratios. Based on past analyses, for a magnetic field strength applied at $\alpha = 0°$ and $\alpha = 90°$, the smaller aspect ratio particles are more susceptible to the magnetic field for the same shear rate [25–27,30]. Therefore, if the magnetic torque and angular velocity influences on a similar particle is greater than the hydrodynamic torque and angular velocity (for a large range of θ_p), then T_0 decreases and τ increases [25–27,30]. As a result, the particles radially migrates toward the channel center.

Next, when the aspect ratio is the same but the average radius increases, T_0 decreases, but $\Delta r_{p\theta}$ increases. When we apply a magnetic field, the value of τ results either in a net positive or negative radial migration and can change, depending on the average radius of the channel. Therefore, for an increasing R_{avg}, a particle with an aspect ratio $AR = 4$ is more susceptible to the magnetic field and has a larger net migration. For smaller R_{avg}, the particle transportation is faster and T_0 increases and thus does not react to a uniform magnetic field. For a larger R_{avg}, the particle is more susceptible to the magnetic field because the particle spends more time in the curved channel and T_0 decreases. Even though T_0 decreased for an increasing R_{avg}, a magnetic field can manipulate the orientation of the particle and its net migration efficiently by increasing the magnetic field strength and τ.

Finally, as r_{p0} increases, T_0 increases since the shear rate is lower towards the channel center. Therefore, it is obvious that since the magnetic field breaks the rotation of symmetry and increases T_0 and τ, the particle orientation is easily manipulated for a larger range of θ_p. By controlling the orientation in the range $0° < \phi'_p < 90°$, the particle migrates toward the channel center or toward the inner channel wall in the range $90° < \phi'_p < 180°$.

4. Conclusions

We numerically investigated the orientation and radial migration of a paramagnetic elliptical particle in a Poiseuille flow in a curved channel and under a uniform magnetic field. We accomplished our numerical computations by using a multi-physics numerical model based on a direct numerical simulation and a finite element method that uses an arbitrary Lagrangian-Eulerian approach. Simulations were used to analyze how the geometry of the channel, particle-wall distance, shear rate, the shape of the particle, and the strength and direction of the magnetic field affect the orientation and net radial migration of the particle. The physics in this paper involve the hydrodynamic force (including wall lift and shear lift forces) and the coupling of the hydrodynamic and magnetic torques.

In the absence of a magnetic field, the particle executes symmetric periodic rotations and results in a zero net migration. The particle dynamics were analyzed by varying three parameters: particle aspect ratio, average channel radius, and initial position. We concluded that the elliptical particle rotates and radially oscillates toward and away from the wall but at different rates.

In the presence of a uniform magnetic field, the periodic rotation of the particle becomes asymmetric, and its angular velocity is modified due to the magnetic torque acting on the particle. As a result, the particle radially migrates toward the channel center since the particle rotates backward in different parts of the channel where the magnetic field is stronger. The particle rotates backward in the first half of the curved channel when the magnetic field is applied at the direction $\alpha = 0°$, whereas a magnetic field applied at $\alpha = 90°$ results in a backward orientation in the second half of the channel. Furthermore, for an increasing magnetic field strength, the magnetic manipulation on an elliptical particle becomes more prominent for a larger range inside the curved channel, resulting in larger positive radial migrations and velocities.

Different from the analyses of a spherical particle in a low Reynolds number flow in a curved channel by Harding et al. [32,33], a spherical particle, in this case, results in a zero net migration. Therefore, a shape-based separation is feasible by applying a uniform magnetic field. Additionally, we can use previous studies on particle separation in straight channels under uniform magnetic fields, in order to combine both channel geometries for a lab-on-a-chip design. This investigation gives researchers another method for an effective particle shape separation technique for industrial and biological applications.

Author Contributions: Conceptualization, C.S.; Methodology, C.S. and J.Z.; Software, C.S. and J.Z.; Formal Analysis, C.S.; Investigation, C.S.; Data Curation, C.S.; Writing–Original Draft Preparation, C.S.; Writing–Review and Editing, C.S.; Visualization, C.S.; Supervision, C.W.; Project Administration, C.S.; Resources, C.S.

Funding: This research received no external funding.

Acknowledgments: The first author gratefully acknowledges the financial support from the Chancellor's Distinguished Fellowship at Missouri University of Science and Technology.

Appendix A. Particle in Upper Curve of the Channel

Additional simulations were computed for a particle transporting near the outer curve of the channel ($R_{avg} < \|(x_p, y_p)\| < R_{out}$). In Figure A1, the effects on the particle dynamics in the absence of a magnetic field and under a uniform magnetic field applied at $\alpha = 0°$ can be seen, whereas Figure A2 shows a magnetic field applied at $\alpha = 90°$. In both figures, we investigate the radial migrations, $\Delta r_{p\pi}$ and $\Delta r_{p\theta}$, in terms of (a) θ_p, (c) t, and the (b) radial velocities, $U_{r\pi}$ and $U_{r\theta}$, and (d) τ in terms of H_0. We see in Figures A1a,c and A2a,c that, in the absence of a magnetic field, there is no radial migration for $\Delta r_{p\pi} = 0$ μm but $\Delta r_{p\theta} > 0$ μm, with radial velocities close to zero and $\tau = 0.5$. Similar to a particle near the inner curve, even though we see a different velocity profile for a fluid in a curved channel than in a straight channel, as seen in Figure A5e, the particle results in a zero net migration after one periodic rotation. From our data, when $\tau = 0.5$, a particle transporting near the lower curve will experience $\Delta r_{p\pi} = 0$ μm and $\Delta r_{p\theta} < 0$ μm, $\Delta r_{p\pi} = 0$ μm and $\Delta r_{p\theta} = 0$ μm for a particle transporting in a straight channel, and $\Delta r_{p\pi} = 0$ μm and $\Delta r_{p\theta} > 0$ μm for a particle transporting near the upper curve.

The effects on a particle radial migration by a magnetic field applied at $\alpha = 0°$ can be seen in Figure A1. In some cases, the particles complete more than one periodic rotation but their net radial migrations, τ values, and radial velocities are different. As an example, we look at the periodic rotations of the particle for the magnetic field strength $H_0 = 4000$ A/m (diamond symbol) in Figure A1a,c. We see that the first periodic rotation results in a radial migration towards the channel wall but then migrates toward the channel center in the second periodic rotation. The triangle symbol in Figure A1b,d only shows the first periodic rotation but the second periodic rotation can be explained as follows. Since τ needs to be equal to 0.5 for a zero net migration (near the value with the magnetic field strength at 2000 A/m [$\tau = 0.49$]), the first periodic rotation for the magnetic field strengths $H_0 = 3000$ A/m and $H_0 = 4000$ A/m are at $\tau = 0.47$ and $\tau = 0.42$, respectively. The τ values will thus result in $\Delta r_{p\pi} < 0$ μm. Therefore, after the second periodic rotation, the τ values are such that $\Delta r_{p\pi} < 0$ μm or $\Delta r_{p\pi} > 0$ μm. A similar conclusion can be made in Figure A1b where the radial velocities for strengths $H_0 = 3000$ A/m and $H_0 = 4000$ A/m are negative. Thus, $U_{r_{p\pi}} > 0$ μm/s after the second periodic rotation for $H_0 = 3000$ A/m, but $U_{r_{p\pi}} < 0$ μm/s for $H_0 = 4000$ A/m.

For the overall transportation of the particle in Figure A1b (square symbol) as the magnetic field increases, $U_{r\theta}$ increases and $\Delta r_{p\theta} > 0$ μm. An additional observation can be made that the particle near the upper curve radially migrates faster towards the channel center than a particle initially near the lower curve.

The effects on particle net radial migration by a magnetic field applied at $\alpha = 90°$ can be seen in Figure A2. We notice that in Figure A2a,c the particle radial migration towards the channel center is greater when a particle is near the upper curve than near the lower curve. The radial velocity, for a particle near the outer wall under all magnetic field strengths at the channel curve exit (square symbol), is thus positive and greater than the radial velocity of a particle near the inner wall. Additionally, we see that, for all magnetic field strengths, the particles radially oscillate towards the channel center since $\tau > 0.5$ for all magnetic field strengths, as shown in Figure A2d, and the radial velocities are positive and increasing for $U_{r\pi}$ (triangle symbol) and $U_{r\theta}$ (square symbol) in Figure A2b.

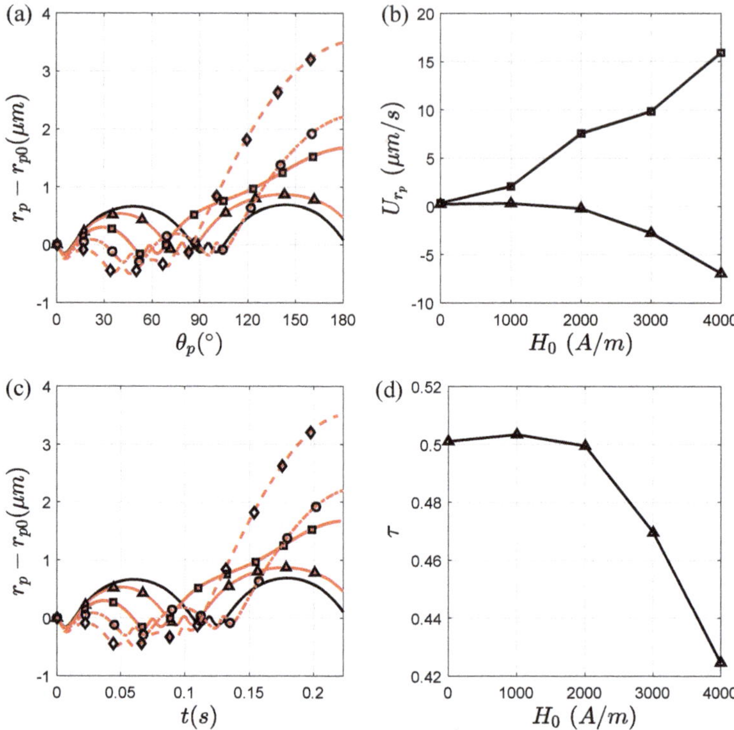

Figure A1. Effect of the magnetic field strength when it is applied at $\alpha = 0°$ on an elliptical particle ($AR = 4$) that is in the upper curve of the channel. The particle is affected by the magnetic field when its center of mass is approximately at $\theta_p = 0$. The radial particle-wall distance change of the particle with respect to (**a**) θ_p, and (**c**) time for, $H_0 = 0$ A/m (solid black line), $H_0 = 1000$ A/m (triangle symbol), $H_0 = 2000$ A/m (square symbol), $H_0 = 3000$ A/m (circular symbol), $H_0 = 4000$ A/m (diamond symbol). (**b**) The radial velocities U_{rp} (triangle symbol) and $U_{r\theta}$ (square symbol) are functions of H_0. (**d**) The dimensionless parameter τ as a function of H_0. We see that as τ decreases, the net migration of the particle decreases.

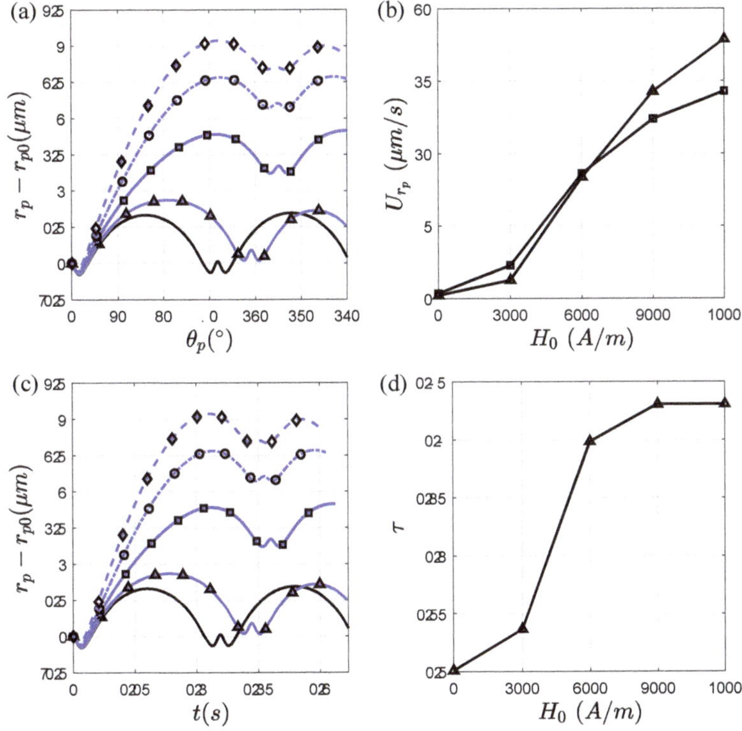

Figure A2. Effect of the magnetic field strength when it is applied at $\alpha = 90°$ on an elliptical particle ($AR = 4$) that is in the upper curve of the channel. The particle is affected by the magnetic field when its center of mass is approximately at $\theta_p = 0$. The radial particle-wall distance change of the particle with respect to (**a**) θ_p, and (**c**) time for, $H_0 = 0$ A/m (solid black line), $H_0 = 1000$ A/m (triangle symbol), $H_0 = 2000$ A/m (square symbol), $H_0 = 3000$ A/m (circular symbol), $H_0 = 4000$ A/m (diamond symbol). (**b**) The radial velocities U_{rp} (triangle symbol) and $U_{r\theta}$ (square symbol) are functions of H_0. (**d**) The dimensionless parameter τ as a function of H_0. We see that as τ increases, the net migration of the particle increases.

Appendix B. Mesh and Time Step Comparison

The accuracy, convergence, and time efficient calculation of our simulation is based on the time step, number of elements in the computational domain Ω, and the particle surface Γ. To ensure that the time step and the number of elements are efficient for this paper, we compare them by establishing a time and grid independent analysis.

The time independence of an elliptical particle in a curved channel was studied for $18,571$ domain elements and 152 elements on the particle boundary as seen in Table A1 and Figure A3. As shown in the plot, the numerical results for the time step $\Delta t = 2 \times 10^{-5}$ s appears to have a large net migration after one periodic rotation. The time step $\Delta t = 1 \times 10^{-5}$ s results in a smaller error. The time steps $\Delta t = 5 \times 10^{-6}$ s and $\Delta t = 2.5 \times 10^{-6}$ s have very similar results and have a very small error, therefore showing that an elliptical particle results in a zero net migration after one periodic rotation, similar to an elliptical particle in a straight channel. However, the computational time for the latter time steps are longer than for $\Delta t = 1 \times 10^{-5}$ s. Thus for accuracy and for computational time for all of our simulations, we use the time step $\Delta t = 1 \times 10^{-5}$ s.

With a time step of $\Delta t = 1 \times 10^{-5}$ s, we compared the simulation results of four different mesh sizes, in the channel domain and around the particle boundary in a Poiseuille flow and in the absence of a uniform magnetic field are shown in Table A2 and Figure A4. As shown in the plot, the numerical results are efficient enough when the channel domain has 18, 571 elements and when there are 152 elements on the particle surface, compared to the lower elements studied. If we increased the number of elements, then there would not be too much difference in the accuracy of the calculations in the simulations. Therefore, in this article, we used 18, 571 elements in the computational domain and 152 elements on the the particle surface, thus giving acceptable accurate results for the particle's transportation and rotation.

Table A1. Four time independence analysis for 18, 571 computational domain elements and 152 particle surface elements.

Time Step for Time Independence Analysis	Time Step Δt
Time 1	2×10^{-5} s
Time 2	1×10^{-5} s
Time 3	5×10^{-6} s
Time 4	2.5×10^{-6} s

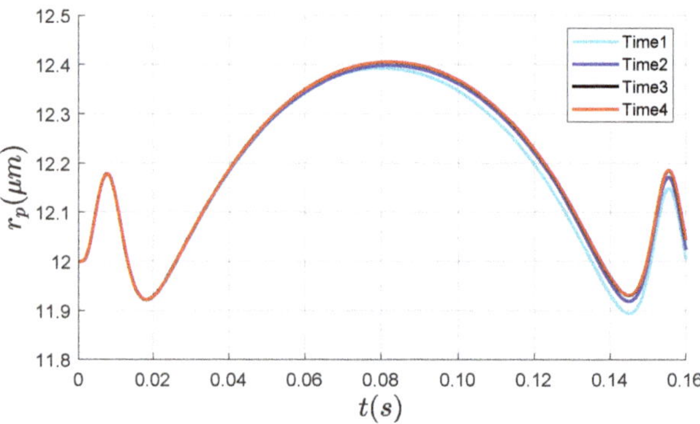

Figure A3. Time independence analysis for a particle's radial migration as a function of time for $AR = 4$, $r_{p0} = 12$ μm, 18, 571 Domain Elements, and 152 Particle Surface Elements. The lines for Time 2, Time 3, and Time 4 are very close, which shows that Time 2 still has a reasonable accuracy and a time efficient computation compared to Time 3 and Time 4.

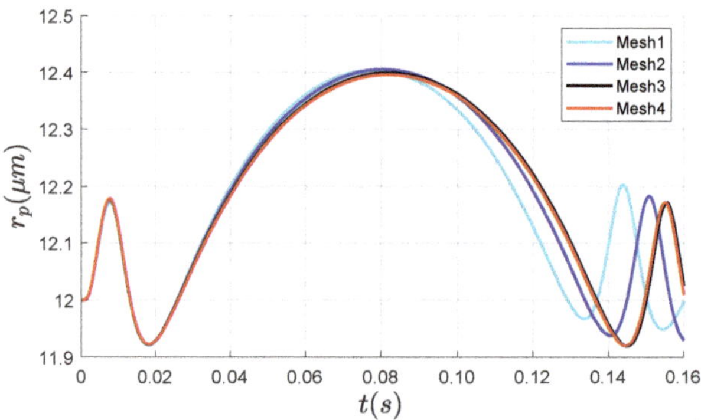

Figure A4. Grid independence analysis for a particle's radial migration as a function of time for $AR = 4$, $r_{p0} = 12$ μm, and a time step $\Delta t = 1 \times 10^{-5}$ s. The lines for Mesh 3 and Mesh 4 are very close, which demonstrates that Mesh 3 still has a reasonable accuracy and a more efficient computational time.

Table A2. Four meshes for grid independence analysis and for $\Delta t = 1 \times 10^{-5}$ s.

Mesh for Grid Independence Analysis	Domain Elements	Particle Surface Elements
Mesh 1	11,321	56
Mesh 2	14,020	72
Mesh 3	18,571	152
Mesh 4	24,208	256

Appendix C. Particle Dynamics Without a Magnetic Field

In this section, we examine the effect of the particle aspect ratio, AR, average radius of the channel, R_{avg}, and initial particle-wall separation distance, r_{p0}, on the particle dynamics in the absence of a magnetic field. For an elliptical particle in a low Reynolds number and in the presence of a channel wall, a particle will oscillate towards and away from the wall according to a non-zero lift velocity perpendicular to the wall as mentioned by Gavze et al. [35]. These oscillatory motions are caused by some of the forces involved as mentioned by Zhou et al. and the inertia mentioned by Zhang et al. [25,27]. We define the wall lift force as a consequence of the coupling of the rotation and translation dynamics of the particle and the particle-wall hydrodynamic interaction. Consequentially, this force pushes the particle towards the channel center. Conversely, the shear gradient lift force, mentioned in this paper, is due to the non-linearity of the fluid velocity profile and its interaction with the particle. As a result this force pushes the particle towards the wall.

For an elliptical particle in a curved channel and in the absence of a magnetic field, whether or not a particle executes a complete periodic rotation and its net radial migration still needs to be analyzed. The main difference between a particle in a straight channel and in a curved channel is based on the velocity profile and the shear rate due to the channel geometries. The shear rate in a straight channel is applied at the same direction, and its magnitude remains the same so long as the channel width and geometry are constant. As shown in previous studies, an elliptical particle in a straight channel spends an equal amount of time in the first half rotation versus the second half rotation, resulting in a zero net migration [25–27,30]. In a curved channel, the shear rate changes directions with respect to the curvature of the channel but the particle net migration still needs to be verified.

Figure A5 shows the dynamics of an elliptical particle with $AR = 4$ initially positioned at $r_{p0} = 12$ μm and $\phi_{p0} = 43°$ and placed in a channel where the average radius is $R_{avg} = 175$ μm. As seen in Figure A5a, the particle transportation throughout the curved channel results in a radial migration towards the inner wall, $\Delta r_{p\theta} < 0$ μm. Similar to previous research articles, we see that, in Figure A5b, the particle oscillates away from the wall in the first half of its periodic rotation ($0° < \phi'_p < 90°$) and towards the wall in the second half ($90° < \phi'_p < 180°$), thus keeping the particle from radially migrating toward the inner channel wall. In Figure A5c, the orientation of the particle, ϕ'_p, is a function of the dimensionless time, t/T_0, where T_0 is the total time for one periodic rotation. The periodic rotation and the angular velocity, seen in Figure A5d, are symmetric with respect to $\phi'_p = 90°$ due to the shear rate. However, even though we have a different channel geometry, the elliptical particle has a net-zero radial migration. In previous works, τ was defined as the ratio between the amount of time the particle spends in the first half of its rotation to its second half [25–27,30]. Thus, a particle with $AR = 4$ transporting in a channel with $R_{avg} = 175$ μm results in $\tau = 0.50$, and the shear gradient lift force acting on the elliptical particle is equal to the second half of its rotation as the wall lift force in the first half, resulting in a zero net radial migration. The radial velocity is also zero, even if we have two different velocity profiles for the curved channel (dashed line) and straight channels (dotted line), as seen in Figure A5e. The wall distance is described as from the inner radius to the outer radius for a curved channel and from the bottom wall to the top wall for a straight channel. We notice that for a straight channel, the velocity profile is symmetric with respect to the channel center (at 25 μm). For the curved channel, however, the velocity profile is asymmetric and the maximum velocity of the flow is between the inner wall and the channel center. However, due to the low inertia acting on the particle, the fluid velocity profile will continue to result in a zero net radial migration.

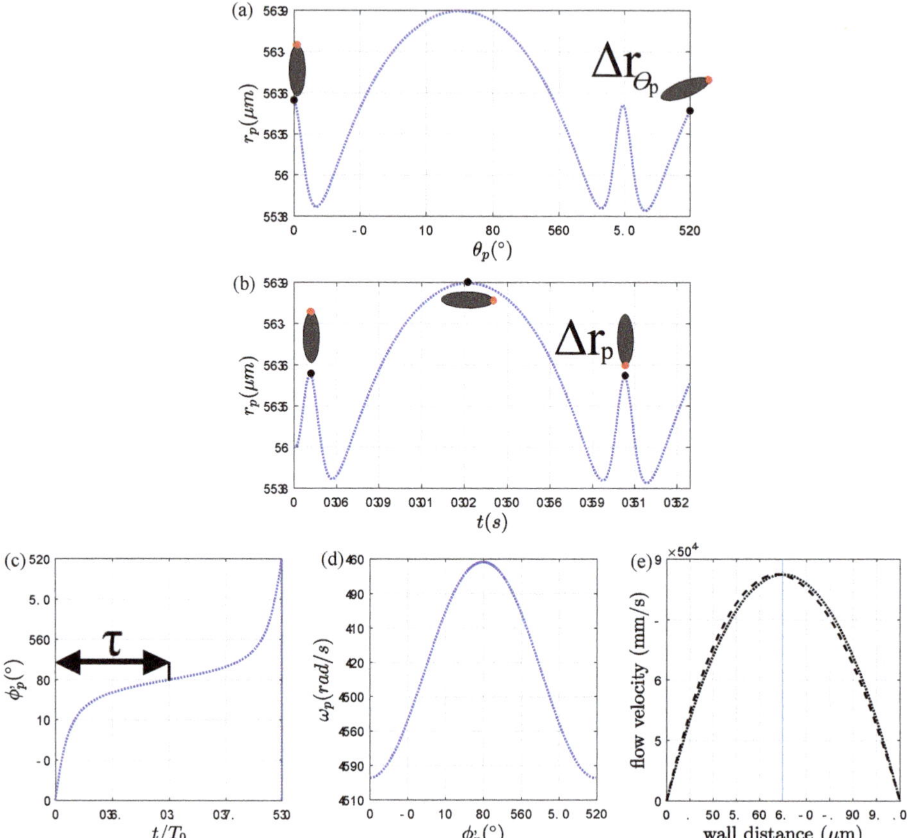

Figure A5. Translation and rotation of an elliptical particle in the absence of a magnetic field. The particle has an aspect ratio $AR = 4$ and the initial radial position is $r_{p0} = 12$ μm, whereas the initial particle direction is $\phi_{p0} = 43°$ in the beginning of the simulation. The channel's average radius is $R_{avg} = 175$ μm. The particle-wall distance is studied with respect to its (**a**) θ_p position; (**b**) time, t, in the curved channel; (**c**) the evolution of the orientation angle ϕ'_p with respect to dimensionless time t/T_0; (**d**) the rotational velocity ω_p versus ϕ'_p; and (**e**) the velocity profiles for a curved channel (dashed line) and straight channel (dotted line).

The net migration of an elliptical particle is shown in Figure A6 for particle aspect ratio, AR, average curve radius, R_{avg}, and initial radial particle-wall distance r_{p0}. With these different parameters, we can establish an outcome for the radial migration of an elliptical particle after one periodic rotation and upon exiting the curved channel.

In Figure A6a,d, we study how the aspect ratio affects the particle dynamics in a curved channel when $r_{p0} = 12$ μm and $R_{avg} = 175$ μm. In Figure A6d, as AR increases from 2 to 4, the particle executes fewer periodic rotations in the curved channel because T_0 increases. Consequentially, for all aspect ratios, $\Delta r_{p\pi} = 0$ μm, but they radially migrate towards and away from the inner wall at different rates and the particle completes its π-rotation at different positions inside the curved channel, θ_p, even though r_{p0} is the same for all cases. For example, for $AR = 2$, the total time for one periodic rotation is

$T_0 = 0.064$ s, $T_0 = 0.096$ s for $AR = 3$, and $T_0 = 0.15$s for $AR = 4$. Likewise, the periodic rotation is executed earlier in the channel as seen in Figure A6a where $\theta_p = 64°$ for $AR = 2$, $\theta_p = 97°$ for $AR = 3$, and $\theta_p = 149°$ for $AR = 4$. Upon exiting the curve, $\Delta r_{p\theta} = -0.09$ μm for $AR = 2$, $\Delta r_{p\theta} = -0.16$ μm for $AR = 3$, and $\Delta r_{p\theta} = -0.023$ μm for $AR = 4$. The particle radial migration velocity throughout the channel can be observed where a particle with $AR = 4$ will oscillate towards the wall but its overall net migration is less than all the other aspect ratios.

In Figure A6b,e, we increase the average radius R_{avg}, while keeping $AR = 4$ and $r_{p0} = 12$ μm. For a larger average radius, the particle will complete more periodic rotations since the arc length has increased. The rotational time, T_0, decreases and the net radial migration after one periodic rotation, $\Delta r_{p\pi}$, is zero for $R_{avg} = 175$ μm and $R_{avg} = 225$ μm. We see that in Figure A6e, the particle cannot execute a full periodic rotation for $R_{avg} = 125$ μm (dash line). In Figure A6b, however, as the particle is exiting the channel curve, $\Delta r_{p\theta}$ increases where $\Delta r_{p\theta} = -0.042$ μm for $R_{avg} = 125$ μm, $\Delta r_{p\theta} = -0.023$ μm for $R_{avg} = 175$ μm, and $\Delta r_{p\theta} = 0.21$ μm for $R_{avg} = 225$ μm. As the average radius increases from $R_{avg} = 175$ μm to $R_{avg} = 225$ μm, T_0 decreases, as well as the θ_p position at the end of the particle periodic rotation, where $\theta_p = 149°$ for $R_{avg} = 175$ μm and $\theta_p = 109°$ for $R_{avg} = 225$ μm.

In Figure A6c,f, r_{p0} increases from 10 μm to 16 μm, while keeping $AR = 4$ and $R_{avg} = 175$ μm. We see in Figure A6c that as r_{p0} increases, T_0 also increases since the shear rate is smaller towards the channel center. The radial migration of the particle, however, will experience a zero net radial migration when the particle is placed further away. In Figure A6f, as r_{p0} increases, the particle will less likely perform a periodic rotation and θ_p increases at $\phi'_p = 90°$. For particles initially further away from the wall, they spend a shorter time in the channel than all other initial positions shown, but the initial particle-wall distance affects the radial migration of the particle. For all initial positions, the wall lift force is stronger than the shear lift force; thus, $\Delta r_{p\pi} > 0$ μm at $\phi'_p = 90°$ and increases when r_{p0} decreases. For $r_{p0} = 12$ μm and $r_{p0} = 10$ μm, the particle can execute a periodic rotation due to the larger shear rate. For $r_{p0} = 12$ μm, $T_0 = 0.15$s and $\theta_p = 149°$ and $T_0 = 0.14$s and $\theta_p = 132°$ for $r_{p0} = 10$ μm for one periodic rotation. At the exit of the channel curve in Figure A6c, the particle will most likely radially migrate towards or away from the inner channel wall but at different rates with $\Delta r_{p\theta} = 0.25$ μm for $r_{p0} = 10$ μm, $\Delta r_{p\theta} = -0.023$ μm for $r_{p0} = 12$ μm, $\Delta r_{p\theta} = -0.095$ μm for $r_{p0} = 14$ μm, and $\Delta r_{p\theta} = 0$ μm for $r_{p0} = 16$ μm. Therefore, in most cases (except at the channel center), the particle experiences a net radial migration towards the inner wall.

The behavior of an elliptical particle migration in a curved channel based on AR, R_{avg}, and r_{p0} on $\Delta r_{p\theta}$, and T_0 can thus be explained as follows. When the particle is in an unbounded fluid domain, the particle freely rotates because there is no inertia acting on the particle [34]. Thus, there is no net lateral migration. When the channel wall is present, there is a particle-wall hydrodynamic interaction that increases the resistance on the particle's rotation [35]. As the particle aspect ratio increases, the periodic rotation of the particle increases because the particle-wall hydrodynamic interaction is more significant for larger aspect ratios than for smaller aspect ratios. Therefore, even if the shear lift force is greater than the wall lift force, a particle with a smaller aspect ratio radially migrates towards and away from the wall faster than particles with a larger aspect ratio. Next, as the average radius increases, the particle completes a periodic rotation faster and at a shorter θ_p. For the first half of its rotation, $\Delta r_{p\pi}$ from $\phi'_p = 0°$ to $\phi'_p = 90°$ increases, indicating that the channel geometry affects the particle dynamics during its transportation throughout the channel curve. Finally, as r_{p0} decreases, the wall lift force on the particle becomes more prominent since the shear rate is also larger near the channel wall. When a particle is initially placed further away from the inner wall, the shear lift force has a larger impact on the radial migration except at $r_{p0} = 25$ μm. Even though the last two parameters have been studied for $AR = 4$, we can make a similar conclusion for $AR = 2$ and $AR = 3$.

We can, therefore, make a few conclusions and predictions for an elliptical particle transporting in a curved channel in the absence of a magnetic field. First, the channel geometry, the particle-wall separation distance, and the particle aspect ratio are important factors that ultimately affect the rotation, the radial oscillatory motion, and the overall channel transportation of the particle in a curved channel.

Second, for a radial oscillatory motion after one periodic rotation, $\Delta r_{p\pi} = 0$ µm at $\phi'_p = 180°$, but the particle-wall distance is positive, $\Delta r_{p\pi} > 0$ µm, at $\phi'_p = 90°$. The transportation of the particle throughout the curved channel is affected since the particle continues to radially oscillate, but the particle migrates either toward or away from the inner wall. In these cases, the shear lift force may be equal to the wall lift force and the radial migration depends on the rotational dynamics of the particle.

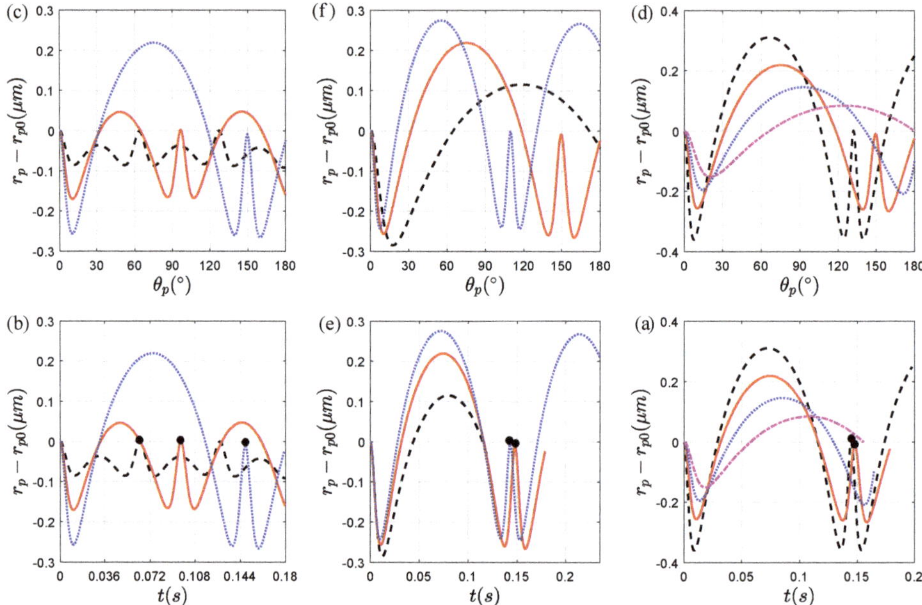

Figure A6. Effects of particle aspect ratio, AR, average radii, R_{avg}, and initial radial position, r_{p0}, values on the radial particle-wall separation, $\Delta r_{p\pi}$ and $\Delta r_{p\theta}$, of elliptical particles for the range $0° \leq \theta_p \leq 180°$ in the absence of a magnetic field. The (**a**) θ_p and (**d**) time dependency on the aspect ratio effect on the particle's radial position ($r_p - r_{p0}$): $AR = 2$ (dash line), $AR = 3$ (solid line), $AR = 4$ (dot line). The initial position is $r_{p0} = 12$ µm and the average channel radius is $R_{avg} = 175$ µm. The (**b**) θ_p and (**e**) time dependency on the average radius effect on $r_p - r_{p0}$: $R_{avg} = 125$ µm (dash line), $R_{avg} = 175$ µm (solid line), $R_{avg} = 225$ µm (dot line). The particle's aspect ratio and position are $AR = 4$, $r_{p0} = 12$ µm, and respectively. The (**c**) θ_p and (**f**) time dependency of the particle's initial radial positions on the effect on the radial position $r_p - r_{p0}$: $r_{p0} = 10$ µm (dash line), $r_{p0} = 12$ µm (solid line), $r_{p0} = 14$ µm (dot line), $r_{p0} = 16$ µm (dash-dot line). The particle's aspect ratio and the average channel radius are $AR = 4$ and $R_{avg} = 175$ µm, respectively. The black dots in (**d,e,f**) represent the end of the particles' π rotation.

References

1. Kang, H.; Kim, J.; Cho, H.; Han, K.-H. Evaluation of Positive and Negative Methods for Isolation of Circulating Tumor Cells by Lateral Magnetophoresis. *Micromachines* **2019**, *10*, 386. [CrossRef] [PubMed]
2. Kim, S.; Ham, S.-I.; Park, M.-J.; Jeon, C.-W.; Joo, Y.-D.; Choi, I.-H.; Han, K.-H. Circulating tumor cell microseparator based on lateral magnetophoresis and immunomagnetic nanobeads. *Anal. Chem.* **2013**, *85*, 2779–2786. [CrossRef] [PubMed]
3. Aviles, M.O.; Chen, H.; Ebner, A.D.; Rosengart, A.J.; Kaminski, M.D.; Ritter, J.A. In vitro study of ferromagnetic stents for implant assisted-magnetic drug targeting. *J. Magn. Magn. Mater.* **2007**, *311*, 306–311. [CrossRef]
4. Lübbe, A.S.; Alexiou, C.; Bergemann, C. Clinical applications of magnetic drug targeting. *J. Surg. Res.* **2001**, *95*, 200–206. [CrossRef] [PubMed]

5. Svoboda, J.; Fujita, T. Recent developments in magnetic methods of material separation. *Miner. Eng.* **2003**, *16*, 785–792. [CrossRef]

6. Ambashta, R.D.; Sillanpää, M. Water purification using magnetic assistance: a review. *J. Hazard. Mater.* **2010**, *180*, 38–49. [CrossRef]

7. Pamme, N. Magnetism and microfluidics. *Lab Chip* **2006**, *6*, 24–38. [CrossRef]

8. Bayat, P.; Rezai, P. Microfluidic curved-channel centrifuge for solution exchange of target microparticles and their simultaneous separation from bacteria. *Soft Matter* **2018**, *14*, 5356–5363. [CrossRef]

9. Schaap, A.M.; Chu, W.C.; Stoeber, B. Continuous size-separation of airborne particles in a microchannel for aerosol monitoring. *IEEE Sens. J.* **2011**, *11*, 2790–2797. [CrossRef]

10. Ozbey, A.; Karimzadehkhouei, M.; Kocaturk, N.M.; Bilir, S.E.; Kutlu, O.; Gozuacik, D.; Kosar, A. Inertial focusing of cancer cell lines in curvilinear microchannels. *Micro Nano Eng.* **2019**, *2*, 53–63. [CrossRef]

11. Di Carlo, D.; Irimia, D.; Tompskins, R.G.; Toner, M. Continuous inertial focusing, ordering, and separation of particles in microchannels. *Proc. Natl. Acad. Sci. USA* **2007**, *104*, 18892–18897. [CrossRef]

12. Wang, L.; Dandy, D.S. High-Throughput Inertial Focusing of Micrometer-and Sub-Micrometer-Sized Particles Separation. *Adv. Sci.* **2017**, *4*, 1700153. [CrossRef] [PubMed]

13. Albagdady, A.; Al-Faqheri, W.; Kottmeier, J.; Meiner, S.; Frey, L.J.; Krull, R.; Dietzel, A.; Al-Halhouli, A. Enhanced inertial focusing of microparticles and cells by integrating trapezoidal microchambers in spiral microfluidic channels. *RSC Adv.* **2019**, *9*, 19197–19204.

14. Lee, J.-H.; Lee, S.-K.; Kim, J.-H.; Park, J.-H. Separation of particles with bacterial size range using the control of sheath flow ratio in spiral microfluidic channel. *Sens. Actuators A* **2019**, *286*, 211–219. [CrossRef]

15. Wu, L.; Guan, G.; Hou, H.-W.; Bhagat, A.A.S.; Han, J. Separation of leukocytes from blood using spiral channel with trapezoid cross-section. *Anal. Chem.* **2012**, *84*, 9324–9331. [CrossRef]

16. Rafeie, M.; Zhang, J.; Asadnia, M.; Li, W.; Warkiani, M.E. Multiplexing slanted spiral microchannels for ultra-fast blood plasma separation. *Lab Chip.* **2016**, *16*, 2791–2802. [CrossRef]

17. Son, J.; Samuel, R.; Raheel, G.; Bruce, K.; Carrell, D.T.; Hotaling, J.M. Separation of sperm cells from samples containing high concentrations of white blood cells using a spiral channel. *Biomicro.* **2017**, *11*, 1. [CrossRef]

18. Syed, M.S.; Rafeie, M.; Vandamme, D.; Asadnia, M.; Henderson, R.; Taylor, R.A.; Warkiani, M.E. Selective separation of microalgae cells using inertial microfluidics. *Bioresour. Technol.* **2018**, *252*, 91–99. [CrossRef]

19. Zhou, Y.; Ma, Z.; Ai, Y. Sheathless inertial cell focusing and sorting with serial reverse wavy channel structures. *Microsys. Nanoeng.* **2018**, *4*, 1–18. [CrossRef]

20. Zhou, Y.; Ma, Z.; Tayebi, M.; Ai, Y. Submicron particle focusing and exosome sorting by wavy microchannel structures within viscoelastic fluids. *Anal. Chem.* **2019**, *91*, 4577–4584. [CrossRef]

21. Zhu, J.; Tzeng, T.-R.J.; Xuan, X. Continuous dielectrophoretic separation of particles in a spiral microchannel. *Electrophoresis* **2010**, *31*, 1382–1388. [CrossRef] [PubMed]

22. Li, M.; Li, S.; Cao, W.; Li, W.; Wen, W.; Alici, G. Continuous particle focusing in a waved microchannel using negative dc dielectrophoresis. *J. Micromech. Microeng.* **2012**, *22*, 095001. [CrossRef]

23. Kim, J.; Park, J.; Müller, M.; Lee, H.-H.; Seidel, H. Uniform magnetic mobility in a curved magnetophoretic channel. *2009 IEEE Sens.* **2009**, 1165–1167.

24. Zhou, R.; Bai, F.; Wang, C. Magnetic separation of microparticles by shape. *Lab Chip* **2017**, *17*, 401–406. [CrossRef]

25. Zhou, R.; Sobecki, C.A.; Zhang, J.; Zhang, Y.; Wang, C. Magnetic control of lateral migration of ellipsoidal microparticles in microscale flows. *Phys. Rev. Appl.* **2017**, *8*, 024019. [CrossRef]

26. Sobecki, C.A.; Zhang, J.; Zhang, Y.; Wang, C. Dynamics of paramagnetic and ferromagnetic ellipsoidal particles in shear flow under a uniform magnetic field. *Phys. Rev. Fluids* **2018**, *3*, 084201. [CrossRef]

27. Zhang, J.; Wang, C. Numerical Study of Lateral Migration of Elliptical Magnetic Microparticles in Microchannels in Uniform Magnetic Fields. *Magnetochemistry* **2018**, *4*, 16. [CrossRef]

28. Matsunaga, D.; Meng, F.; Zöttl, A.; Golestanian, R.; Yeomans, J.M. Focusing and sorting of ellipsoidal magnetic particles in microchannels. *Phys. Rev. Lett.* **2017**, *119*, 198002 . [CrossRef]

29. Matsunaga, D.; Zöttl, A.; Meng, F.; Golestanian, R.; Yeomans, J.M. Far-field theory for trajectories of magnetic ellipsoids in rectangular and circular channels. *IMA J. Appl. Math.* **2018**, *83*, 767–782 . [CrossRef]

30. Zhang, J.; Sobecki, C.A.; Zhang, Y.; Wang, C. Numerical investigation of dynamics of elliptical magnetic microparticles in shear flows. *Micro. Nano.* **2017**, *22*, 83. [CrossRef]

31. Cao, Q.; Li, Z.; Wang, Z.; Han, X. Rotational motion and lateral migration of an elliptical magnetic particle in a microchannel under a uniform magnetic field. *Micro. Nano.* **2018**, *22*, 3. [CrossRef]
32. Harding, B.; Stokes, Y.M.; Bertozzi, A.L. Effect of inertial lift on a spherical particle suspended in flow through a curved duct. *J. Fluid Mech.* **2019**, *875*, 1–43 . [CrossRef]
33. Harding, B. A study of inertial particle focusing in curved microfluidic ducts with large bend radius and low flow rate. In Proceedings of the 21st Australasian Fluid Mechanics Conference, Adelaide, Australia, 10–13 December 2018.
34. Jeffery, G.B. The motion of ellipsoidal particles immersed in a viscous fluid. *Proc. Roy. Soc. Lon.* **1922**, *102*, 161–179. [CrossRef]
35. Gavze, E.; Shapiro, M. Particles in a shear flow near a solid wall: Effect of nonsphericity on forces and velocities. *Int. J. Multiphase Flows* **1997**, *23*, 155–182. [CrossRef]

Article

Accelerated Particle Separation in a DLD Device at Re > 1 Investigated by Means of μPIV

Jonathan Kottmeier [1,*], Maike Wullenweber [2,3], Sebastian Blahout [4], Jeanette Hussong [4], Ingo Kampen [2,3], Arno Kwade [2,3] and Andreas Dietzel [1,2]

[1] Institute for Microtechology, TU Braunschweig, 38124 Braunschweig, Germany; a.dietzel@tu-braunschweig.de

[2] Center of Pharmaceutical Engineering (PVZ), TU Braunschweig, 38106 Braunschweig, Germany; m.wullenweber@tu-braunschweig.de (M.W.); i.kampen@tu-braunschweig.de (I.K.); a.kwade@tu-braunschweig.de (A.K.)

[3] Institute for Particle Technology, TU Braunschweig, 38104 Braunschweig, Germany

[4] Institute for Fluid Mechanics and Aerodynamics, TU Darmstadt, 64287 Darmstadt, Germany; sebastian.blahout@ruhr-uni-bochum.de (S.B.); hussong@sla.tu-darmstadt.de (J.H.)

* Correspondence: j.kottmeier@tu-braunschweig.de; Tel.: +49-(0)531-391-9784

Received: 24 September 2019; Accepted: 9 November 2019; Published: 11 November 2019

Abstract: A pressure resistant and optically accessible deterministic lateral displacement (DLD) device was designed and microfabricated from silicon and glass for high-throughput fractionation of particles between 3.0 and 7.0 μm comprising array segments of varying tilt angles with a post size of 5 μm. The design was supported by computational fluid dynamic (CFD) simulations using OpenFOAM software. Simulations indicated a change in the critical particle diameter for fractionation at higher Reynolds numbers. This was experimentally confirmed by microparticle image velocimetry (μPIV) in the DLD device with tracer particles of 0.86 μm. At Reynolds numbers above 8 an asymmetric flow field pattern between posts could be observed. Furthermore, the new DLD device allowed successful fractionation of 2 μm and 5 μm fluorescent polystyrene particles at *Re* = 0.5–25.

Keywords: microfluidics; deterministic lateral displacement; Reynolds number; particle image velocimetry; size-dependent fractionation

---

1. Introduction

Pharmaceutical drugs with poor water solubility exhibit a size dependent resorption. Therefore, only a small range of particle sizes is suitable to ensure maximal efficacy of a specific particulate pharmaceutical drug. During the process of formulation, a uniform size within a narrow size distribution is sought [1]. Another application of high-throughput microfluidic systems is the size dependent up-concentration of biological materials that behave as particles [2]. To address these tasks, different methods have been implemented in microfluidics, such as fractionation in a spiral microchannel, multi-orifice flow fractionation (MOFF) and deterministic lateral displacement (DLD) [3,4]. In contrast to spiral channels and MOFF devices, DLD arrays allow separating particles of small size differences down to 0.3 μm, in specially designed arrays even down to 10 nm, within a dynamic range of particle sizes around five inside a single array [5,6]. A DLD fractionation device array consists of an array of tilted microposts with a gap between posts (g), post diameter (d) and period (N) as illustrated in Figure 1. The period (N), tilt angle (θ) and row shift fraction (ε) are related as described in Equation (1) [2]. The flow between two posts is divided into a number of streamlines given by the array period each carrying the same volumetric flow rate. Due to the no slip condition at the walls and posts, the streamlines next to the posts are wider compared to the streamlines in the middle between two posts [5]. As long as the particles are smaller than the first streamline (in other words: Are below the critical diameter), they will follow this streamline through the array and will therefore not be displaced. Particles with a diameter higher than this streamline will bump against the posts, thereby the hydrodynamic center of the particles will change the streamline and the particle will move in the so-called displacement mode through the array. The critical diameter (D_c) can be calculated for lower *Re* using Equation (2) as an empirically founded relation [2]. In standard lower throughput operation (*Re* ≤ 1), using spherical particles and cylindrical posts, D_c is only based on the array geometry such as tilt angle and gap between posts [2,5]. To increase the range of particle sizes fractionated inside one array, it can be segmented changing either the gap between posts (*g*) or the tilt angle (*θ*) [2,5]. At high throughput (*Re* > 9) stationary vortices form behind the posts resulting from inertial forces, which change the fractionation properties [5,7–9].

$$\varepsilon = 1/N = \tan(\theta) \tag{1}$$

$$D_c = 1.4 \cdot g \cdot \varepsilon^{0.48} \tag{2}$$

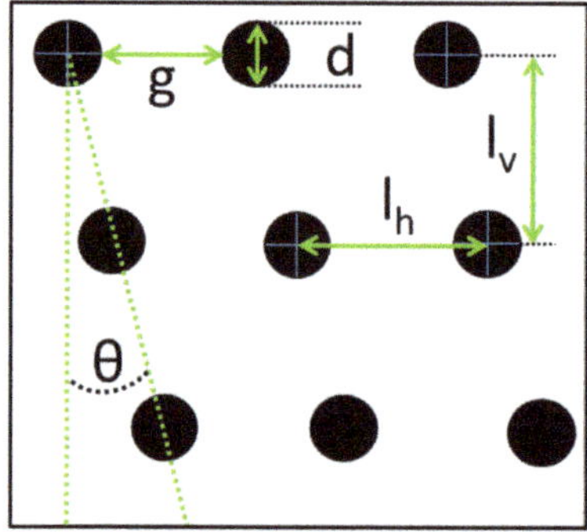

Figure 1. Illustration of the design parameters for a deterministic lateral displacement (DLD) array, with gap between posts (*g*), diameter of posts (*d*), row shift angle (*θ*) defining the period and the row shift fraction, as well as the horizontal and vertical center-to-center distance (I_h and I_v).

In addition to using geometry or flow control, the critical size for fractionation can also be adjusted by electrokinetic effects such as electrophoresis, electro-osmosis or electrostatic effects based on the ion concentration of the buffer solution [10,11]. In addition, a stretchable array was fabricated to alter D_c by mechanical stress applied to the array [12].

High-throughput applications of DLD arrays include the up-concentration of cycling tumor cells from blood samples whilst removing the blood plasma. For this purpose, stacked arrays of triangular posts were operated at a flow rate up to 10 ml/min (*Re* > 40) [13]. In this case, a stable separation was observed up to higher Reynolds numbers (*Re* = 40) [13]. Furthermore, the fractionation characteristics and visualization of flows at higher Reynolds numbers was investigated using 70× up-scaled arrays with circular posts with a diameter of 0.675 mm [7,8,14]. The flow around the posts was visualized using 100 μm particles in a range of 2 ≤ *Re* ≤ 30. With high-speed particle trajectory recording, stationary vortices behind posts have been observed for *Re* > 16 [8]. More recently, a study about miniaturized arrays with a gap between posts (*g*) of 50 μm and 30 μm, operated up to *Re* = 50 (*g* = 50 μm) and *Re* = 35 (*g* = 30 μm) has been reported [9]. By means of high-speed trajectory recording in the smaller array, a shift of the particle positions at the outlets was observed for particles at a diameter around half of the gap (*g* = 30 μm, *d* = 15 μm) explained by emerging vortices behind the microposts which could not be directly visualized [9]. To minimize the vortices behind posts, also airfoil shaped posts were investigated [15]. The experimental results using airfoil shaped posts and circular posts at higher Re were compared to COMSOL multiphysics simulations [9,15]. Up to now, all published experimental works in the field of high-throughput DLDs rely on high-speed particle trajectory recording at the end of the array and observation of the fractionation at the outlets. The fractionation results are typically compared with simulations using computational fluid dynamic (CFD) simulations with ANSYS FLUID and STAR-CCM+, based on the Lattice Boltzmann method; either as a standalone to solve the velocity field or coupled with FEM using the immersed boundary method to account for fluid–particle interactions [5,14,16–19]. However, up to now, there is no direct experimental evidence of the asymmetry of the flow velocity distribution induced by vortices behind the microposts, which would require microparticle image velocimetry (µPIV) inside a DLD array at Re > 1. µPIV is based on picture evaluation using two frames at a time difference adjusted to the particle velocity [20]. The flow field results from cross-correlation of consecutive interrogation windows at identical positions. The amount of details that can be resolved depends on the particle seeding density [20]. To enable such experiments, we fabricated a pressure stable Si/glass DLD system, with a transparent lid allowing µPIV and other optical measurements. The array was segmented to allow fractionation with multiple sizes by changing the tilt angle. Tracer particles must provide sufficient fluorescent signal for µPIV but must be small enough to follow the flow in vortices forming behind the posts. For the first time, µPIV measurements in DLD devices at elevated Re with seed particles smaller than 1 µm could be directly compared with CFD simulations.

2. Materials and Methods

2.1. Segmented DLD Design

To allow a wider range of particle sizes to be fractionated, a segmented array was designed. Either the gap between posts or the tilt angle can be altered between segments to stepwise increase Dc. In this work, we focus on the latter, because changing the gap between posts will lead to a non-uniform fluidic resistance and the segments with a smaller gap could easily clog. We implemented seven segments each with posts of a constant size $d = 5$ µm (defined in the lithographic mask as 10 µm) positioned with a pitch of $l_H = l_V = 20$ µm as illustrated in Figure 2. The tilt angle was changed between the segments according to Table 1 from 1.0° up to 6.7° to increase Dc in steps of 0.8 µm from 3.0 to 7.5 µm. Each segment is designed to displace the particles by 80 µm. The two buffer inlets in Figure 2 are included to provide a focusing of the inlet sample to a defined start position. The widths of the inlet channels (w_i) are different and the individual flow rates of the syringe pumps have to be adjusted to achieve uniform lateral flow. Starting from the desired velocity v inside the array in relation to the simulations, the sum of the flow rates for all three inlet channels Q_{tot} was calculated according to Equation (3). For the calculation, the height (h) and width of the complete array, which is not covered by posts (w_{arr}), is taken into account. Therefore, the ratio of the horizontal center-to-center distance (l_h) and the gap between posts (g) was used to define the free width of the complete array ($w = 730$ µm). The individual volume flow rate for the individual inlet channel Q_i results from the sum of all inlet channel widths and the individual width w_i according to Equation (4) for uniform lateral flow without secondary flows perpendicular to the main flow.

$$Q_{tot} = v \cdot A = v \cdot w_{arr} \cdot h = v \cdot 730 \ \mu m \cdot \frac{g}{l_h} \cdot h \tag{3}$$

$$Q_i = \frac{Q_{tot} \cdot w_i}{\sum w_{i,j}} \tag{4}$$

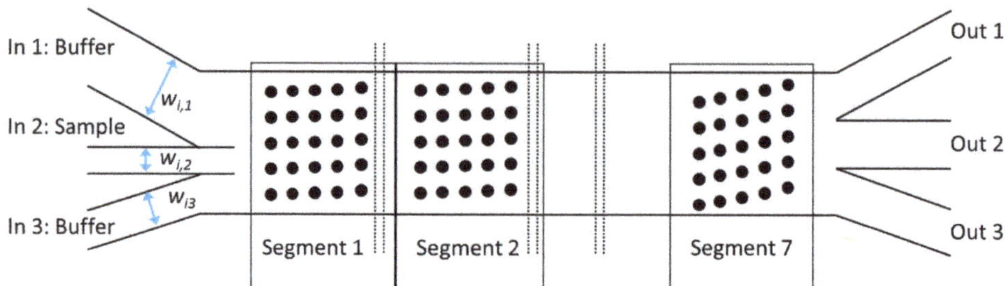

Figure 2. Illustration of the segmented array. The sketch is not to scale and in the real device, there is no gap between segments. The expected particle sizes in each outlet are as follows. Out 1: 0 µm (non-displaced) to 3 µm, Out 2: 3.8–5.3 µm, Out 3: > 6.0 µm.

Table 1. Variables within each segment to allow 80 µm displacement, tilt angle, length of each segment and resulting critical particle size for fractionation.

Segment	1	2	3	4	5	6	7
Tilt Angle [°]	1	1.6	2.3	3.2	4.2	5.4	6.7
Length [mm]	4.62	2.90	1.98	1.44	1.08	0.86	0.68
D_c	3.0	3.8	4.5	5.3	6.0	6.8	7.5

2.2. Fabrication of the DLD Devices

In order to allow for high throughput, the arrays were made of silicon and glass, sealed by anodic bonding as described earlier [21] to resist high pressures. The microfabrication steps for high pressure DLD devices are illustrated in Figure 3. The microstructures for the post arrays and fluidic channels are transferred into photoresist by photolithography before dry etching the silicon in a Bosch process (using STS Multiplex ICP, Surface Technology Systems, Newport, UK) to a depth of 20 µm. Figure 4a shows the scanning electron micrograph of a single post. The fluidic vias in silicon for fluidic connections were fabricated using a fs-laser induced plasma cutting process that can produce almost perpendicular via sidewalls [22]. The wafer was placed inside a holder and immersed in a constant water flow to generate a moving thin water film on top of the surface to facilitate plasma agitation and heat transport. Before cutting, the arrays were covered with a photoresist to prevent destruction of the microposts and uptake of particles generated by laser cutting. A scanning electron micrograph (Figure 4b) shows a fluidic via with a sidewall angle of 88°. To lower agglomeration around posts and improve the filling of the device, a 200 nm PECVD (plasma enhanced chemical vapor deposition using STS 310 PC from STS instruments) oxide layer was applied which lowers the surface tension. The arrays were finally sealed using anodic bonding at increased voltage (700 V instead of the typically used 400 V) to allow a strong bond even in the presence of the insulating oxide layer.

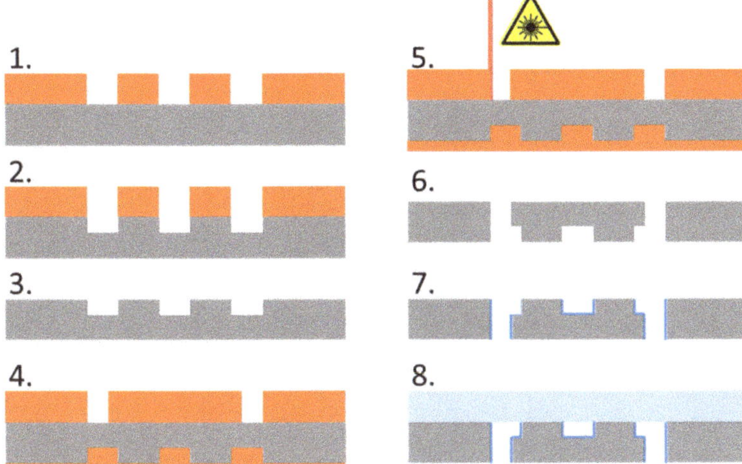

Figure 3. Schematic illustration of the microfabrication steps for a Si/glass DLD array. The steps are: 1. Photolithographic structuring of the resist for micropost arrays and connection channels; 2. Dry etching of silicon (Bosch Process); 3. Stripping of the resist; 4. Filling of the arrays with resist, structuring the resist for fluidic vias; 5. Fs-laser cutting of the vias; 6. Stripping the resist; 7. Deposition of 200 nm oxide layer; 8. Anodic bonding of glass and silicon.

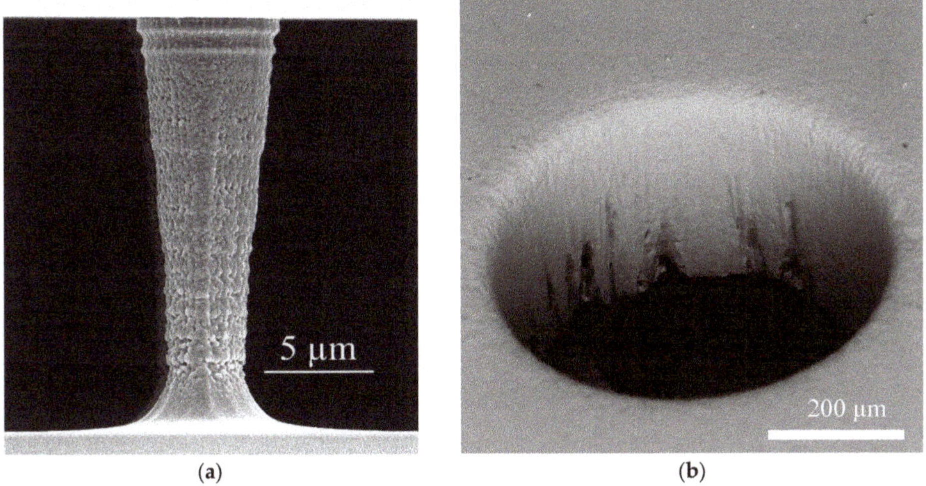

Figure 4. Scanning electron micrographs showing the post arrays and fluidic connections. (a) Detailed view of the micropost array showing the tapered profile. (b) Via for fluidic connections created by fs-laser induced plasma cutting.

To further analyze the dimensions of an array used in μPIV measurements in detail, we broke one of the used arrays with a diamond pen to define the breaking point. A SEM micrograph of two posts is depicted in Figure 4a and shows the tapered shape of a single post ($d_{min} = 3.59 \pm 0.33$ μm; $d_{max} = 6.56 \pm 0.43$ μm; $d_{avg} = 5.08 \pm 1.54$ μm and $h = 19$ μm). In the further considerations, we therefore assumed an averaged post diameter of $d = 5$ μm, which leads to a gap of $g = 15$ μm and a height of h = 19 µm. Figure 4b shows defects at the bottom of the vias after laser induced plasma cutting. These are a result of the non-uniform resist coating caused by the hydrophobic nature of the photoresist.

2.3. Flow Control Setup

As described in Section 2.1, the array is fed using three inlet channels, which were connected to three NEMESYS syringe pumps. Inlet 1 was connected to a mid-pressure pump equipped with a 10 mL stainless steel syringe. The two low-pressure modules for inlets 2 and 3 feeding the sample and smaller buffer channels were each equipped with a 2.5 mL glass syringe. The DLD chip was placed between an aluminum plate (for fixation) and a PMMA holder (with fluidic inlets) to connect the array using standard high-pressure liquid chromatography (HPLC) connectors.

Table 2 shows the flow rates in the experiments ($Q_1, \ldots, Q_3$) and the resulting average velocities in the inlets (v_{in}) which are equal for all inlets thus avoiding secondary flows perpendicular to main flow direction. Furthermore, in Table 2 the average velocity inside the array (v_{arr}) as well as the estimated Reynolds number (Re) is shown. The inlet velocity inside the inlet channels was calculated according to Equation (5) using Q_{tot}, the height of the array (h) and the sum of the widths of the inlet channels ($w_{i,1} + w_{i,2} + w_{i,3}$) as described in Section 2.1. v_{arr} was calculated according to Equation (6). The effective width of the array not covered by posts is given as w_{arr}. Therefore, the designed width of the complete array (w = 730 µm) was multiplied with the ratio of the horizontal center-to-center distance of the posts (l_h) and the gap between the posts (g). The Reynolds number was calculated, using v_{arr} and standard values for water at 25 °C for the density (ρ = 1000 kg/m^3) and viscosity ($\eta = 10^{-3}$ Pa·s) according to Equation (7). The hydrodynamic length was taken as the gap between the posts.

For particle fractionation experiments at different Reynolds numbers, 2 µm and 4 µm fluorescent green labeled polystyrene particles (exc. 470 nm, emm. 505 nm, FluoroGreen Thermo Scientific, bought via Distrilab Particle Technology, Leusden, The Netherlands) were suspended in a 1% (w/v) Pluronic F-127 solution and injected through the sample inlet.

$$v_{in} = \frac{Q_{tot}}{h \cdot w_{in}} = \frac{Q_1 + Q_2 + Q_3}{h \cdot (w_{i,1} + w_{i,2} + w_{i,3})} \tag{5}$$

$$v_{arr} = \frac{Q_{tot}}{h \cdot w_{arr}} = \frac{Q_{tot}}{h \cdot 730 \ \mu m \cdot g / l_h} \tag{6}$$

$$Re = \frac{\rho \cdot v_{arr} \cdot g}{\eta} \tag{7}$$

Table 2. Input flow rates and velocities in the inlet channels and between the post and Reynolds number.

Q1: Buffer [1]	Q2: Sample [1]	Q3: Buffer [1]	V_{in} [2]	V_{post} [2]	Re
20.01	8.30	146.89	0.194	0.25	3.87
40.02	16.59	293.79	0.387	0.51	7.62
100.04	41.48	734.47	0.968	1.27	19.05
160.07	66.37	1175.16	1.549	2.03	30.48

[1] All Flow rates in µL/min. [2] All velocities in m/s.

2.4. µPIV Setup

µPIV measurements were performed utilizing an upright EPI-fluorescence microscope (Nikon Eclipse LV100) with an objective lens of M = 50 magnification and a numerical aperture of NA = 0.6 (50× Nikon CFI60 TU Plan Epi ELWD, Nikon, Düsseldorf, Germany) leading to a correlation depth of 7.03 µm according to the calculation described by Olsen and Adrian [23]. Since µPIV measurements utilize volume illumination, the fluorescence signal of defocused particles has a contribution to the cross-correlation result. The correlation depth indicates the distance relative to the focal plane at which this contribution occurs and is therefore a measure for the spatial averaging of the velocity field along the channel height [20,23]. Images are recorded with a double-frame CCD camera (LaVision Imager pro SX, LaVision GmbH, Göttingen, Germany). Laser and camera are triggered and monitored with the commercial software DaVis 8.4 (LaVision GmbH, Göttingen, Germany). The camera has a resolution of 2058 pixel (px) × 2456 px, which results in a field of view of 0.227 mm × 0.190 mm. The minimal size of tracer particles for PIV measurements is limited because sufficient light from the particle has to reach the camera. Fluorescent (emm. 532 nm, exc. 605 nm) polystyrene particles of 0.86 μm ± 0.04 μm in diameter (Fluoro-Max R900 from Thermo Scientific) were still providing sufficient light. Prior to injecting, particles were suspended in the carrier fluid, which is a mixture of water and 1% (w/v) Pluronic F-127. Before injecting particle suspensions, the array was flushed for 30 min with the carrier fluid. Tracer particles are excited by the beam of a double-pulsed, dual-cavity Nd:YAG laser (Litron Nano S 65-15 PIV, Litron Lasers, Rugby, UK, emm. 532 nm), which is coupled into the microscope. Images for μPIV measurements were taken using the double frame exposure mode with a short time interval (*dt*) between *dt* = 0.6 μs and 50 μs depending on the flow rate. The time interval was manually set to ensure a particle shift between the double frames of ~10 px. Double frame images ware taken at a rate of 6.5 fps. Image processing steps applied to the raw fluorescence image are depicted in Figure 5. Image processing (tools provided by DaVis 8.4) is used to isolate the particle fluorescence from background noise of the camera. Furthermore, stationary agglomerations around posts were removed before vector evaluation. Background noise was further reduced by subtracting the sliding Gaussian average of 19 frames. Second, the salt and pepper noise was reduced using a band pass filter removing everything below 5 px and above 12 px isolating the particle size signal which is in the range of 7 to 11 px. To completely isolate the particle signal, it was necessary to set all intensities below a defined value (3 counts) as provided by the software, to zero, as illustrated in Figure 5b. In a next step, the areas with posts obtained from white light images were manually masked out.

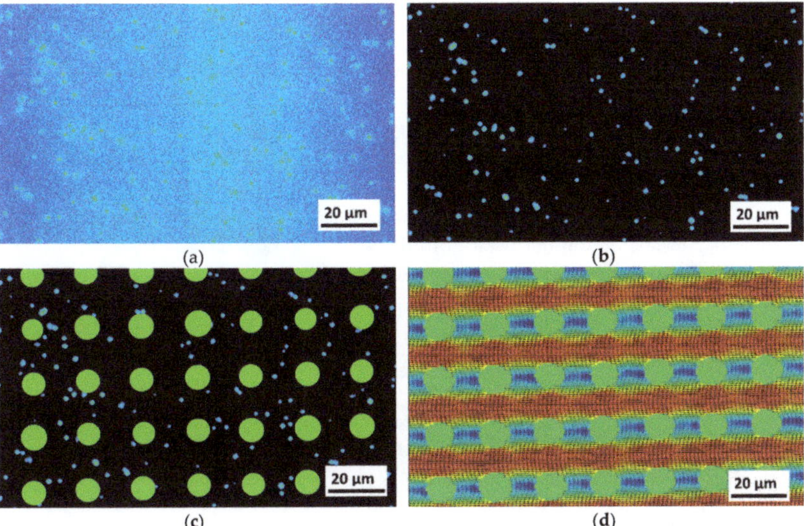

Figure 5. Illustration of the image processing steps (example image taken for *Re* = 7.62): (**a**) The raw camera fluorescence image. In the center of the image, the brightness level slightly but abruptly changes, which is an artifact induced by the camera technology and does not cause problems with the image correlation procedures. (**b**) Image after post processing. (**c**) Image with post areas masked. (**d**) The final μPIV result of the measurement set consisting of 1500 frames. All Images show precisely the same area.

Vector calculation is achieved using the shift of a correlation peak inside interrogation windows in two subsequent images, as described elsewhere [20]. In this work, we chose a so-called multi-pass procedure with calculation passes using each a different interrogation window size and start position. For the first pass, the interrogation window (96 px × 96 px) covered the size of a micropost to get a first fast overview. The interrogation windows of the second and third pass was chosen to be at 32 px, to resolve details around the posts. Between the second and third pass, the start position of the interrogation windows was changed. Each measurement contained 1500 images and was evaluated using the sequential sum of correlation, as provided by the software, which summarizes the average of all cross-correlation results in the correlation plane to compensate for the relatively low particle seeding density.

2.5. Simulation

Computational fluid dynamic (CFD) simulations were performed using the open-source software package OpenFOAM. The simulations were carried out in a three-dimensional fragment of the microsystem, consisting of seven post rows with three posts per row at a height of 21 μm (Figure 6). The microposts have a conical shape with a diameter of 6.5 μm at the top and 3.5 μm at the bottom. Density (ρ) and kinematic viscosity (v) of the Newtonian fluid were assumed as $\rho = 1$ g/cm^3 and $v = 0.01$ cm^2/s respectively. Fluid flow is achieved by defining an inlet velocity and a pressure drop in the x-axis direction. The inlet velocities v_{inlet} (free approaching flow) were calculated according to Equation (8) by multiplying the average velocity v between posts by the ratio of the gap g to the center-to-center distance I_h.

$$v_{inlet} = v \cdot \frac{g}{I_h} \tag{8}$$

Inlet velocities are listed with their corresponding Reynolds numbers in Table 3. In order to obtain uniform velocity profiles normal to the flow direction (y-axis), cyclic boundary conditions were assigned to the upper and lower patches. Because of the pressure drop it was not possible to define cyclic boundary conditions in x-axis direction. This problem was solved by using a function that maps the fully developed velocity profile, which is generated at the outlet, to the inlet patch. In z direction, no-slip boundary conditions were assumed. The Navier-Stokes equations are solved using the PISO algorithm which has the ability to solve transient, incompressible flows. The reason for using this transient solver is that it is also used in the CFD-DEM (computational fluid dynamics-discrete element method) coupling, in order not to get differences in the flow field due to a solver change when introducing particles into the system. The flow field was discretized using Finite Volume Discretization. A laminar flow was assumed. The time step was chosen small enough that a Courant number of 0.5 was not exceeded. The calculation period was sufficient for the flow to fully develop and not change over time. At the velocities given here, the state of the flow field remains in the stationary range.

As a post-processing step, velocity and vector fields at the middle-z-position of the channel were visualized using the software ParaView. Since the correlation depth in the μPIV measurements was 7.03 μm, a layer of the simulation domain with a thickness of 7 μm with layer center at $z = 0$ (being the middle position in the channel) was cut and averaged over the layer thickness to enable a comparison to experimental results. For $Re = 1.52$ this process was repeated for different z-positions in 2.5 μm steps in order to compare the flow profile over the height of the microsystem. For getting the maximum velocities, the global maximum of these averaged layers was determined. For the average velocities between the posts, the mean value of the velocity profile between two adjacent posts was taken.

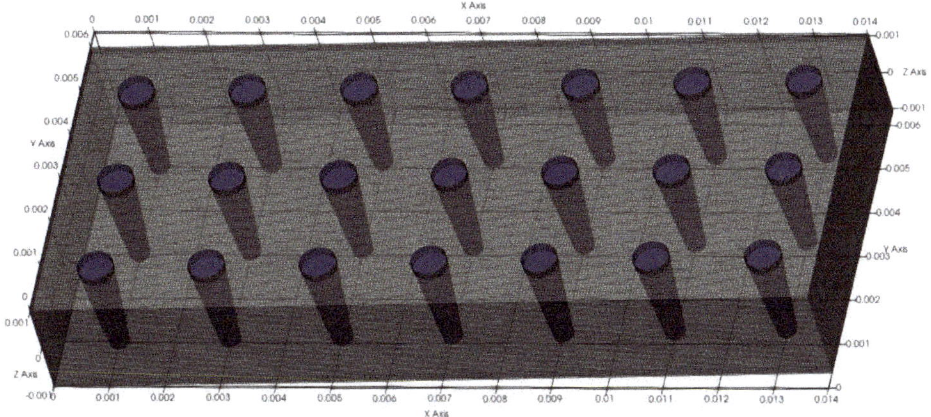

Figure 6. Simulation domain with tapered microposts (dimensions in 10^{-2} m).

Table 3. Inlet velocities at different Reynolds numbers.

V_{in} (cm/s)	3.8	7.6	19.0	38.1	57.1	76.2	95.2	114.3	133.3	152.4	175.0	200.0	225.0	250.0
Re	0.76	1.52	3.81	7.62	11.43	15.24	19.05	22.86	26.67	30.48	35.00	40.00	45.00	50.00

3. Results and Discussion

3.1. Simulations

The wake behind the posts observed in the simulations was compared for different Reynolds numbers. The flow around the DLD microposts can be compared to the flow around a cylinder, which has often been investigated in the literature [24,25]. However, adjacent microposts can also influence the flow. In literature, attached flow behind a cylinder was proven for $Re < 5$, while for larger Re up to 40 a fixed pair of vortices was detected [24]. With further increase in Reynolds number, it was found that the vortices became unstable and began to detach alternately (Karman vortex street). The transition range to turbulence is $150 \leq Re < 300$ [24]. Considering the flow around the posts in the microsystem, a similar behavior can be observed. Whilst at Reynolds number up to $Re = 1$ the wake behind the posts is nearly symmetrical to the stagnation point flow in front of the posts, with growing Reynolds numbers the wake elongates which is expressed in a growing asymmetry (when mirrored with the flow in front of the posts) as shown in Figure 7. The relative velocities in different z-planes did not change during the simulation of tapered posts and are therefore not displayed. At Reynolds numbers of about 50 the wake of the trailing edge passes almost directly into the leading edge flow of the following post. This means that the microsystem is continuously traversed along the direction of flow by lines in which the fluid and thus the particle velocity is strongly decelerated. Vector fields show that at Reynolds numbers up to around 1 the fluid stream splits up in front of the posts and flows back together directly behind the posts (Figure 8a). At Reynolds numbers around 20, a static vortex begins to form behind the posts, which grows with the Reynolds number and causes a stall (Figure 8b–d). The flow remains stationary at $Re \leq 50$ and no vortex shedding was observed.

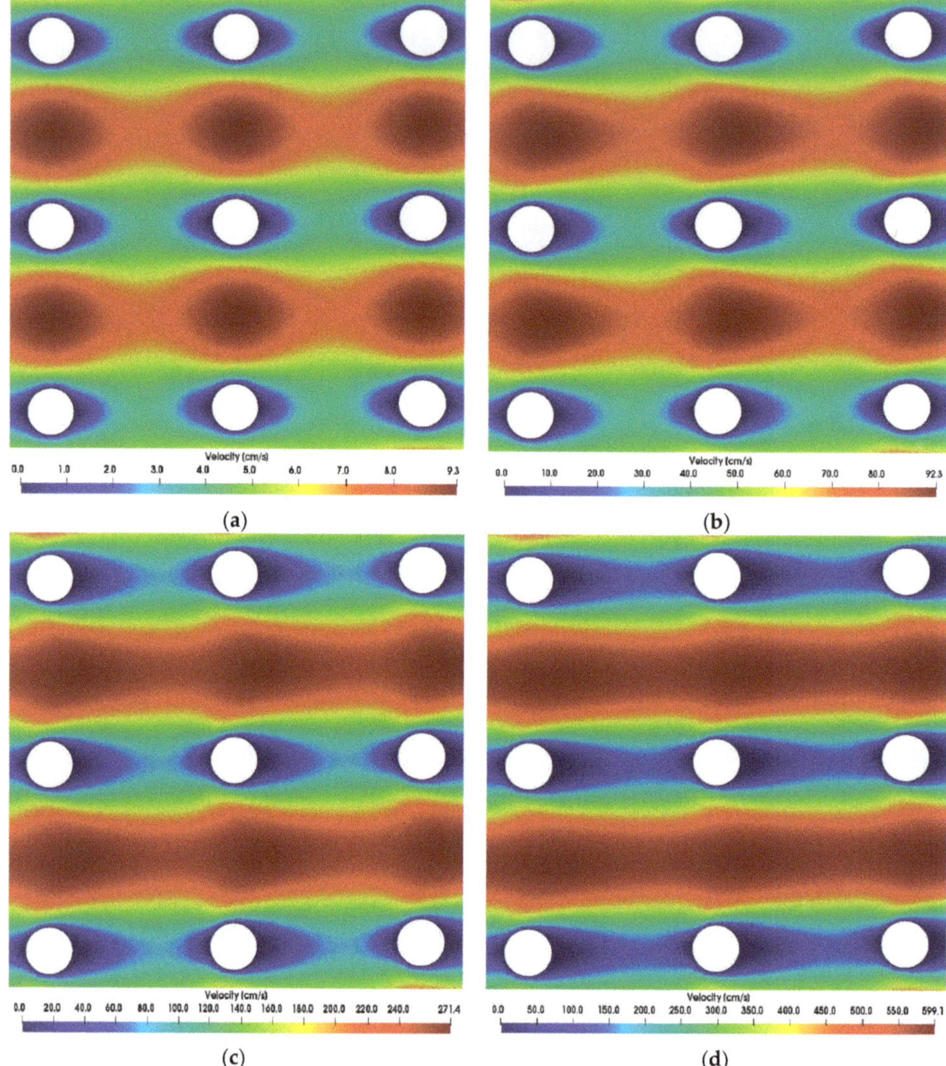

Figure 7. Simulations show emerging asymmetric flow between posts at different Reynolds numbers. The direction of flow is from left to right. Color scales refer to the velocity magnitude. (**a**) $Re = 0.76$, (**b**) $Re = 7.63$, (**c**) $Re = 22.86$, (**d**) $Re = 50$.

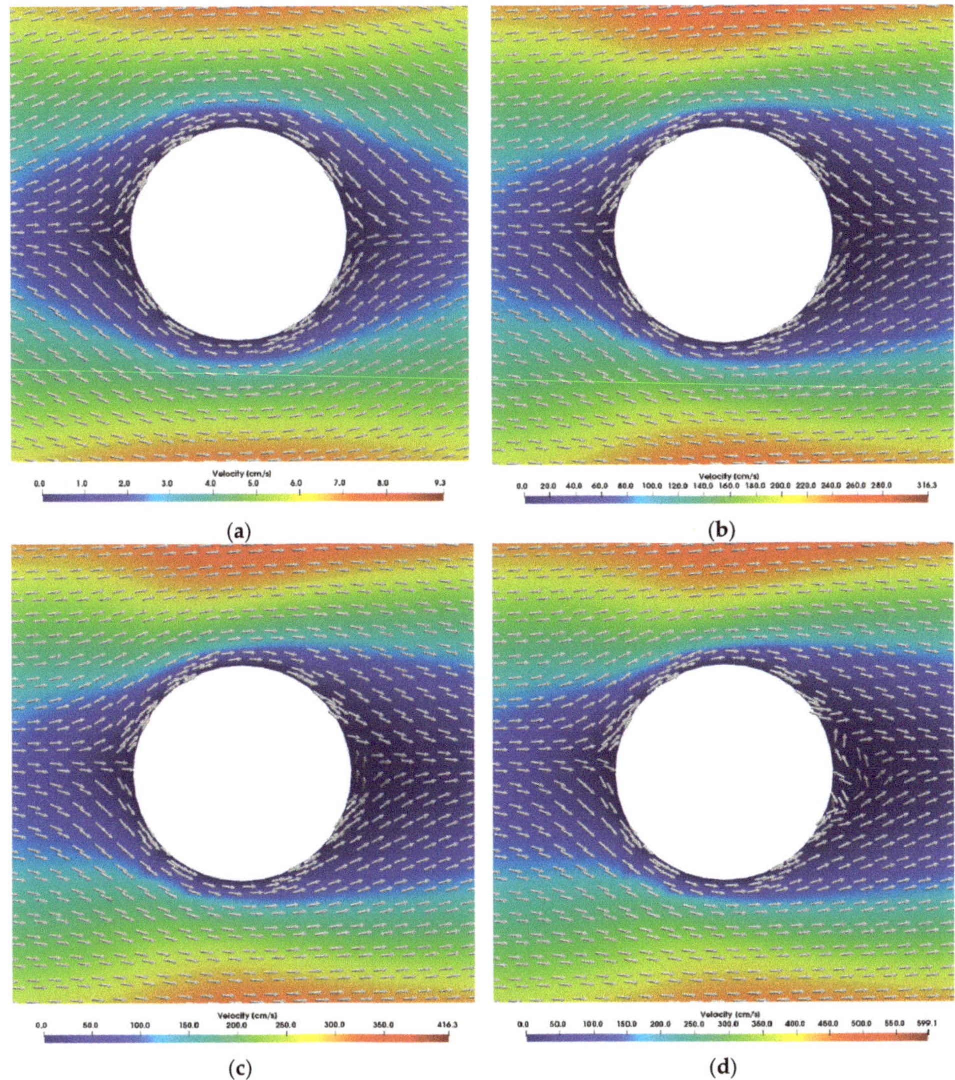

Figure 8. Simulated vector fields between posts at different Reynolds numbers. Static vortices begin to arise with a growing Reynolds number. Color scales refer to the velocity magnitude. The direction of flow is from left to right. (**a**) $Re = 0.76$, (**b**) $Re = 26.67$, (**c**) $Re = 35$, (**d**) $Re = 50$.

3.2. Comparison between Simulations and μPIV Results

The simulation results for different Reynolds numbers were compared to the μPIV results at position $z = 0$ (Figures 9–13). For each μPIV measurement, a white light image was taken to obtain the position of the microposts. Therefore, the posts were manually marked and masked out. Asymmetries between the area in front of the post and behind the post appear in the simulations as well as in the experiments at $Re > 1$. These asymmetries grow with increasing Reynolds number, which is especially visible in the simulations. At Reynolds numbers up to around 10, the simulation velocity fields resemble the μPIV results. At higher Reynolds numbers, a difference is notable between μPIV and simulations (see the color code in Figures 11–13). The μPIV velocities are slightly lower when compared to the ones obtained in the simulations. In addition, the flow asymmetries in Figures 12 and 13 are difficult to detect. Therefore, we have highlighted the differences before and after the posts. It is possible that vortices form also in z-direction, but our measurement equipment is not capable to detect vorticities along the channel height. The maximum and average velocities in flow direction at the median z layer between posts that are direct neighbors in y direction (perpendicular to the flow direction) as obtained from μPIV experiments and from CFD simulations are shown in Figure 14. The velocities obtained in the simulations linearly increase with the Reynolds number but the velocities from μPIV are slightly lower at Reynolds numbers higher than 15. We assume that the major reasons for the deviations are agglomerations of particles around posts in sections of the array located outside the field of view, which could not be monitored simultaneously. Other sources of deviation could be slight variations in the masking during post-processing of μPIV measurements that affect the average velocities between posts. Furthermore, the average velocity from the simulation is obtained inside a 7.03 μm thick layer at the median z-position, in which posts are assumed with perpendicular sidewalls and gaps are therefore constant. The layer thickness was adjusted to the experimental correlation length of 7.03 μm. The real silicon microposts are tapered, as could be seen in the SEM micrographs, we expect a varying gap and flow velocities changing with depth.

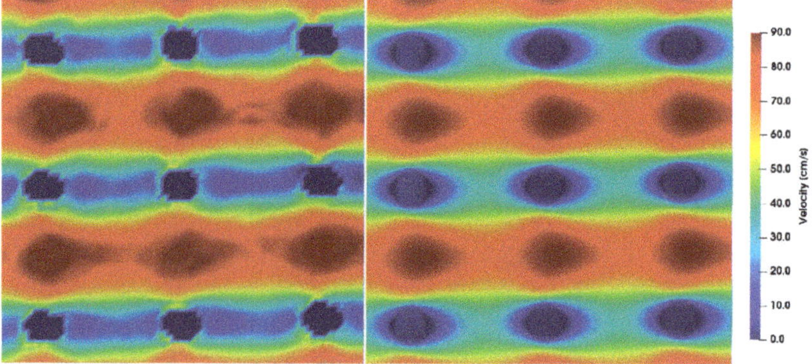

Figure 9. Microparticle image velocimetry (μPIV) measurements (**left**) compared to simulations (**right**) showing the velocity field around posts at *Re* = 7.62. The direction of flow is from left to right, the pitch between the posts is 20 μm at a time difference between double frames of *dt* = 3 μs.

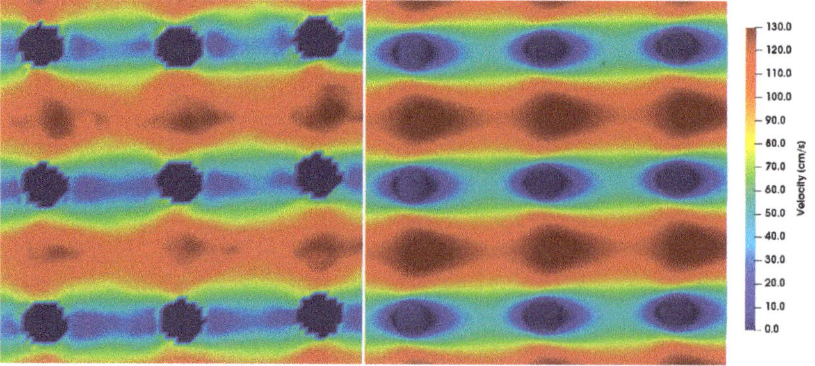

Figure 10. μPIV measurements (**left**) compared to simulations (**right**) showing the velocity field around posts at *Re* = 11.43. The direction of flow is from left to right, the pitch between the posts is 20 μm and the time difference between double frames is *dt* = 1 μs.

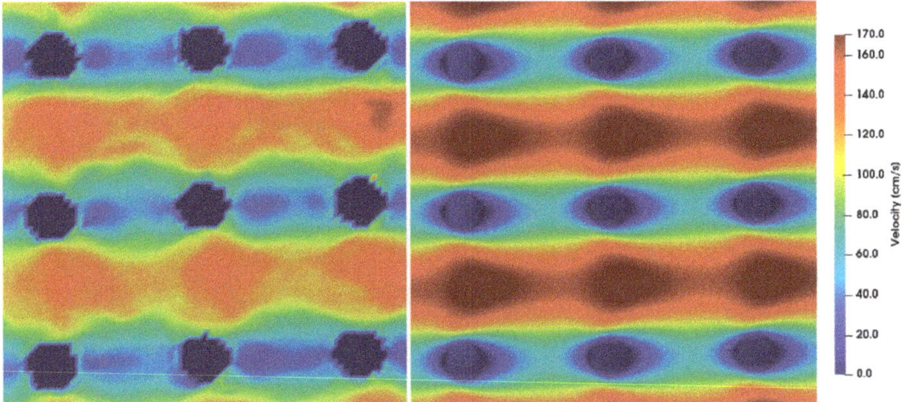

Figure 11. µPIV measurements (**left**) compared to simulations (**right**) showing the velocity field around posts at $Re = 15.24$. The direction of flow is from left to right, the pitch between the posts is 20 µm and the time difference between double frames is $dt = 1$ µs. The µPIV measurements do not attain the maximum velocities achieved by the simulations.

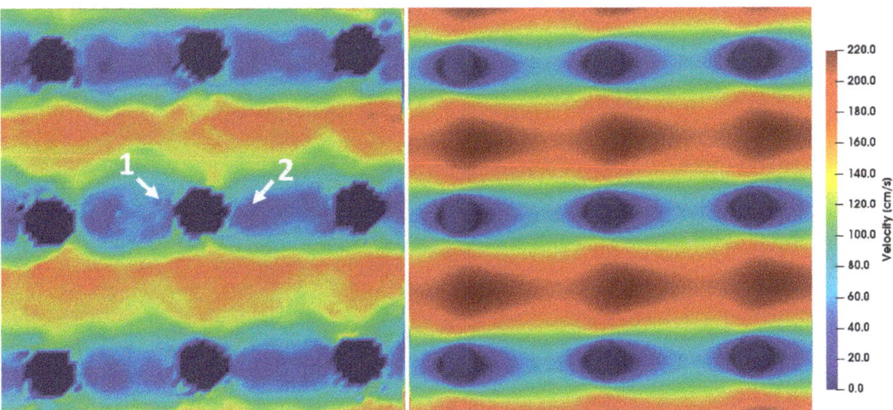

Figure 12. µPIV measurements (**left**) compared to simulations (**right**) showing the velocity field around posts at $Re = 19.05$. The direction of flow is from left to right and the pitch between the posts is 20 µm and the time difference between double frames is $dt = 0.8$ µs. For a better highlighting of the asymmetry, two arrows are inserted into the picture. The velocity before the post is higher (arrow 1) than the velocity after the post (arrow 2). Furthermore, the µPIV measurements do not attain the maximum velocities achieved by the simulations.

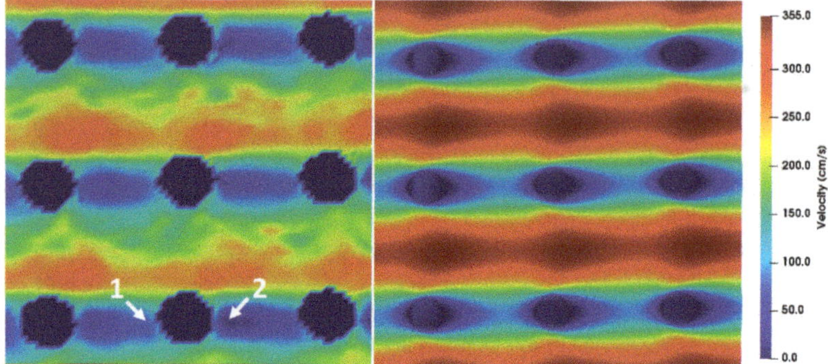

Figure 13. µPIV measurements (**left**) compared to simulations (**right**) showing the velocity field around posts at $Re = 30.48$. The direction of flow is from left to right and the pitch between the posts is 20 µm and the time difference between double frames is $\Delta t = 0.6$ µs. For a better highlighting of the asymmetry, two arrows are inserted into the picture. The velocity before the post is higher (arrow 1) than the velocity after the post (arrow 2). Furthermore, the µPIV measurements do not attain the maximum velocities achieved by the simulations.

To gain information about the flow profile along the z-direction, we conducted µPIV measurements at different z-positions in the area between neighboring posts. Five areas between posts were manually selected in which the average and maximum velocities were calculated for different z-heights and compared to the simulation of tapered posts. The resulting parabolic flow profiles at $Re = 1.52$ are depicted in Figure 15. It is notable that the velocity maximum in the µPIV experiments is shifted 2.5–5 µm in direction of the top cover of the microsystem. Since the posts are tapered as shown in Figure 4a, deviations of the velocity maximum value and the maximum position between µPIV and simulation scan be explained by the height dependent gap between the posts. The channel bottom position ($z = -10$) was obtained by focusing particles after allowing them to settle for 30 min (density particles 1.05 g/cm^3, density water 1 g/cm^3). We believe that the velocity shift and higher values of maximal velocities are caused by the tapered profile of the posts, which leads to a changed partial pressure along the z-direction of the array. The flow counteracts the pressure and thereby the velocity is modified in comparison to the simulation.

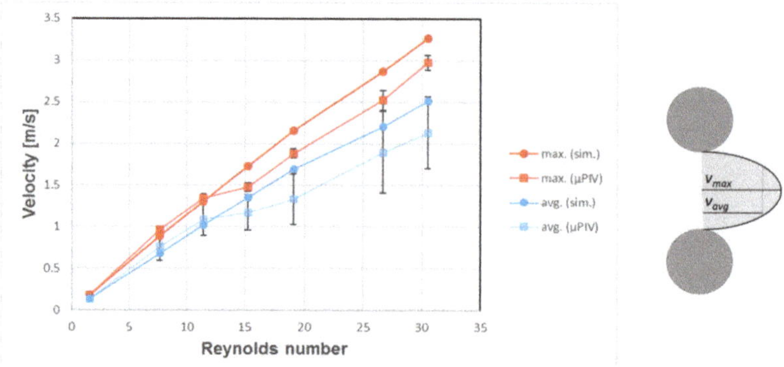

Figure 14. Velocities (maximum and average between posts) depending on Reynolds number at median z layer. The simulation data shows a linear correlation between Reynolds number and both maximum and average velocity. The µPIV data, however, show that velocities do not continue to grow in proportion at Reynolds numbers higher than 15.

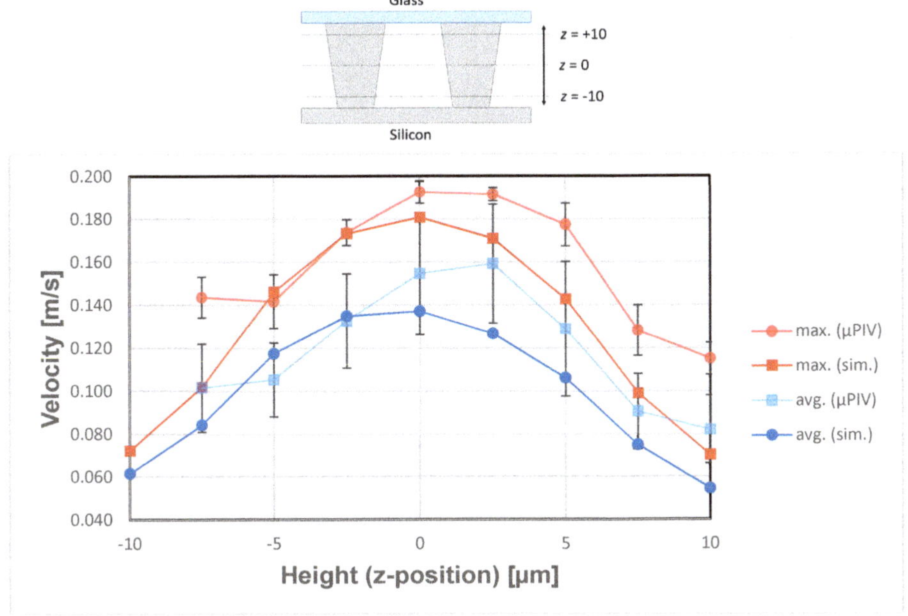

Figure 15. Maximum and averaged velocities at $Re = 1.52$ between neighboring posts obtained by µPIV and simulations for varied focal depth (z-axis within the array). For this evaluation, for every layer, four rectangle areas between posts were chosen out of the µPIV velocity fields and the velocity maxima and mean values of these four areas were averaged.

3.3. Fractionation Results

Due to the emerging and growing wake behind the microposts at higher Re, the fractionation behavior of an array will change compared to an operation at lower *Re*. Therefore, particles at a size of 2 µm and 4.8 µm were fed through the sample inlet channel of the array. The fractionation behavior was observed at different *Re* (*Re* = 0.5–25). Due to the design of the array (see also Table 1), we expect the 2 µm particles not to be displaced and therefore remain at the inlet position. Whereas the 4.8 µm particles would be displaced until Section 3 (120 µm wider compared to 2 µm particles) and reach a position at the end of outlet 1 and the beginning of outlet 2. At higher Reynolds numbers, we expect an increase in particle size for each segment caused by the asymmetric flow behind the posts as shown by our PIV measurement and by Dincau et al. [8]. Therefore, the 4.8 µm particles should be displaced further at higher *Re* until they reach outlet 3. Figure 16 shows the sum of 200 consecutive video frames at different Reynolds numbers overlaid with the white light background. The images were taken at 20× magnification. The white dots around the posts in Figure 16 are caused by spontaneous agglomeration around the microposts in individual frames. The probability of bigger particles agglomerating around microposts is lower, therefore posts appear less decorated by particle agglomeration in the fluorescence image. The smaller particles were not visibly affected by the increase of Reynolds numbers and thus remained in Outlet 1. Due to the fluorescence label of the particles, a DLS (Dynamic Light Scattering) measurement of the different outlets was not possible. However, it is visible that the 4.8 µm particles were strongly affected for $Re > 10$, increasing the displacement until they reach outlet 3 at $Re = 25$.

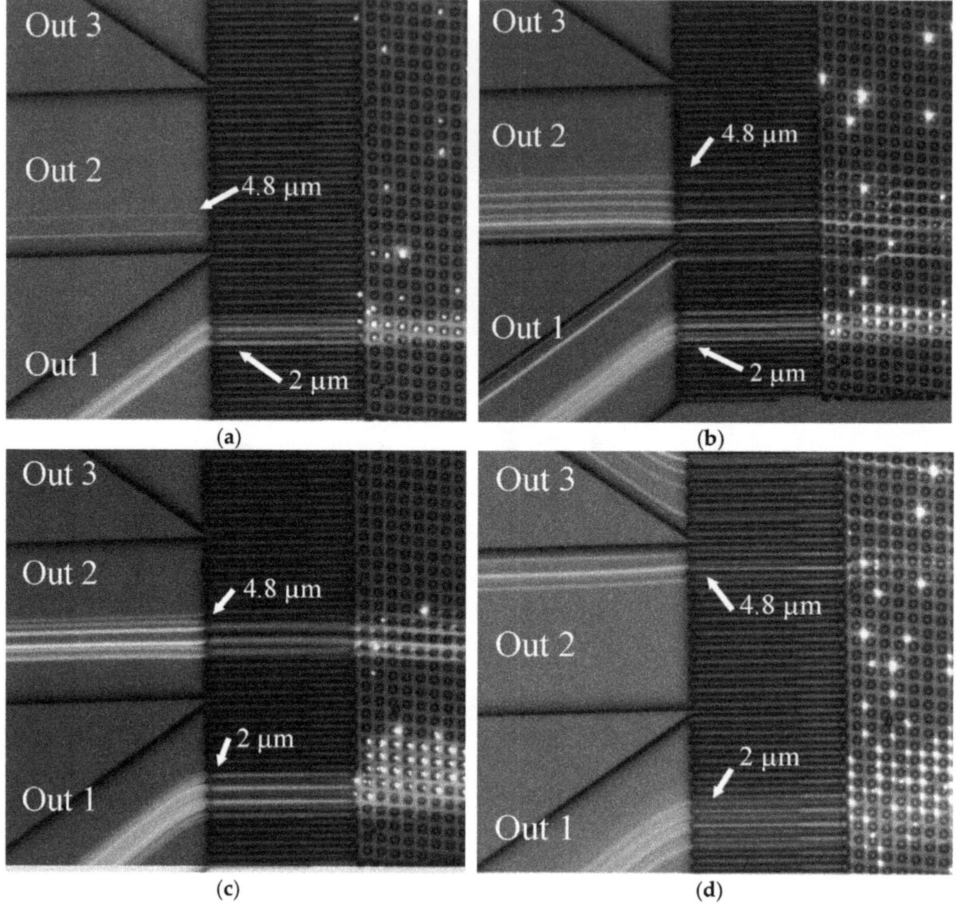

Figure 16. Overlay of 200 video frames with the white light image of the DLD chip areas showing the fractionation position at the end of the array as obtained with fluorescent particles at a size of 2 μm and 4.8 μm for different velocities (Reynolds numbers). (**a**) *Re* = 0.5; (**b**) *Re* = 10; (**c**) *Re* = 15; (**d**) *Re* = 25.

4. Conclusions and Outlook

In this work, we successfully designed and fabricated a DLD array consisting of 7 segments to allow fractionation of particles in sub-micrometer steps by changing the tilt angle with a constant gap between the posts. In order to enable operation at higher *Re*, pressure stable systems were realized using processes of dry etching and anodic silicon to glass bonding. CFD simulations predicted growing wakes behind microposts (asymmetric flow) at *Re* > 1. The flow field asymmetry was confirmed for the first time with μPIV measurements using 0.86 μm fluorescent polystyrene particles. The influence of the wakes on particle fractionation was demonstrated using particles at a size of 2 μm and 4.8 μm, which were displaced to slightly different positions at higher *Re*. This confirms that D_c is not only based on array geometry, but can also be influenced by flow velocity The obtained results will be important to understand and optimize high-throughput fractionation using pressure stable DLD devices, which may find interesting applications in post processing of pharmaceutical particulate formulations.

Author Contributions: Conceptualization: A.D. and A.K.; methodology: M.W., J.K.; software: M.W.; validation, M.W., J.K.; formal analysis: M.W., J.K.; investigation: J.K., M.W. and S.B.; SEM images: I.K.; resources: A.D.,

A.K. and J.H.; data curation: M.W.; writing—original draft preparation: J.K., M.W. and S.B.; writing—review and editing: A.D., I.K., A.K. and J.H.; visualization: J.K. and M.W.; supervision: A.K., A.D. and J.H.; project administration: A.D., A.K. and J.H.; funding acquisition, A.D., A.K. and J.H.

Funding: This research was funded by Deutsche Forschungsgemeinschaft (DFG) SPP 2045 "MehrDimPart".

Conflicts of Interest: The authors declare no conflict of interest. The funders had no role in the design of the study; in the collection, analyses, or interpretation of data; in the writing of the manuscript, or in the decision to publish the results.

References

1. Shekunov, B.Y.; Chattopadhyay, P.; Tong, H.H.Y.; Chow, A.H.L. Particle size analysis in pharmaceutics: Principles, methods and applications. *Pharm. Res.* **2007**, *24*, 203–227. [CrossRef] [PubMed]
2. Davis, J. Microfluidic Separation of Blood Components through Deterministic Lateral Displacement. Ph.D. Thesis, Princeton University, Princeton, NJ, USA, 2008.
3. Sajeesh, P.; Sen, A.K. Particle separation and sorting in microfluidic devices: A review. *Microfluid Nanofluid* **2014**, *17*, 1–52. [CrossRef]
4. Albagdady, A.; Al-Faqheri, W.; Kottmeier, J.; Meinen, S.; Frey, L.J.; Krull, R.; Dietzel, A. Enhanced inertial focusing of microparticles and cells by integrating trapezoidal microchambers in spiral microfluidic channels. *RSC Adv.* **2019**, *9*, 19197–19204. [CrossRef]
5. McGrath, J.; Jimenez, M.; Bridle, H. Deterministic lateral displacement for particle separation: A review. *Lab Chip* **2014**, *14*, 4139–4158. [CrossRef] [PubMed]
6. Wunsch, B.H.; Smith, J.T.; Gifford, S.M.; Wang, C.; Brink, M.; Bruce, R.L.; Austin, R.H.; Stolovitzky, G.; Astier, Y. Nanoscale lateral displacement arrays for the separation of exosomes and colloids down to 20 nm. *Nat. Nanotechnol.* **2016**, *11*, 936–940. [CrossRef] [PubMed]
7. Lubbersen, Y.S.; Schutyser, M.A.I.; Boom, R.M. Suspension separation with deterministic ratchets at moderate Reynolds numbers. *Chem. Eng. Sci.* **2012**, *73*, 314–320. [CrossRef]
8. Lubbersen, Y.S.; Dijkshoorn, J.P.; Schutyser, M.A.I.; Boom, R.M. Visualization of inertial flow in deterministic ratchets. *Sep. Purif. Technol.* **2013**, *109*, 33–39. [CrossRef]
9. Dincau, B.M.; Aghilinejad, A.; Hammersley, T.; Chen, X.; Kim, J.-H. Deterministic lateral displacement (DLD) in the high Reynolds number regime: High-throughput and dynamic separation characteristics. *Microfluid Nanofluid* **2018**, *22*, 869. [CrossRef]
10. Zeming, K.K.; Thakor, N.V.; Zhang, Y.; Chen, C.-H. Real-time modulated nanoparticle separation with an ultra-large dynamic range. *Lab Chip* **2016**, *16*, 75–85. [CrossRef] [PubMed]
11. Calero, V.; Garcia-Sanchez, P.; Honrado, C.; Ramos, A.; Morgan, H. AC electrokinetic biased deterministic lateral displacement for tunable particle separation. *Lab Chip* **2019**, *19*, 1386–1396. [CrossRef] [PubMed]
12. Beech, J.P.; Tegenfeldt, J.O. Tuneable separation in elastomeric microfluidics devices. *Lab Chip* **2008**, *8*, 657–659. [CrossRef] [PubMed]
13. Loutherback, K.; D'Silva, J.; Liu, L.; Wu, A.; Austin, R.H.; Sturm, J.C. Deterministic separation of cancer cells from blood at 10 mL/min. *AIP Adv.* **2012**, *2*, 42107. [CrossRef] [PubMed]
14. Lubbersen, Y.S.; Fasaei, F.; Kroon, P.; Boom, R.M.; Schutyser, M.A.I. Particle suspension concentration with sparse obstacle arrays in a flow channel. *Chem. Eng. Process. Process Intensif.* **2015**, *95*, 90–97. [CrossRef]
15. Dincau, B.M.; Aghilinejad, A.; Chen, X.; Moon, S.Y.; Kim, J.-H. Vortex-free high-Reynolds deterministic lateral displacement (DLD) via airfoil pillars. *Microfluid Nanofluid* **2018**, *22*, 869. [CrossRef]
16. Krüger, T.; Holmes, D.; Coveney, P.V. Deformability-based red blood cell separation in deterministic lateral displacement devices—A simulation study. *Biomicrofluidics* **2014**, *8*, 054114. [CrossRef] [PubMed]
17. Kulrattanarak, T.; van der Sman, R.G.M.; Lubbersen, Y.S.; Schroën, C.G.P.H.; Pham, H.T.M.; Sarro, P.M.; Boom, R.M. Mixed motion in deterministic ratchets due to anisotropic permeability. *J. Colloid Interface Sci.* **2011**, *354*, 7–14. [CrossRef] [PubMed]
18. Kulrattanarak, T.; van der Sman, R.G.M.; Schroën, C.G.P.H.; Boom, R.M. Analysis of mixed motion in deterministic ratchets via experiment and particle simulation. *Microfluid Nanofluid* **2011**, *10*, 843–853. [CrossRef]
19. Liu, Z.; Huang, F.; Du, J.; Shu, W.; Feng, H.; Xu, X.; Chen, Y. Rapid isolation of cancer cells using microfluidic deterministic lateral displacement structure. *Biomicrofluidics* **2013**, *7*, 11801. [CrossRef] [PubMed]

20. Lindken, R.; Rossi, M.; Grosse, S.; Westerweel, J. Micro-Particle Image Velocimetry (microPIV): Recent developments, applications, and guidelines. *Lab Chip* **2009**, *9*, 2551–2567. [CrossRef] [PubMed]

21. Temiz, Y.; Lovchik, R.D.; Kaigala, G.V.; Delamarche, E. Lab-on-a-chip devices: How to close and plug the lab? *Microelectron. Eng.* **2015**, *132*, 156–175. [CrossRef]

22. Kruusing, A. Underwater and water-assisted laser processing: Part 2—Etching, cutting and rarely used methods. *Opt. Lasers Eng.* **2004**, *41*, 329–352. [CrossRef]

23. Olsen, M.G.; Adrian, R.J. Out of focus effects on particle image visibility and correlation in microscopic particle image velocimetry. *Exp. Fluids* **2000**, *29*, S166–S174. [CrossRef]

24. Blevins, R.D. *Flow-Induced Vibration*, 2nd ed.; repr. ed. w/updating and new preface; Krieger: Malabar, FL, USA, 1994; ISBN 9781575241838.

25. Kawamura, T.; Takami, H. Computation of high Reynolds number flow around a circular cylinder with surface roughness. *Fluid Dyn. Res.* **1986**, *1*, 145–162. [CrossRef]

Article

Nozzle-Shaped Electrode Configuration for Dielectrophoretic 3D-Focusing of Microparticles

Salini Krishna [1], Fadi Alnaimat [1,2,*] and Bobby Mathew [1,2]

[1] Mechanical Engineering Department, United Arab Emirates University, Al Ain, UAE
[2] National Water Center, United Arab Emirates University, Al Ain, UAE
* Correspondence: falnaimat@uaeu.ac.ae

Received: 11 July 2019; Accepted: 26 August 2019; Published: 31 August 2019

Abstract: An experimentally validated mathematical model of a microfluidic device with nozzle-shaped electrode configuration for realizing dielectrophoresis based 3D-focusing is presented in the article. Two right-triangle shaped electrodes on the top and bottom surfaces make up the nozzle-shaped electrode configuration. The mathematical model consists of equations describing the motion of microparticles as well as profiles of electric potential, electric field, and fluid flow inside the microchannel. The influence of forces associated with inertia, gravity, drag, virtual mass, dielectrophoresis, and buoyancy are taken into account in the model. The performance of the microfluidic device is quantified in terms of horizontal and vertical focusing parameters. The influence of operating parameters, such as applied electric potential and volumetric flow rate, as well as geometric parameters, such as electrode dimensions and microchannel dimensions, are analyzed using the model. The performance of the microfluidic device enhances with an increase in applied electric potential and reduction in volumetric flow rate. Additionally, the performance of the microfluidic device improves with reduction in microchannel height and increase in microparticle radius while degrading with increase in reduction in electrode length and width. The model is of great benefit as it allows for generating working designs of the proposed microfluidic device with the desired performance metrics.

Keywords: dielectrophoresis; focusing; microchannel; microfluidics; microparticles; modeling

---

1. Introduction

Microfluidic devices employ channels with hydraulic diameters smaller than 1 mm and this brings about certain advantages such as low sample and reagent requirement, small footprint, portability, and low power consumption [1,2]. Microfluidic devices are used for several applications including type- and size-based separation of microparticles [3]. For achieving separation, it is important to investigate the characteristics of each microparticle and subsequently move it to the appropriate group. This requires arranging randomly dispersed microparticles in an orderly fashion. Conventional biomedical devices such as flow cytometers (separating and counting of cells) and Coulter counters (counting of cells) are being miniaturized to avail several of the aforementioned benefits [4,5]. For these devices to function efficiently, it is important that cells pass through the sensing zone of the same individually, i.e., arrangement of randomly dispersed cells in an orderly fashion. Arranging of randomly dispersed microparticles is referred to as focusing [6]. Three-dimensional (3D) focusing refers to arranging randomly ordered microparticles in a single file. Several non-invasive actuation forces, including dielectrophoresis (DEP), are employed in microfluidic devices for 3D-focusing [6]. DEP refers to the movement of dielectric microparticles, suspended in an electrically conductive medium when subjected to a non-uniform electric field [7]. When the movement of the microparticle is towards the maxima of the electric field then, DEP is referred to as a positive-DEP (pDEP). On the other hand, DEP is referred to as negative-DEP (nDEP) when the movement of the microparticles is towards the minima of the electric field [7]. Employing DEP in microfluidic devices requires neither sheath flow nor specialized wafers and these represent the merits of DEP over other types of actuation forces. Additionally, DEP scales well with miniaturization, i.e., the electric field required for realizing DEP can be achieved with low voltages. The time averaged magnitude of the force associated with DEP is provided in Equation (1) [7].

$$\begin{bmatrix} F_{DEP,x} \\ F_{DEP,y} \\ F_{DEP,z} \end{bmatrix} = 2\pi\varepsilon_m r_p^3 Re[f_{CM}] \begin{bmatrix} \frac{\partial}{\partial x} \\ \frac{\partial}{\partial y} \\ \frac{\partial}{\partial z} \end{bmatrix} E_{RMS}^2 \tag{1}$$

The real part of the Clausius–Mossotti factor (f_{CM}) is dependent on the permittivity and conductivity of the medium and microparticle as shown in Equation (2); Equation (3) presents the relationship of conductivity of microparticle to its bulk and surface conductivities [8]. The polarity of $Re[f_{CM}]$ determines whether a microparticle will experience pDEP or nDEP; when $Re[f_{CM}]$ is positive, then the microparticle will experience pDEP and if $Re[f_{CM}]$ is negative, then the microparticle will experience nDEP.

$$Re[f_{CM}] = \frac{\left(\varepsilon_p + 2\varepsilon_m\right)\left(\varepsilon_p - \varepsilon_m\right) + \frac{\left(\sigma_p + 2\sigma_m\right)\left(\sigma_p - \sigma_m\right)}{\omega^2}}{\left(\varepsilon_p + 2\varepsilon_m\right)^2 + \frac{\left(\sigma_p + 2\sigma_m\right)^2}{\omega^2}} \tag{2}$$

$$\sigma_p = \sigma_{bulk,p} + 2\frac{K_{s,p}}{r_p} \tag{3}$$

It can be noticed from Equation (2) that as $\omega \to 0$, ε_m and ε_p do not influence $Re[f_{CM}]$. On the other hand, the influence of σ_m and σ_p on $Re[f_{CM}]$ is non-existent as $\omega \to \infty$. $Re[f_{CM}]$ of microparticles such as polystyrene, latex, and silica microparticles is positive and negative as $\omega \to 0$ and $\omega \to \infty$, respectively. In most microfluidic devices, the electrodes are located near the sidewalls and it is desired to 3D focus microparticles at the center of the microchannel; for this, it is necessary to subject microparticles to nDEP which in turn requires operating the microfluidic device at high operating frequency. Figure 1 provides the variation of $Re[f_{CM}]$ with operating frequency with respect to the radius and conductivity of the medium for polystyrene ($\varepsilon_p = 2.55\varepsilon_o$ F/m, $\rho_p = 1055$ kg/m^3, $\sigma_{bulk,p} = 0$, and $K_{s,p} = 2.85 \times 10^{-9}$ S) and silica microparticles ($\varepsilon_p = 3.8\varepsilon_o$ F/m, $\rho_p = 2000$ kg/m^3, $\sigma_{bulk,p} = 0$, and $K_{s,p} = 0.82 \times 10^{-9}$ S) [8]. Figure 1 reveals that these microparticles experience nDEP at high operating frequencies as mentioned earlier. Additionally, it can be noticed that these microparticles experience nDEP at high operating frequencies irrespective of the conductivity of the medium; this implies that the performance of any microfluidic device that employs nDEP for purposes of focusing is independent of the conductivity of the medium.

Figure 2 shows the microfluidic device analyzed in this article. The electrode configuration consists of two electrodes on the top surface and two electrodes on the bottom surface; all electrodes are right-triangle shaped and each electrode on the top surface aligns with a similar electrode on the bottom surface. For this electrode configuration, the minima of the gradient of the electric field is located at the center of the microchannel. Thus, any microparticle subjected to nDEP using the electrode configuration shown in Figure 2 will be pushed towards the center of the microchannel thereby achieving 3D-focusing of microparticles; the electrode configuration functions like a nozzle with regard to the movement of microparticles.

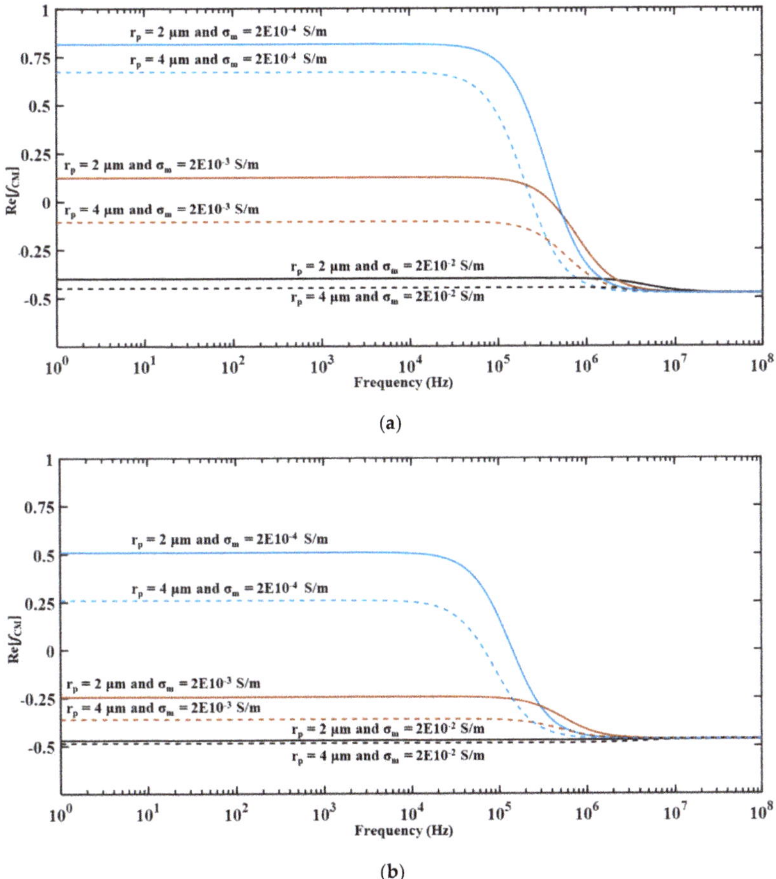

(a)

(b)

Figure 1. Variation of $Re[f_{CM}]$ with frequency for (**a**) polystyrene and (**b**) silica microparticles.

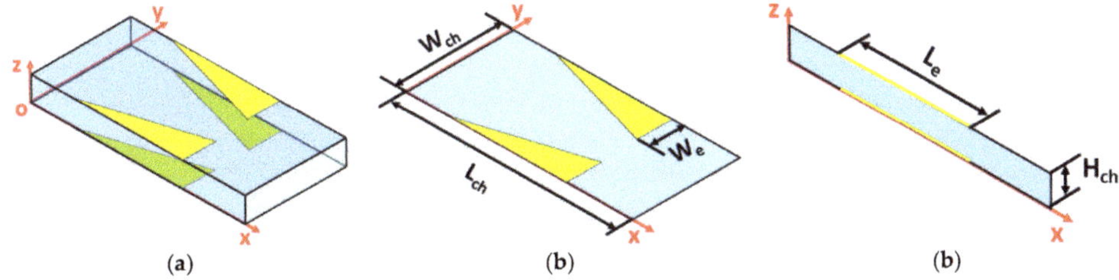

(a)   (b)   (b)

Figure 2. Schematic of the electrode configuration for dielectrophoresis (DEP)-based three dimensional (3D) focusing analyzed in this article (**a**) perspective view, (**b**) top view, and (**c**) side view.

Figure 3 depicts the top view of the trajectory of a typical microparticle along with the prominent forces acting on it inside the microfluidic device. In the regions away from the electrodes, the prominent force acting on the microparticle is that due to drag while in the regions near the electrodes, the prominent forces acting on the microparticle are those associated with drag and nDEP. In the presence of all forces, it is important that the microparticle is subjected to nDEP force for sufficient duration to be deflected along the desired path.

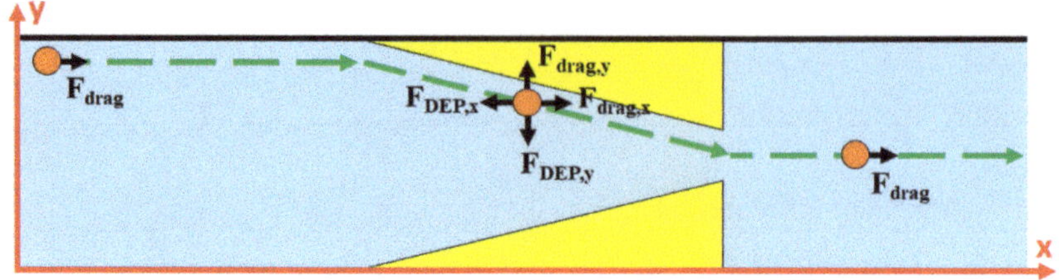

Figure 3. Schematic of the top-view of the trajectory of microparticle.

Regarding the electrode configuration shown in Figure 2, the starting width of the electrodes is zero and this allows for focusing microparticles that are originally located very close to the sidewalls; this is the primary merit of the proposed electrode configuration. Holmes et al. [9] previously showed that microfluidic device employing electrodes with non-zero starting width might not always focus microparticles that start from regions close to the sidewalls. Moreover, the chip-to-world electrical connection is easier for this electrode configuration compared with the electrode configuration consisting of multiple finite-sized interdigitated transducer (IDT) electrode pairs [10]. Additionally it is possible to apply nDEP over a greater length with the electrode configuration in Figure 1 compared with finite-sized IDT electrodes pairs. The ability to apply nDEP over a greater length allows for either handling greater throughput for a specific applied electric potential or reducing the applied electric potential for a specific throughput [10]. The proposed electrode configuration can be incorporated along with other phenomena, such as acoustophoresis, magnetophoresis, and inertial microfluidics, as required in the same microfluidic device [11–13].

Researchers have developed DEP based microfluidic devices for 3D-focusing of microparticles. Morgan et al. [14] modeled and built a microfluidic device, with two planar electrodes each on the top and bottom surfaces of the microchannel, for purposes of 3D-focusing. Each planar electrode has an upstream curved section and a downstream rectangular section. The four curved sections function together to direct all microparticles, by subjecting them to nDEP, to the central region of the microfluidic device between the four rectangular sections. Subsequently, the microparticles are subjected to nDEP by the rectangular section of the electrodes to 3D-focus them at the center of the microchannel. Yu et al. [15] created a microfluidic device with interdigitated transducer (IDT) electrodes, running over the entire circumference of a circular microchannel, for 3D-focusing of a microparticle. In this microfluidic device, an electric field is established between every electrode and its two neighboring electrodes. The minima of the gradient of the electric field is at the center of the microchannel due to which microparticles subjected to nDEP are focused at the center of the microchannel. Holmes et al. [9] constructed a microfluidic device with two pairs of trapezoid-shaped electrodes. Electrodes of each pair protrude into the microchannel from the sides; one electrode of a pair is located on the top surface and aligns with the other electrode, of the same pair, located on the bottom surface. In this device, microparticles experience nDEP in the vertical and horizontal directions to focus the same at the center of the microchannel. Alnaimat et al. [10] developed the mathematical model of a microfluidic device with multiple pairs of IDT electrodes on both sides of the bottom surface of the microchannel for DEP-based 3D-focusing. The analyzed microfluidic device can 3D-focus microparticles at any lateral location along the width of the microchannel; for equal applied voltages, the microparticle is focused at the center of the microchannel and for unequal applied voltages the microparticle is focused at locations other than the center of the microchannel. Alnaimat et al. [16] also modeled a microfluidic device with two continuous electrodes each on the top and bottom surfaces of the microchannel; the electrodes on each side formed a pair. The minima of the gradient of the electric field is located at the center of the microchannel when both pairs of electrodes have equal applied voltages, thereby allowing for 3D-focusing of microparticles at the same. On the other hand, for unequal applied voltages the minima of the gradient of the electric field is located away from the center of the microchannel thereby allowing 3D-focusing microparticles at locations other than the center of the microchannel, specifically at the location of the minimum gradient of electric field.

Several electrode configurations have been developed in the past for DEP-based focusing [4]; however, most do not work well for microparticles very close to the sidewalls and can handle only low throughout. As already detailed, the electrode configuration presented in this article can overcome these drawbacks. Regarding the model developed in this article, it is a dynamic model, thereby allowing for determining the axial distance as well as duration required for reaching steady state conditions unlike static models. Knowledge of the axial distance required for reaching steady state dictates the actual size of the microfluidic device.

2. Theoretical Model

The model corresponding to the microfluidic device shown in Figure 2 is developed in this section. The purpose of the model is to calculate the trajectory of the microparticle in the microfluidic device and in turn determine the vertical and horizontal focusing parameters. The trajectory of the microparticle is described by Equation (4); it is based on Newton's 2nd law and relates the acceleration of the microparticle to the net external force acting on the microparticle. External forces acting on the microparticle are drag, gravity, virtual mass, buoyancy, and DEP [10,16,17].

$$\begin{bmatrix} \ddot{x}_p \\ \ddot{y}_p \\ \ddot{z}_p \end{bmatrix} + 9\frac{\mu_m}{(2\rho_p+\rho_m)r_p^2}\begin{bmatrix} \dot{x}_p \\ \dot{y}_p \\ \dot{z}_p \end{bmatrix} - \frac{\rho_m}{(2\rho_p+\rho_m)}\begin{bmatrix} \dot{u}_{m,x} \\ \dot{u}_{m,y} \\ \dot{u}_{m,z} \end{bmatrix} - 9\frac{\mu_m}{(2\rho_p+\rho_m)r_p^2}\begin{bmatrix} u_{m,x} \\ u_{m,y} \\ u_{m,z} \end{bmatrix} + 2\frac{(\rho_p-\rho_m)}{(2\rho_p+\rho_m)}\begin{bmatrix} 0 \\ 0 \\ g \end{bmatrix} - \frac{3}{2}\frac{\varepsilon_m Re[f_{CM}]}{(2\rho_p+\rho_m)}\begin{bmatrix} \frac{\partial}{\partial x} \\ \frac{\partial}{\partial y} \\ \frac{\partial}{\partial z} \end{bmatrix}E_{RMS}^2 = 0 \quad (4)$$

The initial conditions associated with the Equation (4) include the initial displacements and velocities as represented in Equations (5) and (6), respectively [10,16,17]. The microparticle can occupy any location along the cross-section of the microchannel at the start of the electrode configuration and this is presented in Equation (5). Moreover, it is medium that is dragging the microparticle to the start of the electrode configuration due to which the velocity of the microparticle at the start of the electrode configuration is same as that of the medium as shown in Equation (6).

$$\begin{bmatrix} x_p \\ y_p \\ z_p \end{bmatrix}\Bigg|^{t=0} = \begin{bmatrix} 0 \\ Y \\ Z \end{bmatrix} \quad (5)$$

$$\begin{bmatrix} \dot{x}_p \\ \dot{y}_p \\ \dot{z}_p \end{bmatrix}\Bigg|^{t=0} = \begin{bmatrix} u_{m,x}|_{x_p=0,y_p=Y,z_p=Z} \\ u_{m,y}|_{x_p=0,y_p=Y,z_p=Z} \\ u_{m,z}|_{x_p=0,y_p=Y,z_p=Z} \end{bmatrix} \quad (6)$$

From Equation (4) it is evident that information on the electric field and velocity of the medium are required for solving the same. The information on electric field can be determined only if information on electric potential inside the microchannel is available. Information on electric potential inside the microchannel will be available upon solving Equation (7). Afterwards, solving Equation (8) will provide information on electric field inside the microchannel [10,16,17].

$$\left(\frac{\partial^2}{\partial x^2} + \frac{\partial^2}{\partial y^2} + \frac{\partial^2}{\partial z^2}\right)V = 0 \quad (7)$$

$$\begin{bmatrix} E_x \\ E_y \\ E_z \end{bmatrix} = -\begin{bmatrix} \frac{\partial}{\partial x} \\ \frac{\partial}{\partial y} \\ \frac{\partial}{\partial z} \end{bmatrix}V \quad (8)$$

The boundary conditions associated with Equation (7) include applied electric potential on the electrodes and electrical insulation elsewhere. All governing equations, i.e., Equations (4), (7) and (8), are numerically solved. The finite difference method (FDM) is used for solving Equations (4) and (7). For implementing FDM, second order central difference scheme replaces the first order differential term of Equation (4) while the second order central difference scheme of Equations (4) and (7) replaces second order central differential term. In order to evaluate the electric field, the first order differential terms of Equation (8) are replaced by second order difference schemes. The time step for Equation (4) is held at 0.00001 s. The node-to-node distance in the x-and z-directions are set at 25 µm and 1 µm, respectively. The node-to-node distance in the y-direction is determined based on the node-to-node in the x-direction and the electrode dimensions; this approach allows for having a structured grid inside the computational domain even though the electrode has a triangular shape. The details of implementing FDM to solve Equations (7) and (8) can be found in Alnaimat et al. [10,16] and Mathew et al. [17]. The implementation of FDM in Equation (4) and its subsequent rearrangement results in Equation (9) which can be used for finding displacements at the next time step. Equation (9) is evaluated until the axial displacement of the microparticle becomes equal to the length of the microfluidic device.

$$
\begin{bmatrix} x_p^{n+1} \\ y_p^{n+1} \\ z_p^{n+1} \end{bmatrix} = \left(\frac{1}{\frac{1}{\Delta t^2} + \frac{9}{2\Delta t} \frac{\mu_m}{(2\rho_p + \rho_m) r_p^2}} \right) \left(\frac{2}{\Delta t^2} \begin{bmatrix} x_p^n \\ y_p^n \\ z_p^n \end{bmatrix} - \left(\frac{1}{\Delta t^2} - \frac{9}{2\Delta t} \frac{\mu_m}{(2\rho_p + \rho_m) r_p^2} \right) \begin{bmatrix} x_p^{n-1} \\ y_p^{n-1} \\ z_p^{n-1} \end{bmatrix} + \frac{\rho_m}{(2\rho_p + \rho_m)} \begin{bmatrix} \dot{u}_{m,x} \\ \dot{u}_{m,y} \\ \dot{u}_{m,z} \end{bmatrix} +
$$
$$
9 \frac{\mu_m}{(2\rho_p + \rho_m) r_p^2} \begin{bmatrix} u_{m,x} \\ u_{m,y} \\ u_{m,z} \end{bmatrix} - 2 \frac{(\rho_p - \rho_m)}{(2\rho_p + \rho_m)} \begin{bmatrix} 0 \\ 0 \\ g \end{bmatrix} + \frac{3}{2} \frac{\varepsilon_m Re[f_{CM}]}{(2\rho_p + \rho_m)} \begin{bmatrix} \frac{\partial}{\partial x} \\ \frac{\partial}{\partial y} \\ \frac{\partial}{\partial z} \end{bmatrix} E_{RMS}^2 \right)
\tag{9}
$$

The velocity of the medium is dictated by the Navier–Stokes equations and the continuity equation. Additionally, the flow is usually fully developed at the start of the electrode configuration and this allows for reducing the governing equations to a simplified version of one of the components of the Navier–Stokes equation. The velocity of flow under these conditions is provided in Equation (10) [7]. Since the flow is fully developed, the medium has velocity only in the axial direction as stated in Equation (10).

$$
\begin{bmatrix} u_{m,x} \\ u_{m,y} \\ u_{m,z} \end{bmatrix} = \begin{bmatrix} \dfrac{48 Q_m \sum_{i=1,3,5}^{\infty} \left(\frac{(-1)^{\left(\frac{i-1}{2}\right)}}{i^3} \right) \cos\left[i \frac{\pi}{W_{ch}} \left(\frac{W_{ch}}{2} - y \right) \right] \left\{ 1 - \dfrac{\cosh\left[i \frac{\pi}{W_{ch}} \left(\frac{H_{ch}}{2} - z \right) \right]}{\cosh\left(\frac{i\pi}{2} \frac{H_{ch}}{W_{ch}} \right)} \right\}}{\pi^3 W_{ch} H_{ch} \left[1 - \dfrac{192 W_{ch}}{\pi^5 H_{ch}} \sum_{i=1,3,5}^{\infty} \frac{\tanh\left(i \frac{\pi}{2} \frac{H_{ch}}{W_{ch}} \right)}{i^5} \right]} \\ 0 \\ 0 \end{bmatrix}
\tag{10}
$$

The performance of the microfluidic device used for 3D-focusing is quantified in terms of Equations (11) and (12). Equations (11) and (12) represent the degree of 3D-focusing of microparticles. Mathematically, Equations (11) and (12) represent the standard deviation associated with the final position of the microparticle from the ideal 3D-focused position of the microparticle; performance of the microfluidic device studied by Morgan et al. [14] for 3D-focusing was quantified in a similar manner. Based on the working of the microfluidic device detailed above, the ideal 3D-focused position of the microparticle is the center of the microparticle. Reduction in magnitude of the vertical and horizontal focusing parameters indicates the closeness of the microparticles to the horizontal and vertical planes passing through the center of the microchannel, respectively. It is stressed here that smaller the focusing parameters, better the performance of the microfluidic device in achieving 3D-focusing. Both focusing parameters need to be small for the proposed device to realize 3D-focusing.

$$\Delta w = \sqrt{\frac{\sum_{i=1}^{N}\left(y_{p,i} - \frac{W_{ch}}{2}\right)^2}{N-1}} \tag{11}$$

$$\Delta d = \sqrt{\frac{\sum_{i=1}^{N}\left(z_{p,i} - \frac{H_{ch}}{2}\right)^2}{N-1}} \tag{12}$$

3. Model Validation

Holmes et al. [9] developed a microfluidic device with two trapezoid shaped electrodes located on the top and bottom surfaces of the microchannel; the device is schematically shown in Figure 4. It can be noticed that the electrode configuration of Holmes et al. [9] is similar to that considered in this article; the difference being the non-zero starting width for the electrode configuration of Holmes et al. [9]. Holmes et al. [9] provided trajectories of microparticles in their device and this information is used for validating the model developed in this article. The width and height of the microchannel of the microfluidic device constructed by Holmes et al. [9] is 250 and 40 μm, respectively; the electrode length is 500 μm with the starting and ending widths of the electrodes being 55 and 105 μm, respectively. Holmes et al. [9] conducted tests for different applied electric potentials (0, 5, 10, 15, and 20 V_{pp}) and documented the trajectories for all electric potentials. Holmes et al. [9] conducted tests using 6 μm (diameter) latex beads suspended in water. Information on volumetric flow rate (s) is not available in Holmes et al. [9]; however, they mention that for applied electric potential of 20 V_{pp} the axial velocity of microparticles is 1.05 mm/s after achieving cent percent 3D-focusing. For applied electric potential of 20 V_{pp}, the microparticles are 3D-focused near the center of the microchannel and this coupled with the fact the microparticles are transported axially by the medium allows for assuming the centerline velocity of the medium to be same as the velocity of the microparticles. Based on the assumption that the centerline velocity of the medium to be 1.05 mm/s, the volumetric flow rate is determined and employed in the model. Figure 4 compares the trajectories of Holmes et al. [9] and the model; Holmes et al. [9] provided information on the trajectories at the end of the electrodes as shown in Figure 4. The model is used to calculate the trajectory of microparticles starting from several locations along the width of the microchannel as depicted in Figure 4. Upon comparison of the trajectories, it can be noticed that there is good match between the two and this validates the model.

The model detailed here is based on several assumptions all of which are detailed in Mathew et al. [18] and includes negligible interaction between microparticles as well as between the microparticle and medium. These assumptions are satisfactorily achieved in all microfluidic devices handling dilute samples. On the other hand, when handling dense suspensions these assumptions do not hold leading to the need for modifying the coefficient associated with the drag equation. Knoerzer et al. [19] developed an algorithm, called dynamic drag force based on iterative density mapping, for determining the trajectory of microparticles while including the effect of these interactions; their model has been experimentally validated.

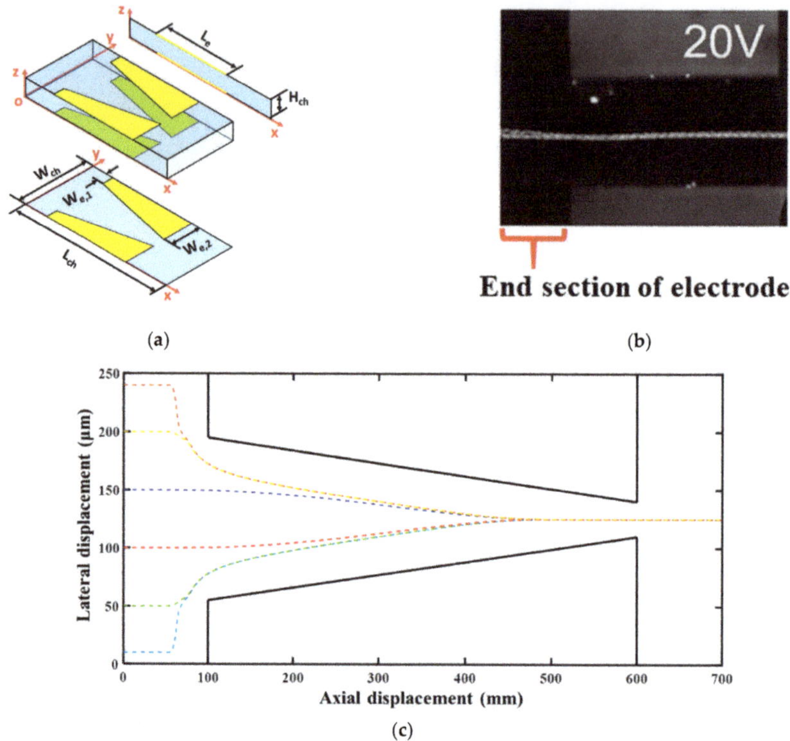

Figure 4. (**a**) Microfluidic device of Holmes et al. [9], (**b**) trajectories in the microfluidic device of Holmes et al. [9], and (**c**) trajectories based on the model (r_p = 6 µm, W_{ch} = 250 µm, H_{ch} = 40 µm, L_e = 500 µm, L_{ch} = 700 µm, $W_{e,1}$ = 55 µm, $W_{e,2}$ = 105 µm). Figure 4b is reproduced with permission from Elsevier, Copyright (2006) from Holmes et al. [9].

4. Results and Discussions

The ability of the microfluidic device in achieving 3D-focusing is investigated in this section. Studies are carried out using polystyrene microparticles suspended in water (ε_m = 78.5ε_0, ρ_m = 998 kg/m^3, μ_m = 10^{-3} Pa·s) [8]. The trajectories of 81 microparticles (uniformly distributed) originating from the inlet of the microchannel are shown in Figure 5 along with projections of each trajectory on to vertical and horizontal planes; similar simulations with 100 microparticles were carried out by Morgan et al. [9]. It can be noticed that the proposed microfluidic device can achieve 3D-focusing at the center of the microchannel irrespective of the initial position of the microparticle. For this electrode configuration, the horizontal DEP force in the vertical plane passing through the center of the microchannel is zero and this leads to all microparticles being horizontally pushed to this vertical plane. Similarly, the vertical component of the nDEP force is zero in the horizontal plane passing through the center of the microchannel which causes all microparticles to experience vertical motion towards this horizontal plane. The vertical and horizontal motions of the microparticles happen simultaneously, thereby leading them to be almost 3D-focused at the center of the microchannel.

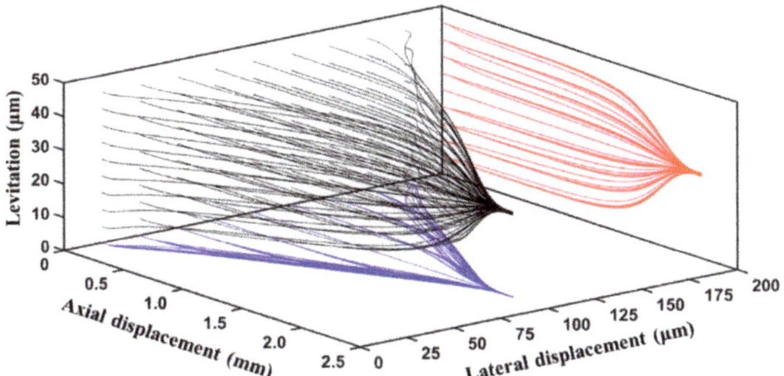

Figure 5. Trajectories of microparticles in the microfluidic device subjected to negative-DEP (nDEP) force (r_p = 2 μm, $V_1 = V_2$ = 20 V_{pp}, L_e = 2000 μm, L_{ch} = 2200 μm, W_e = 80 μm, Q_m = 100 μL/h, W_{ch} = 200 μm, H_{ch} = 50 μm).

From the equations presented in the previous section of the article it is evident that the performance of the microfluidic device depends on parameters such as microchannel height, electrode width and length, volumetric flow rate, and applied electric potential. In the following part of this article, the influence of these parameters are investigated.

Figure 6 presents the variation of horizontal and vertical focusing parameters with respect to the applied electric potential. It can be noticed that with increase in applied electric potential the focusing parameters decrease indicating the enhancement in performance of the microfluidic device in achieving 3D-focusing. An increase in applied electric potential increases the electric potential inside the microchannel which in turn increases all components of the nDEP force inside the microchannel. This increase in components of the nDEP force causes the microparticles to move closer to the horizontal and vertical planes passing through the center of the microchannel, while passing through the region of the electrodes, thereby leading to the observed increase in performance of the microfluidic device. At low values of applied electric potential, the focusing parameters is almost equal to the focusing parameters in the absence of applied electric potential. The horizontal and vertical focusing parameters in the absence of applied electric potential are ca. 52 and ca. 13 μm, respectively. This indicates that there exists a threshold applied electric potential below which no appreciable degree of focusing can be expected when all other parameters are maintained constant.

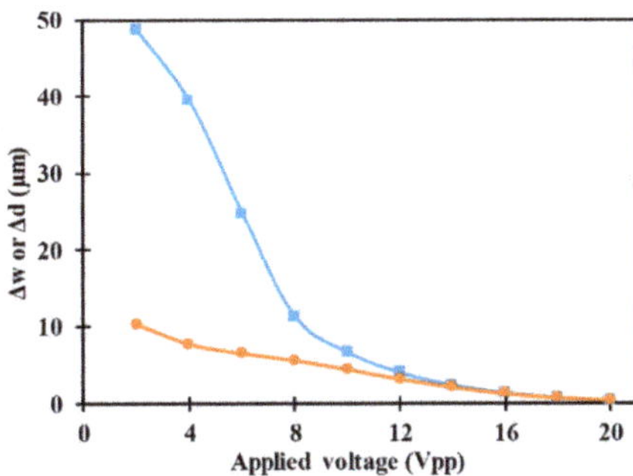

Figure 6. Variation of Δw (■) and Δd (●) with applied voltage (r_p = 2 μm, L_e = 2000 μm, L_{ch} = 2200 μm, W_e = 80 μm, Q_m = 100 μl/h, W_{ch} = 200 μm, H_{ch} = 50 μm).

Additionally, it can also be noticed from Figure 6 that the vertical focusing parameter is significantly smaller than the horizontal focusing parameter for low applied electric potentials. The width of the microchannel is much greater than the height of the microchannel; the small height implies that the electric field, and subsequently the nDEP force, is stronger along the height than along the width. This, coupled with the fact that the average displacement required to reach the horizontal plane passing through the center of the microchannel is smaller than the average displacement required to reach the vertical plane passing through the center of the microchannel, leads to vertical focusing parameter being smaller than horizontal focusing parameter. At high applied voltages, the electric field and the nDEP forces are strong in both directions to realize excellent degree of focusing in the vertical and horizontal directions thereby leading to both focusing parameters assuming low values.

Figure 7 shows the influence of volumetric flow rate on the horizontal and vertical focusing parameters. Increase in volumetric flow rate increases the horizontal focusing parameter indicating degradation in the ability of the microfluidic device to achieve 3D-focuisng. Increase in volumetric flow rate increases the x-component of the force associated with drag at every location inside the microchannel while maintaining the x-component of the nDEP force constant. Thus, with an increase in volumetric flow rate, the velocity of the microparticle in the axial direction increases. An increase in velocity of the microparticle in the axial direction reduces the duration for which the same is subjected to y-component of nDEP force. This reduction in duration leads to deterioration of the degree of deflection of the microparticle towards the interior of the microchannel which ultimately causes the microparticle to move into the region between the top and bottom electrodes. Once the microparticle moves into this space, between the electrodes, it experiences weakening nDEP force due to which there is no further chance of being focused in the horizontal direction. Increase in volumetric flow rate increases the vertical focusing parameter as well for the same reason, i.e. reduction in residence time for which the z-component of nDEP force acts on the microparticle. It can be noticed form Figure 7 that the variation of the vertical focusing parameter with volumetric flow rate is much smaller than that of horizontal focusing parameter with respect to volumetric flow rate and it is due to the fact that the width of the microchannel is much greater than the height of the microchannel. Additionally, it can be concluded from Figure 7 that an increase in volumetric flow rate will cause the focusing parameters to increase till they become equal to the focusing parameters of a device without any focusing ability, i.e., horizontal and vertical focusing parameters of ca. 52 µm and ca. 13 µm, respectively. This indicates that there is an upper limit for volumetric flow rate beyond which no focusing is possible when all other parameters are maintained constant.

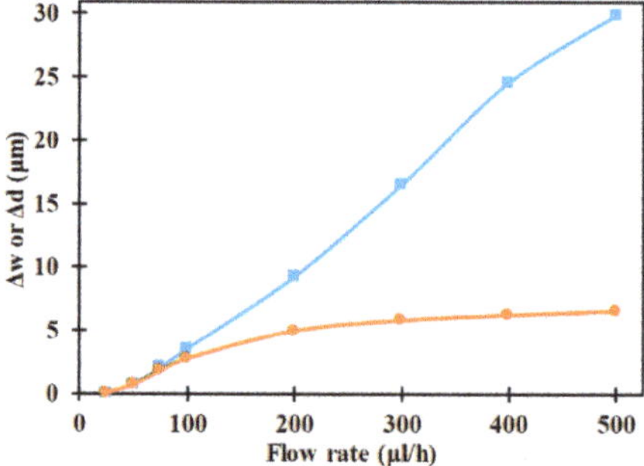

Figure 7. Variation of Δw (■) and Δd (●) with flow rare (r_p = 2 µm, L_e = 2000 µm, L_{ch} = 2200 µm, W_e = 80 µm, V_1 = V_2 = 12 V_{pp}, W_{ch} = 200 µm, H_{ch} = 50 µm).

Figure 8 represents the influence of electrode length on the vertical and horizontal focusing parameters. It can be noticed that with an increase in electrode length the horizontal focusing parameter reduces which in turn depict enhancement in performance of the microfluidic device in achieving 3D-focusing. An increase in length of the electrodes reduces the orientation of the electrodes, with respect to the sidewalls, which in turn enhances the y-component of the nDEP force acting on the microparticle. Additionally, an increase in electrode length reduces the x-component of the nDEP force which along with the x-component of the drag remaining constant leads to an increase in velocity of the microparticle. The increase in the y-component of the nDEP force with the increase in electrode length is greater than the increase in velocity of the microparticle, in the axial direction, under the same conditions and these lead to the reduction in horizontal focusing parameter as observed in Figure 7. Regarding the vertical focusing parameter, it reduces with the increase in electrode length as well and this can be attributed to the increase in the residence time for which the microparticles are subjected to the nDEP force in the vertical direction. Additionally, it can be noticed that the variation of the vertical focusing parameter with electrode length is smaller than that associated with horizontal focusing parameter and it is due to the microchannel height being smaller than the microchannel width.

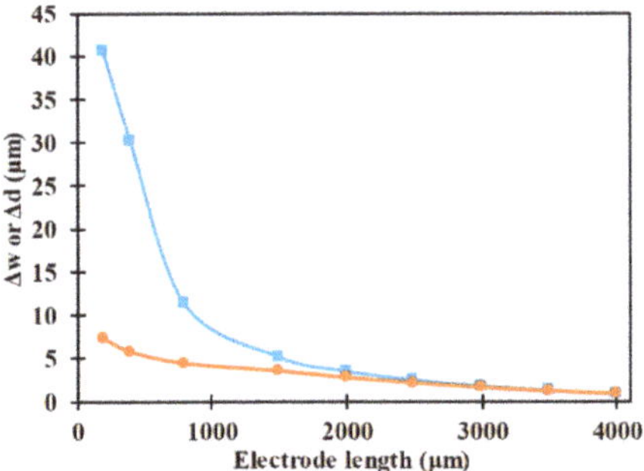

Figure 8. Variation of Δw (■) and Δd (●) with electrode length ($r_p = 2$ μm, $W_e = 80$ μm, $L_{ch} = L_e + 200$ μm, $V_1 = V_2 = 12$ V_{pp}, $W_{ch} = 200$ μm, $H_{ch} = 50$ μm).

The influence of the electrode width on the vertical and horizontal focusing parameters is analyzed in Figure 9; focusing parameters decrease with the increase in electrode width, which is indicative of the improvement in the performance of the microfluidic device in achieving 3D-focusing. Increase in electrode width increases the lateral displacement of the microparticle towards the center of the microchannel which in turn leads to it ending up closer to the center of the microchannel after passing over the length of the electrodes. With an increase in electrode width, the orientation of the electrode with respect to the sidewall increases, thereby leading to increase and decrease in the x- and y-components of the nDEP force, respectively. The increase in the x-component of the nDEP forces leads to reduction in the velocity of the microparticles in the axial direction. The reduction in the y-component of the nDEP force, with change in electrode width, is smaller than the reduction in the axial velocity, with change in electrode width, and this allows for the microparticle to be subjected to nDEP force for duration required to achieve the desired deflection.

Regarding the vertical focusing parameter, it remains almost constant with an increase in electrode width before significantly reducing at high electrode width; the vertical focusing parameters at low electrode widths is almost equal to the vertical focusing parameter in the absence of nDEP forces, i.e., ca. 13 μm. With an increase in electrode width the number of microparticles occupying the space between the top and bottom electrodes increase along with the increase in degree of non-uniformity of the electric field. These are the reasons for the observed trend in the variation of vertical focusing parameter with electrode width. Since microparticles are uniformly distributed over the cross-section of the microchannel, the distance between each sidewall and the nearest column of microparticles is

30 µm, while that between each sidewall and the nearest second and third columns of microparticles are 60 µm and 90 µm, respectively. Thus, for electrode width of 25 µm no microparticles are present in the space between the top and bottom electrodes while for electrode widths of 50 µm and 75 µm the number of columns of microparticles present in the space between the electrodes are one and two, respectively. For an electrode width of 125 µm, the first four columns of microparticles from each sidewall occupy the space between the electrodes and this coupled with enhancement z-component of nDEP force, associated with the increase in electrode width, leads to strong vertical focusing as depicted in Figure 9.

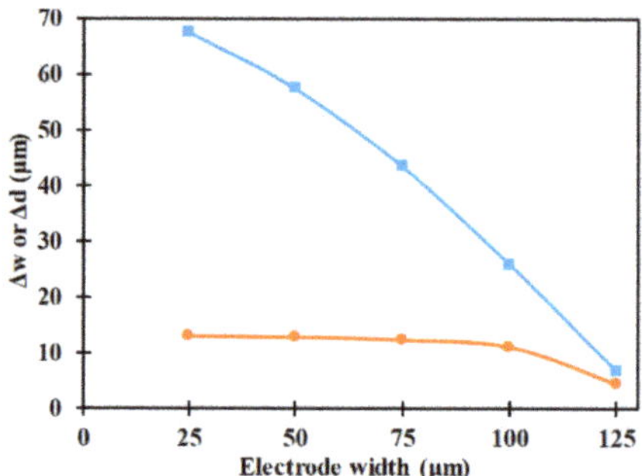

Figure 9. Variation of Δw (■) and Δd (●) with electrode width (r_p = 2 µm, L_e = 2000 µm, L_{ch} = 2200 µm, V_1 = V_2 = 12 V_{pp}, W_{ch} = 200 µm, H_{ch} = 50 µm).

Figure 10 shows the variation of focusing parameters with respect to microchannel height. It can be noticed that an increase in microchannel height initially decreases the focusing parameters before increasing them. This indicates the existence of an optimal microchannel height for purposes of 3D-focusing using the electrode configuration shown in Figure 2. The initial increase in microchannel height leads to an increase in non-uniformity of the electric field inside the microchannel and this leads to an increase in the nDEP force in all directions thereby leading to the decrease in focusing parameters. On the other hand, with a further increase in microchannel height the electric field inside the microchannel decreases leading to the reduction of the nDEP force in all directions which in turn leads to the degradation in performance of the microfluidic device, i.e., increase in focusing parameters. It can also be noticed that as the microchannel height becomes equal to the microchannel width, the focusing parameters become comparable; this is due to the nDEP forces in both directions becoming comparable.

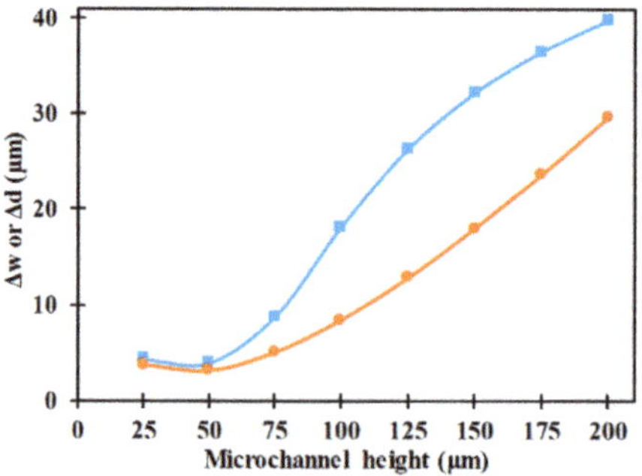

Figure 10. Variation of Δw (■) and Δd (●) with electrode height (r_p = 2 µm, L_e = 2000 µm, L_{ch} = 2200 µm, W_e = 80 µm, V_1 = V_2 = 12 V_{pp}, W_{ch} = 200 µm).

Figure 11 depicts the influence of microparticle radius on the vertical and horizontal focusing parameters associated with the microfluidic device. It can be noticed that increase in microparticle radius reduces the focusing parameters which in turn indicates that enhancement in performancve of the microfluidic device. This is because the nDEP force depends on the cubic power of the radius, as can be observed from Equation (1).

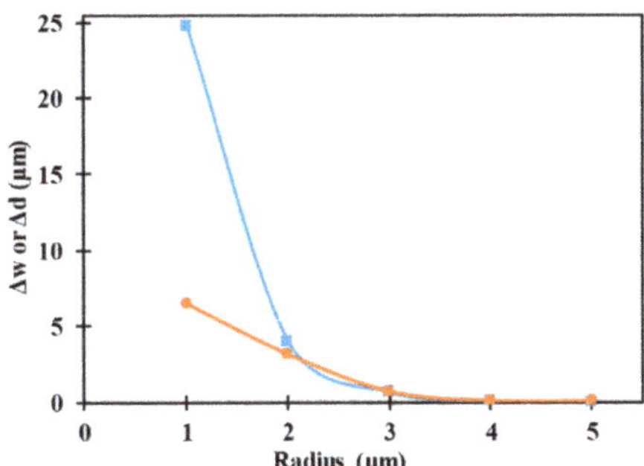

Figure 11. Variation of Δw (■) and Δd (●) with microparticle radius ($W_e = 80$ μm, $L_e = 2000$ μm, $L_{ch} = 2200$ μm, $V_1 = V_2 = 12$ V$_{pp}$, $W_{ch} = 200$ μm, $H_{ch} = 50$ μm).

5. Conclusions

This article presents a nozzle-shaped electrode configuration for dielectrophoresis-based 3D-focusing of microparticles. A mathematical model of this microfluidic device is developed and used for studying its feasibility in achieving 3D-focusing. Additionally, the mathematical model is used for parametric study to understand the influence of operating/geometric parameters on the 3D-focusing efficiency of the device. The parameters analyzed include applied electric potential, volumetric flow rate, electrode width and length, and microchannel height. The performance of the microfluidic device is quantified in terms of horizontal and vertical focusing parameters which are mathematically equivalent to the standard deviation of the horizontal and vertical positions of the microparticle evaluated at the exit section on the microfluidic device. The focusing parameters decrease with an increase in applied electric potential, thereby indicating enhancement in performance of the microfluidic device. The focusing parameters increase with an increase in volumetric flow rate, which is indicative of the degradation in performance of the microfluidic device. An increase in electrode width and length improves the focusing parameters, thereby indicating the enhancement in performance of the microfluidic device. An increase in microchannel height generally increases the focusing parameters, thereby leading to degradation in performance of the microfluidic device. The focusing parameters are influenced by the radius of the microparticle such that an increase in radius of microparticle reduces the focusing parameters, thereby indicating improvement in the performance of the microfluidic device. The model is particularly useful in selecting the geometric and operating parameters of the proposed microfluidic device so as to operate it at the desired performance metrics, i.e., focusing parameters.

Author Contributions: Formal analysis, S.K.; Conceptualization, F.A. and B.M.; Funding acquisition, F.A.; Investigation, S.K., F.A. and B.M.; Methodology, S.K., F.A. and B.M.; Software, S.K. and B.M.; Writing—original draft, S.K., F.A. and B.M.; Writing—review and editing, S.K., F.A. and B.M.

Funding: Authors acknowledge funding from National Water Center of UAEU under grant #31R153.

Conflicts of Interest: Authors report no conflict of interest.

Nomenclature

Δd	*horizontal focusing parameter (m or µm)*
E	*electric field (V/m)*
F	*force (N)*
H	*height (m or µm)*
K_s	*surface conductance (S)*
N	*number of microparticles considered for determining focusing parameters*
$Re[f_{CM}]$	*real part of Clausius–Mossotti factor*
r	*radius (m or µm)*
u	*velocity (m/s)*
V	*applied electric voltage (V)*
W	*width (m or µm)*
Δw	*vertical focusing parameter (m or µm)*
x	*displacement in the x-direction (m or µm)*
Y	*initial position of microparticles along the width of the microchannel (m or µm)*
y	*displacement in the y-direction (m or µm)*
Z	*initial positon of microparticle along the height of the microchannel (m or µm)*
z	*displacement in the z-direction (m or µm)*

Greek alphabet

E	permittivity (F/m)
ε_0	permittivity of free space (F/m)
Σ	conductivity (S/m)
P	density (kg/m^3)
Ω	operating frequency (rad/s)

Subscripts

1	*electrode pair on the right-hand side with respect to the flow direction*
2	*electrode pair on the left-hand side with respect to the flow direction*
bulk	*bulk*
DEP	*dielectrophoresis*
m	*medium*
p	*microparticle*

References

1. Mathew, B.; Weiss, L. MEMS Heat Exchangers. In *Materials and Failures in MEMS and NEMS*; Tiwari, A., Raj, B., Eds.; Wiley- Scrivener: New York, NY, USA, 2015; pp. 63–120.
2. Sackmann, E.K.; Fulton, A.L.; Beebe, D.J. The present and future role of microfluidics in biomedical research. *Nature* **2014**, *507*, 181–189. [CrossRef] [PubMed]
3. Chen, J.; Chen, D.; Xie, Y.; Yuan, T.; Chen, X. Progress of microfluidics for biology and medicine. *Nano-Micro Lett.* **2013**, *5*, 66–80. [CrossRef]
4. Yang, S.-Y.; Hsiung, S.-K.; Hung, Y.-C.; Chang, C.-M.; Liao, T.-L.; Lee, G.-B. A cell counting/sorting system incorporated with a microfabricated flow cytometer chip. *Meas. Sci. Technol.* **2006**, *17*, 2001–2009. [CrossRef]
5. Wu, Y.; Benson, J.D.; Critser, J.K.; Almasri, M. MEMS-based Coulter counter for cell counting and sizing using multiple electrodes. *J. Micromech. Microeng.* **2010**, *20*, 085035. [CrossRef]
6. Xuan, X.; Zhu, J.; Church, C. Particle focusing in microfluidic devices. *Microfluid. Nanofluid.* **2010**, *9*, 1–16. [CrossRef]
7. Alazzam, A.; Mathew, B.; Khashan, S. Microfluidic platforms for bio-applications. In *Advanced Mechatronics and MEMS Devices II*; Zhang, D., Wei, B., Eds.; Springer: Cham, Switzerland, 2017; pp. 253–282.
8. Alnaimat, F.; Ramesh, S.; Adams, S.; Parks, N.; Lewis, C.; Wallace, K.; Mathew, B. Model based performance study of dielectrophoretic flow separator. *IEEE Sens. Lett.* **2019**, *6*, 1–4. [CrossRef]
9. Holmes, D.; Morgan, H.; Green, N.G. High throughput particle analysis: Combining dielectrophoretic particle focusing with confocal optical detection. *Biosens. Bioelectron.* **2006**, *21*, 1621–1630. [CrossRef] [PubMed]

10. Alnaimat, F.; Ramesh, S.; Alazzam, A.; Hilal-Alnaqbi, A.; Waheed, W.; Mathew, B. Dielectrophoresis-based 3D-focusing of microscale entities in microfluidic device. *Cytom. Part A* **2018**, *93*, 811–821. [CrossRef] [PubMed]

11. Urbansky, A.; Ohlsson, P.; Lenhsof, A.; Garofalo, F.; Scheding, S.; Laurell, T. Rapid and effective enrichment of mononuclear cells from blood using acoustophoresis. *Sci. Rep.* **2017**, *7*, 17161. [CrossRef] [PubMed]

12. Mekkaoui, S.; Le Roy, D.; Audry, M.-C.; Lachambre, J.; Dupuis, V.; Desgouttes, J.; Deman, A.-L. Arrays of high aspect ratio magnetic microstructures for large trapping throughput in lab-on-chip systems. *Microfluid. Nanofluid.* **2018**, *22*, 119. [CrossRef]

13. Nguyen, N.; Thurgood, P.; Arash, A.; Pirogova, E.; Baratchi, S.; Khoshmanesh, K. Inertial microfluidics with integrated vortex generators using liquid metal droplets as fugitive ink. *Adv. Funct. Mater.* **2019**, *29*, 1901998. [CrossRef]

14. Morgan, H.; Holmes, D.; Green, N.G. 3D focusing of nanoparticles in microfluidic channels. *IEE Proc.-Nanatechnol.* **2003**, *150*, 76–80. [CrossRef] [PubMed]

15. Yu, C.; Vykoukal, J.; Vykoukal, D.M.; Schwartz, J.; Shi, L.; Gascoyne, P.R.C. A three-dimensional dielectrophoretic particle focusing channel for microcytometry applications. *J. Microelectromech. Syst.* **2005**, *14*, 480–487.

16. Alnaimat, F.; Ramesh, S.; Hilal-Alnaqbi, A.; Alazzam, A.; Dagher, S.; Mathew, B. 3D focusing of micro-scale entities in dielectrophoretic microdevice. *Med. Devices Sens.* **2019**, *2*, e10028. [CrossRef]

17. Mathew, B.; Alazzam, A.; Kashan, S.; El-Khasawneh, B. Path of microparticles in a microfluidic device employing dielectrophoresis for hyperlayer field-flow fractionation. *Microsyst. Technol.* **2016**, *22*, 1721–1732. [CrossRef]

18. Mathew, B.; Alazzam, A.; Abutayeh, M.; Gawanmeh, A.; Khashan, S. Modeling the trajectory of microparticles subjected to dielectrophoresis in a microfluidic device for field flow fractionation. *Chem. Eng. Sci.* **2015**, *138*, 266–280. [CrossRef]

19. Knoerzer, M.; Szydzlk, C.; Tovar-Lopez, F.J.; Tang, X.; Mitchell, A.; Khoshmanesh, K. Dynamic drag force based on iterative density mapping: A new numerical tool for three-dimensional analysis of particle trajectories in a dielectrophoretic system. *Electrophoresis* **2016**, *37*, 645–657. [CrossRef] [PubMed]

Article

Real Time Electronic Feedback for Improved Acoustic Trapping of Micron-Scale Particles

Charles P. Clark [1], Vahid Farmehini [2], Liam Spiers [1], M. Shane Woolf [1], Nathan S. Swami [2] and James P. Landers [1,3,*]

1   Department of Chemistry, University of Virginia, Charlottesville, VA 22903, USA
2   Department of Electrical and Computer Engineering, University of Virginia, Charlottesville, VA 22903, USA
3   Departments of Mechanical Engineering and Pathology, University of Virginia, Charlottesville, VA 22903, USA
*   Correspondence: jpl5e@virginia.edu; Tel.: +1-434-243-8658

Received: 6 June 2019; Accepted: 11 July 2019; Published: 21 July 2019

Abstract: Acoustic differential extraction has been previously reported as a viable alternative to the repetitive manual pipetting and centrifugation steps for isolating sperm cells from female epithelial cells in sexual assault sample evidence. However, the efficiency of sperm cell isolation can be compromised in samples containing an extremely large number of epithelial cells. When highly concentrated samples are lysed, changes to the physicochemical nature of the medium surrounding the cells impacts the acoustic frequency needed for optimal trapping. Previous work has demonstrated successful, automated adjustment of acoustic frequency to account for changes in temperature and buffer properties in various samples. Here we show that, during acoustic trapping, real-time monitoring of voltage measurements across the piezoelectric transducer correlates with sample-dependent changes in the medium. This is achieved with a wideband peak detector circuit, which identifies the resonant frequency with minimal disruption to the applied voltage. We further demonstrate that immediate, corresponding adjustments to acoustic trapping frequency provides retention of sperm cells from high epithelial cell-containing mock sexual assault samples.

Keywords: acoustic differential extraction; feedback; frequency

1. Introduction

The forensic community is hampered by a significant backlog of evidence that requires DNA analysis, with a sizable subset of this consisting of sexual assault samples [1–3]. This stockpile of evidence awaiting testing exceeds 100,000 samples in the United States, with many laboratories requiring a minimum of four months for analysis of each sample [4,5]. A contributing factor to this backlog is the current methodology used for obtaining DNA identification from sexual assault evidence. DNA analysis to generate a short tandem repeat (STR) profile from the sperm cells in sexual assault evidence involves sample extraction, amplification of target DNA, and electrophoretic separation of DNA fragments to conclusively identify an individual. However, before DNA analysis can be completed, sperm cells must be isolated from the rest of the genetic material in the sample. To achieve this, each sample undergoes differential extraction, a technique that relies on multiple centrifugation and washing steps to pellet sperm cells after female epithelial cells (E-cells) are lysed, allowing for extraction of only male DNA [6]. This widely accepted approach is effective, but is labor-intensive, requires a highly trained analyst and can take multiple hours per sample [7]. Hence, there is a significant need for a faster, more automated technique to isolate sperm cells from sexual assault samples. By removing the need for a skilled analyst to perform manual separation steps, cell loss is circumvented, more efficient processing of samples results, which ultimately leads to faster resolution of investigations.

Acoustic differential extraction (ADE) is an attractive alternative to conventional differential extraction for sexual assault samples. As described in our previous work [8–10], ADE offers an automated approach for separating sperm cells from the E-cell lysate and other cellular debris, and can be incorporated with existing chemical protocols in forensic laboratories. The ADE approach generates a standing acoustic wave within a microfluidic chamber, using sound waves to capture and purify sperm cells. This is made possible by performing a differential lysis, which ruptures all non-sperm cells in a sexual assault sample, leaving sperm as the largest (and only intact) particles in solution. When subjected to a standing acoustic wave, those intact sperm cells will experience an acoustic radiation force, and can be held in place under moderate flow conditions while all other cellular debris and free DNA is washed away. The ADE technique works effectively with mock sexual assault samples [10], and compares favorably to conventional DE for authentic sexual assault samples [11]. The crucial component of ADE is the acoustic radiation force, which causes particles to aggregate at low-pressure nodes due to their acoustic contrast with the surrounding medium, with larger particles experiencing a stronger primary acoustic radiation force [12]. As shown in Equation (1), the primary radiation force (F_r) is dependent on the acoustic pressure amplitude (p_0), the volume of the particle (V_c), the compressibility of the fluid (β_w), the wavelength of applied sound (λ), the wavenumber (k), the distance from a pressure node (x), and the acoustic contrast factor (ϕ). The acoustic contrast factor itself can be expanded (Equation (2)) and is a function of the compressibility and density of both the particle (ρ_c and β_c) and the surrounding fluid (ρ_w and β_w).

$$F_r = -\left(\frac{\pi p_0^2 V_c \beta_w}{2\lambda}\right) \cdot \phi(\beta, \rho) \sin(2kx) \tag{1}$$

$$\phi(\beta, \rho) = \frac{5\rho_c - 2\rho_w}{2\rho_c + \rho_w} - \frac{\beta_c}{\beta_w} \tag{2}$$

A factor that is not included in these equations, but which is crucial for understanding the effect described, is the sound velocity. Changes to density and compressibility of a fluid will alter the speed of sound in that liquid, as shown in Equation (3) (v = speed of sound, K = compressibility, ρ = density). Changes to sound velocity, in turn, will alter the wavelength generated by an applied frequency, as seen in Equation (4) (c = speed of sound, λ = wavelength, f = frequency). Thus, shifts in fluid density or compressibility, which impact sound velocity, will necessitate a corresponding shift in applied frequency to generate a resonant acoustic wave.

$$v = (K\rho)^{1/2} \tag{3}$$

$$\lambda = \frac{c}{f} \tag{4}$$

In theory, this suggests that significant changes to the compressibility or density of the liquid medium would result in a similar shift in the required frequency, essentially changing the resonance condition. What makes these parameters crucial for changing the resonance condition is that they are variable with each sample, unlike the particle volume, compressibility, or density, which remain roughly constant for sperm cells across human males [13]. The fluid density and compressibility may be altered by drastic changes in cell concentration before and after lysis, in turn causing a shift in the frequency of sound required to generate a resonant acoustic wave. The resonance condition of an acoustic oscillator is dependent on the sound velocity, frequency, and wavelength, as well as the cavity dimensions. Changes to any of these parameters will impact the resonance condition, which in turn affects the pressure amplitude of the standing wave. The degree of change to the pressure amplitude depends on the Q-factor of the resonator, a measure of a system's oscillation at resonance compared to its bandwidth [14]. For a high-Q material, like the glass resonator cavity in this application, the amplitude–frequency relationship has a narrow range, where deviation from the resonant frequency will result in drastic loss of acoustic pressure amplitude [15]. That loss manifests as weaker acoustic trapping, and failure to retain particles. Thus, changes in resonance condition due to alteration in fluid compressibility or density, must be addressed by shifting the applied frequency.

The fluidic properties of each sample can vary with different pieces of sexual assault evidence, which can subsequently impact acoustic trapping performance. This effect was made clear during testing of the ADE prototype in an external forensic laboratory, where sperm cells were easily captured in some cases, but acoustic trapping suffered with some other samples. Some sexual assault samples will contain extremely high concentrations of E-cells (<50,000 cells), which upon differential lysis will significantly change the physical properties of the sample media. We hypothesized that, with potentially hundreds of thousands of E-cells present in a single sample, lysis alters the compressibility and density of the sample, thereby impacting the resonance condition, which is to say the necessary wavelength and frequency of sound required will shift. This phenomenon has been described more generally in the literature, with multiple studies showing that when lysis occurs, the cellular content perfuses into the surrounding liquid and causes an increase in viscosity [16–19]. There is an empirical dependence between viscosity, density, and compressibility of a liquid [20,21], and more specifically to acoustic trapping applications, groups have shown that changes in viscosity of the surrounding fluid impacts the acoustic radiation force experienced by particles in a standing wave [22,23]. With a proper understanding of the relationship between these parameters, and the knowledge that cellular lysis and variation among samples will impact fluid properties, we assume that significant changes in density and compressibility due to sample characteristics will adversely affect the ability to trap sperm cells due to a shift in optimal resonant frequency.

Sexual assault samples can vary dramatically in their makeup, containing anywhere from thousands to hundreds of thousands of cells. This wide range of possible samples means that the optimal acoustic trapping frequency will change on a sample-to-sample basis, which could have a devastating impact: It has been shown that a deviation of as little as 0.05 MHz from the ideal trapping frequency can result in a 10-fold loss of trapping efficiency [24]. In terms of our application, which uses between 7–8 MHz, this change is only a shift of less than 1% of the applied frequency. As it impacts ADE, this would mean that if the applied acoustic frequency is not precisely regulated for each sample, loss of trapping efficiency would occur and the majority of the sperm cells could be lost. An elegant solution is required to accommodate the bandwidth of samples encountered, specifically, adjust the acoustic trapping parameters to the character of each sample. Inspired by pioneering work from the Nilsson/Laurell group at Lund University [24], we are developing a feedback system that can automatically adjust the applied acoustic frequency through rapid electronic measurements. Previous literature suggested that impedance can be used to predict the optimal trapping frequency by monitoring output voltage [25–27], and a similar approach was adapted by the Lund group. Their work showed that, by calculating a root mean square power spectrum from the measured impedance and output impedance of their function generator, the optimal trapping frequency could be updated continuously during acoustic trapping. This allowed for automated adjustment of the trapping frequency as the surrounding medium was alternated between two very different solutions—a wash buffer and human plasma. The change in media led to a shift of 0.03 MHz in optimal trapping frequency, but the frequency tracking system successfully adjusted and continued to effectively trap 12 μm diameter particles. We postulated that this approach might be effective for acoustic differential extraction. During the ADE protocol, a variety of solutions traverse the trap zone—water, sexual assault sample cell lysate, bead solution, and wash buffer. Each of these constitutes a significant change in the acoustic trapping environment. Hence, a feedback system for sensing those changes, and subsequently adjusting the optimal trapping frequency, is needed to prevent unacceptable loss of sperm cells during the trapping process.

The main aim of this work was to modify the hardware in the ADE prototype [10] to facilitate real-time feedback that reflects a change in the medium conditions during trapping, ultimately enabling efficient acoustic trapping of sperm cells from sexual assault samples. By measuring the voltage drop across the piezoelectric transducer during the acoustic trapping event, we sought to identify reproducible trends that can inform real-time changes. From these relationships between output voltage and optimal trapping frequency, applied parameters could be adjusted and improved during the ADE process. The specific parameter, which is most crucial to precisely tune is the applied acoustic frequency, which is unique for each test depending on both the microchannel height and the liquid properties of that specific ample. Our hypothesis was that, by monitoring the voltage output from the piezoelectric transducer, we would be able to rapidly identify the optimal trapping frequency as it changes during sample trapping due to differences in liquid properties between each solution. This information could then be used to adjust the applied frequency to avoid suboptimal trapping and, hence, prevent loss of sperm cells. To summarize the big picture, our previous work implemented visual measurements of bead trapping to account for chip-to-chip variation and tune the applied frequency. This work seeks to account for sample-to-sample variation, and adjust in real-time based on rapid voltage output measurements.

2. Materials and Methods

2.1. Instrumentation for Real Time Electronic Feedback

The acoustic differential extraction prototype has been described in depth by our previous publications [9,10]. Precise fluidic control is achieved via two solenoid valves (LabSmith AV201, LabSmith Inc. Livermore, CA, USA) and three syringe pumps (LabSmith SPS01, LabSmith Inc. Livermore, CA, USA), which interface with the microchip pneumatically, preventing any liquid contact or contamination of the hardware. The electronic components of the instrument include a waveform generator (Hantek DDS-3X25, Hantek, Shandong, China), oscilloscope (Hantek 6052BE, Hantek, Shandong, China), and home-built amplifier, which provide control over the applied acoustic frequency. The same previously reported instrument was used for this study, however there were significant additions to the electronic domain, namely the incorporation of a custom printed circuit board (PCB) that measured the voltage output of the piezoelectric transducer during an acoustic trapping test.

In the classic peak detector, used in previous works (24, ESI), the holding capacitor is charged with a current (I_c), which is highly dependent of capacitor voltage (V_c). This current decreases as the capacitor voltage increases. The capacitor voltage should ideally become equal to V_a that is the amplitude of attenuated signal after the voltage divider unit. However, when the capacitor is charging toward the desired level of V_a, the charging current is not high like when the capacitor is empty. This reduced current in higher frequencies may be unable to completely charge the capacitor, therefore the amount of error (relative difference between V_c and V_a) may increase accordingly. The trapping frequency of biological samples in our experiments is around 7–8 MHz where the classic peak detector does not provide the sufficient accuracy (>5% error) that is required for discerning subtle variations of signal around resonant frequency.

As a solution to this limitation, we used an upgraded design for the classic peak detector, called the wideband peak detector. The details of operation have been fully explained in an online publication [28], and a simplified schematic of the wideband peak detector is included below. Figure 1 shows the simplified schematics of the circuit used for precise measurement of the AC voltage amplitude across the piezo transducer. The incoming voltage ($V_p \approx 18$ Vpp) was attenuated by the factor of three through the voltage divider unit. The maximum amplitude at the output of this unit (V_a) was around 6 Vpp that lies within the allowable input range of the peak detector unit. The DC voltage corresponding to the peak amplitude (V_c) was stored in the holding capacitor (C). A fast comparator (Ad8561, Analog Devices, MA) continuously compared the already held value of the peak with the input AC voltage (V_a). Whenever the present value of input on the positive input of the comparator went higher than the DC voltage on its negative input (V_c), the comparator output swung high (+5V). It gave rise to a charging current (I_c) through R3 and the diode (D) increases the capacitor voltage. When the input AC voltage (V_a) went below the capacitor voltage (V_c), the comparator output went down ($\approx 0\ V$) and the capacitor retained its voltage due to the unidirectional behavior of diode. In steady state, the capacitor holds the peak level of the input signal. The red-colored circuitry in Figure 1 was added to convert the classic peak detector to the wideband peak detector. In short, the amplifier (Gain = G) acts as an additional current source (I_2) for the capacitor in a way that the sum of I_1 and I_2 is relatively a constant value equal to I_c, independent of the capacitor voltage (V_c) as confirmed by Equation (5). By choosing appropriate values for R_3, R_5, and G, the second term in Equation (5) becomes zero and I_c can be expressed as Equation (6), independent of V_c.

$$I_c = I_1 + I_2 = \left(\frac{5 - 2V_D}{R_3} + \frac{-2V_D}{R_5} \right) + V_c \left(\frac{-1}{R_3} + \frac{G - 1}{R_5} \right) \tag{5}$$

$$I_c = \frac{5 - 4V_D}{R_3} \tag{6}$$

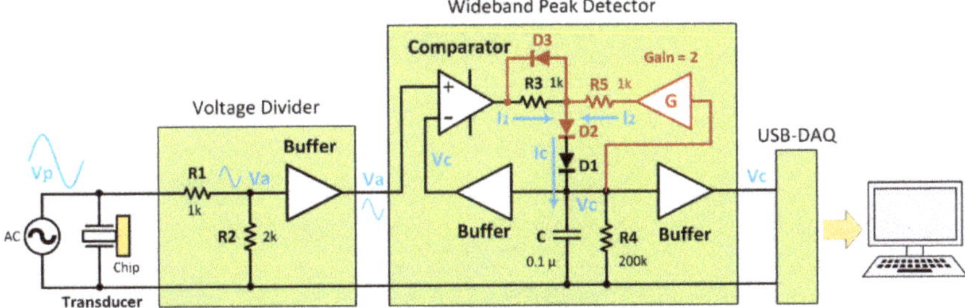

Figure 1. The simplified schematics of the wideband peak detector and the key voltage points.

The peak detector should be able to track the variation of input amplitude with an acceptable time lag (<0.1 s), therefore the resistor R4 was added to deplete the capacitor in case the input voltage drops to lower values. However, R4 should be large enough to avoid voltage ripple on V_c. The DC voltage buffers in the peak detector unit were implemented with very high input impedance operational amplifiers (OPA4170, Texas Instruments, Dallas, TX, USA) to minimize the charge loss in the capacitor. However, a wide bandwidth operational amplifier (LM6171, Texas Instruments, Dallas, TX, USA) used for the voltage buffer in the voltage divider unit to preserve the signal amplitude at frequencies around 8 MHz. The wideband peak detector collected voltage data and transfers it via a data acquisition card (National Instruments, Austin, TX, USA) to an external laptop. The voltage data was plotted in real time through LabView software, which displays the output voltage as a function of time. This information was then used by the operator to select a new acoustic frequency during sample trapping.

2.2. Chemical Protocol for Cellular Lysis and Mock Samples

Buccal swabs were collected from healthy donors to be used in this study. The cotton swabs were reconstituted in 600 μL H_2O and agitated for 60 s to remove cells from the substrate. Cellular concentration was determined via hemocytometry with the green fluorescent nucleic acid stain SYTO-11. Sperm cells were collected from anonymous healthy donors, and cell concentration also determined via hemocytometry. E-cells were lysed using the prepGEM reagent (ZyGEM NZ Ltd, Hamilton, New Zealand) and incubated according to manufacturer instructions.

3. Results

3.1. Microchip Design for Acoustic Capture of Sperm Cells

A glass-PDMS-glass microchip was used for the resonator domain, with a poly(methyl)methacrylate (PMMA) layer affixed for sample reservoirs and downstream collection chambers. Microfluidic architecture was designed in computer-assisted drawing software, and features ablated using CO_2 laser-etching. Layers were adhered through solvent bonding, plasma oxidation, and pressure-sensitive adhesive. In the schematic shown in Figure 2, reservoirs 1, 2, and 3 were filled with fluorescent beads, the sexual assault sample lysate, and a wash buffer, respectively. The first step for sample processing is to flow fluorescent beads from reservoir 1 through the trap site, while scanning through eight different frequencies covering 0.14 MHz. By using both electronic measurements and visual camera-based monitoring of the trap site, the optimal trapping frequency can be determined for that specific microchip. The optimal trapping frequency is specific to the height of the resonant chamber and, thus, will shift slightly due to variation in material thickness between each microchip. Samples containing lysed E-cells and intact sperm cells were mobilized (pneumatically) from reservoir 2 through the trap site, while the previously defined optimal trapping frequency was applied. During the trapping process, the applied frequency was shifted up by 0.01 MHz every 2 s, and real-time voltage data from the piezoelectric transducer used to maintain the optimal trapping frequency regardless of any shift in sample composition. Finally, the piezo frequency was maintained at the previously determined optimal frequency (typically between 7.4–7.8 MHz) to retain all trapped cells in the trap zone, while wash buffer flowed from reservoir 3 to remove any residual free DNA or cellular debris from the pellet.

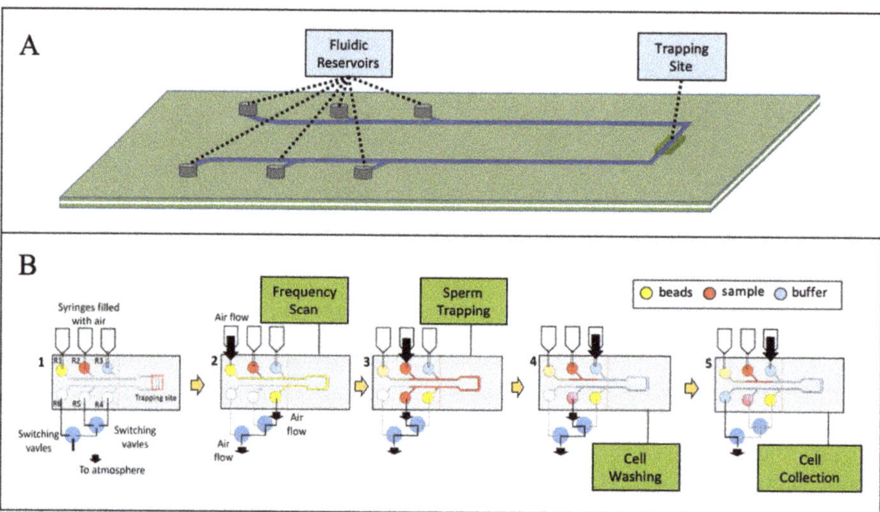

Figure 2. Acoustic differential extraction (ADE) microchip and workflow. (**A**) The multilayer chip composed of glass, polydimethylsiloxane (PDMS), and poly(methyl)methacrylate (PMMA) contains six reservoirs and one acoustic trapping site. (**B**) Workflow for the trapping, washing, and capture of sperm cells from a sexual assault sample. The "upstream" reservoirs 1, 2, and 3 contain fluorescent beads, sample, and wash buffer, respectively. The "downstream" reservoirs 4, 5, and 6 are for the collection of the sperm fraction, non-sperm fraction, and waste fraction, respectively.

3.2. Effect of Increased E-cell Concentration on Acoustic Trapping Efficiency

In order to experimentally confirm our observed (during prototype development) and anecdotal (from forensic labs who evaluated the prototype) findings, a study was conducted to measure the impact of lysed E-cell concentration on acoustic trapping frequency. By trapping fluorescent beads in solutions of varying E-cell concentration, using the same microfluidic chip and hardware, any change in the optimal trapping frequency should theoretically be attributed to the E-cell concentration. Figure 3 shows the visual aggregation of fluorescent beads in the acoustic trap site, which is used to determine optimal trapping frequency. Solutions were tested with beads suspended either in water, lysate from 75,000 cells, lysate from 100,000 cells, or lysate from 150,000 female epithelial cells. Each solution was exposed to six different applied frequencies. Optimal trapping frequency was determined by visual identification of the largest single aggregate of fluorescent particles, as determined by an algorithm that measured the number of yellow pixels in each frame. The data shows that this optimal frequency is 7.76 MHz for beads in water, but shifts to 7.78 MHz for 75,000 and 100,000 cells, and up to 7.80 MHz for 150,000 cells. This may seem like a miniscule change in frequency, but note that if 7.76 MHz was used, the vast majority of the cells would fail to trap and, hence, be lost from 100,000 and 150,000 cell samples. This simple experiment clearly demonstrates that the sample E-cell concentration dramatically influences the optimal trapping frequency, and that when an overwhelming number of E-cells are present, trapping could be obliterated. Given the sample-to-sample differences common in forensic evidence, it became clear that a feedback system was required, one that could adjust the applied frequency in real-time to match the sample constituency in terms of E-cell concentration. This forces a new standard, one that uses beads in water to define the chip-specific frequency (correlating to chamber height [9]), but in addition, uses a sample-specific adjustment in real-time to assure that optimal sperm cell capture ensues. Stated differently, the bead-determined initial trapping frequency defines the general frequency region for trapping under ideal conditions, while sample-specific voltage measurements determine the real-time trapping frequency for optimal sperm cell retention.

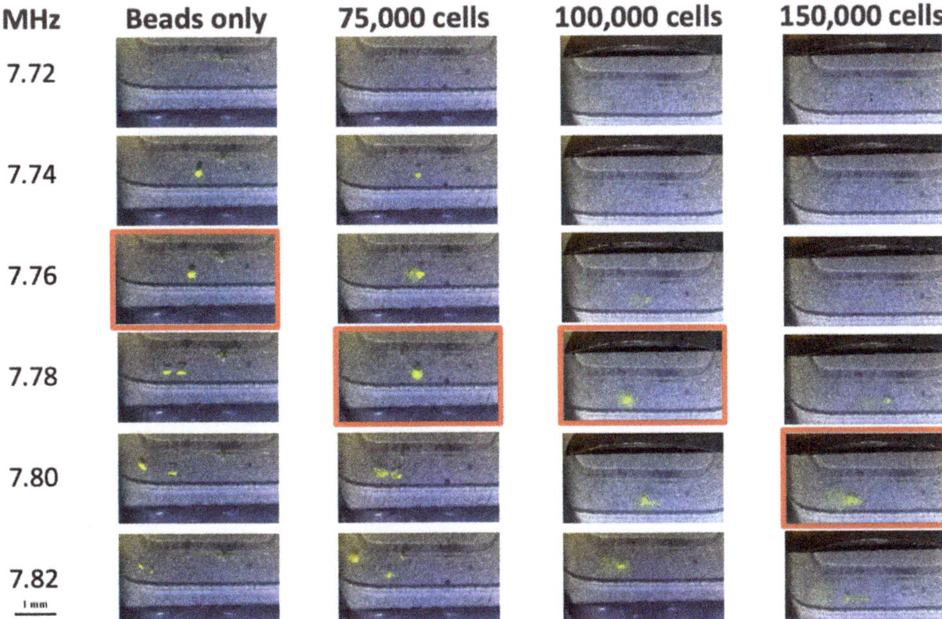

Figure 3. Shift in optimal trapping frequency with increasing epithelial cell concentration. Over a range of six applied frequencies (*y*-axis) different acoustic trapping was achieved, visualized by the aggregation of yellow fluorescent beads. In samples with increasing concentration of lysed epithelial cells (*x*-axis) there was a clear increase in the optimal trapping frequency.

3.3. Output Voltage Measurements Can Identify Optimal Trapping Frequency

During acoustic trapping, the resonance frequency occurs when the piezoelectric transducer most efficiently converts input electrical energy into mechanical energy [29]. As the piezo approaches its resonant frequency there was less resistance to vibration, meaning that electrical impedance would be at a minimum when resonance was achieved. This relationship is how Nilsson et al. successfully determined the optimal frequency for their acoustic trapping system, using impedance measurements to identify when resistance to vibration was at a minimum. Our approach uses this same principle, seeking to identify the piezo's resonance by determining at what frequency the minimum opposition to vibration occurs. However, instead of using an impedance analyzer, we directly measure the 'output voltage' of the piezoelectric transducer. Impedance itself is defined as the opposition a circuit presents to a current when voltage is applied, and Ohm's Law (for AC) states that V = IZ (V = voltage, I = current, Z = impedance) [30,31]. Therefore, by monitoring the output voltage of our acoustic trapping circuit, any changes in the output voltage represents a change in impedance and, thus, changes in the resistance to vibration of the piezo. Although the piezo is bonded to a glass coupling layer, the high-Q value of glass allows for efficient accumulation of acoustic energy [32], so any detectable changes in output voltage can be accurately attributed to piezo resonance rather than thermal effects or energy loss from the glass layer.

This principle was demonstrated by trapping fluorescent beads in a glass microchip at eight different acoustic frequencies. By simultaneously monitoring the voltage output of the piezo at each frequency, and visually monitoring aggregation of the beads, we can determine when optimal trapping occurs and decipher whether there is any concurrent trend in voltage out. Figure 4 shows the results of this experiment, plotting the output voltage (y-axis) against the applied frequency (x-axis), overlayed with still frames from video monitoring of the trapping at each frequency. Two conclusions can be drawn from this data; first, a clear minimum in the output voltage can be identified at 7.54 MHz. Second, this correlates with the strongest aggregation of fluorescent beads (identified as a single, dense aggregate in the middle of the channel) at 7.54 MHz. This suggests that the optimal trapping frequency can be determined from measuring the output voltage, rather than relying on visual monitoring of the bead aggregate size. The implication of this result is that rapid voltage measurements carried out during sample testing should be effective in identifying the optimal trapping frequency. Given this correlation between these two parameters, it is reasonable to expect that real-time adjustments could be made to the applied frequency to retain the maximum number of sperm cells despite changes in the medium (liquid) environment.

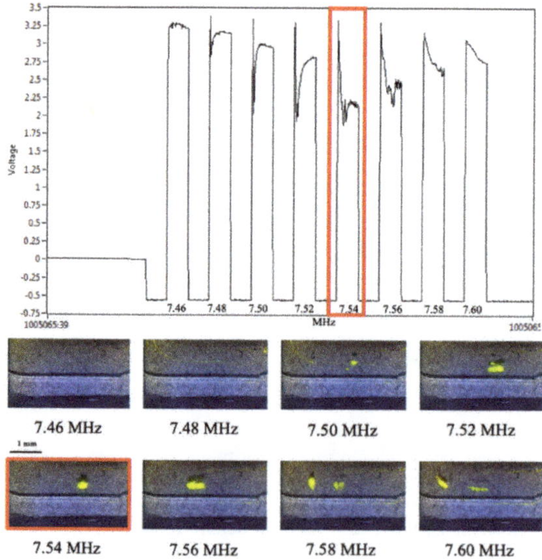

Figure 4. Voltage response to acoustic trapping. Electronic measurements of the piezo during acoustic trapping can be matched to visual data gathered via video monitoring of the trapping event. The lowest voltage output occurred at an applied frequency of 7.54 MHz, matching the frequency at which the optimal visual trapping of fluorescent beads was observed.

3.4. Shift in Optimal Trapping Frequency with Changes to Fluidic Properties

If the problem at the core of ADE inefficiency with a broad range of sexual assault samples was differences in liquid properties as the number of epithelial cells varies, it was imperative to understand the relationship between those differences and output voltage. The use of a pristine solution of glycerol varying in concentration to alter viscosity was the simplest approach that evaluates changes in output voltage during rapidly changing fluidic conditions. An ADE microchip was loaded with three solutions: (A) Fluorescent beads in water (reservoir 1), (B) fluorescent beads in 5% glycerol (reservoir 2), and (C) fluorescent beads in 10% glycerol (reservoir 3). By first conducting a frequency scan from 7.52 MHz to 7.76 MHz with the water solution, then rapidly switching flow to the 5% and 10% glycerol solutions, respectively, abrupt changes in optimal trapping frequency were measured solely via voltage output from the piezo. Figure 5A shows the voltage output data during a frequency scan of fluorescent beads in water where a clear minimum was observed at 7.58 MHz, indicative of the optimal trapping frequency. However, when a 5% glycerol solution flows through the trapping zone, the minimum output voltage shifts to 7.64 MHz (Figure 5B). Similarly, when this is followed by trapping beads in 10% glycerol, the minimum output voltage (indicating optimal trapping frequency) increased to 7.68 MHz (Figure 5C). This confirms the hypothesis that changing viscosity of the sample will directly impact acoustic trapping ability. Samples with higher percent glycerol were tested, further shifting the minimum output voltage. At a 20% glycerol solution some focusing of particles was observed, and at 30% glycerol the limit of acoustic trapping was reached, with no particle aggregation.

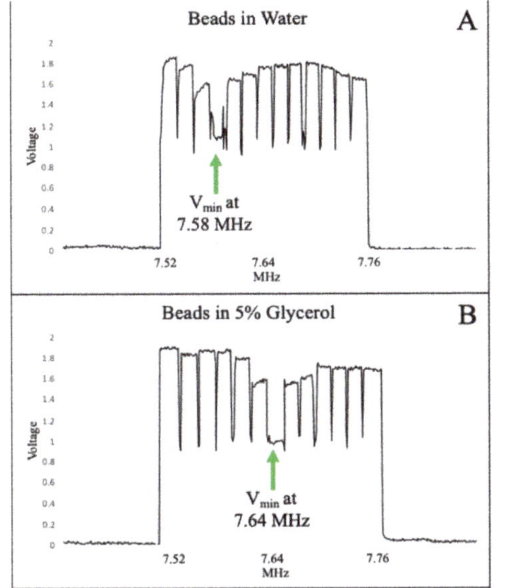
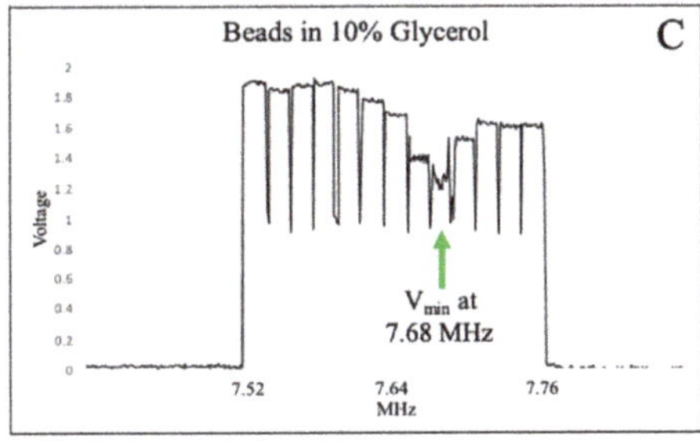
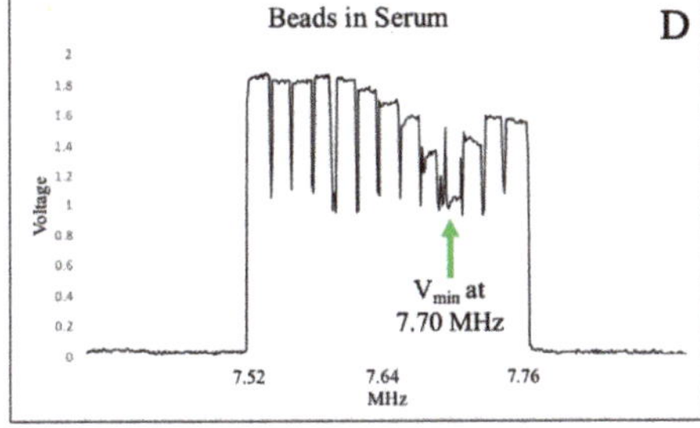

Figure 5. Shift in voltage minimum with changing viscosity. (**A**) A clear minimum in voltage output was observed at 7.58 MHz, indicating the optimal trapping frequency. (**B**) When flow was switched to 5% glycerol, the optimal trapping frequency shifted up to 7.64 MHz. (**C**) In 10% glycerol, the optimal trapping frequency shifted to 7.68 MHz. (**D**) In human serum, the minimum output voltage was 7.70, most similar to 10% glycerol.

3.5. Application of Real Time Feedback to a Mock Sexual Assault Sample

Having demonstrated the relationship between voltage output of the piezo and optimal trapping frequency, mock sexual assault samples were prepared to test the feedback system. A sample containing 290,000 total cells, with a cell ratio of 5:1 female E-cells:male sperm cells, was prepared with a standard differential lysis procedure [6], and the resultant cell lysate exposed to acoustic differential extraction. Using the previously described feedback system deemed "ResFinder" a rapid scan of eight different frequencies was conducted using a 'scanning solution' consisting of fluorescent beads suspended in water. With the purpose of identifying the optimal trapping frequency for this specific microchip, the scanning process was executed in <20 s, and the data shown in Figure 6Ai clearly indicates that a minimum in voltage output occurred at 7.59 MHz—the optimal trapping frequency for that ADE chip. With a frequency 'starting point' of 7.59 MHz, the mock sample was then flowed through the trapping site and with the piezo activated. Figure 6Aii shows the data collected during sample trapping, where the applied frequency was increased by 0.01 MHz while monitoring the voltage output of the piezo. When a minimum voltage output was identified as 7.61 MHz, scanning was terminated and 7.61 MHz was utilized for the remainder of the sample trapping. This resulted in the capture of a large aggregate of sperm cells over the 30 s trapping event, which can be visualized by the increasing size of the sperm cell clump seen in Figure 6, panels B and C. Panel B shows the brightfield microscopic view of the trapping site, as sperm cells are collected in a flow stream over a period of 30 s. The part of the trap site highlighted by the red ellipse shows no cells in Figure 6Bi while the 'blurred' area in Figure 6Bii–iv represents the growing aggregate of cells. While a trained eye can easily discern the cell aggregate in Panel B, Panel C shows the same images after 'color adjustment'. Color manipulation and image analysis are increasingly commonplace, and can be driven by simple smartphone 'apps' or by open source image processing software [33,34]. RGB is a common color space for image capture and for reporting image analysis results [35,36]. HSB (hue/saturation/brightness) color space transforms raw RGB values into a more perceptual color space, and has been routinely employed in our lab for qualitative and quantitative purposes; this is particularly useful for the investigation of colorimetric chemical reactions [37,38]. However, manipulation of color in RGB and HSB color spaces can be cumbersome and unintuitive. Here we exploit L*a*b* color space, which more closely approximates human vision and, as such, can be more intuitive, thus, easing the color manipulation process. In fact, L*a*b* is one of the more powerful and instinctive photo editing color modes used, for example, in Adobe Photoshop. Briefly, two key color adjustments were applied in L*a*b* color space using the ImageJ/Fiji 3D Color Inspector plugin (v2) [39–42]. Color contrast was boosted x14.09 and the color was rotated −90°. This color manipulation approach permits better visualization of the shadow associated with the trapped cell aggregate, which grows over time as more cells are acoustically-trapped, and this is easily identified as the black region identified by the white ellipse in Figure 6Cii–iv. The color adjustment is essentially pulling out the shadow of the sperm pellet, and contrasting it to the background of the channel. This effect is illustrated by Panel Di,ii where two trapping events, beads, and sperm cells with beads, each generate a clearly visible shadow on the bottom of the channel. Following this demonstration of real-time feedback with a mock sexual assault sample, the pellet of sperm was successfully mobilized from the trapping site and captured from the microchip.

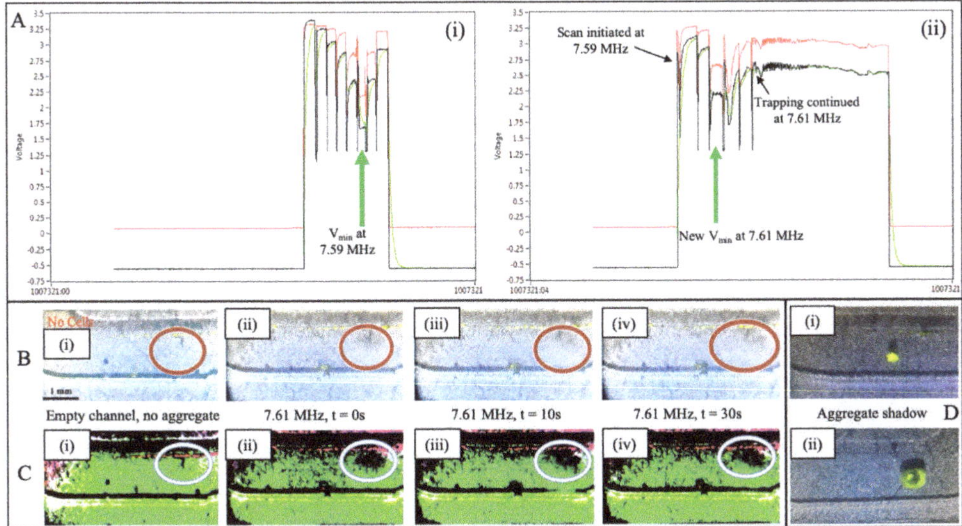

Figure 6. Real-time feedback during mock sample trapping. (**Ai**) Trapping fluorescent beads in water across eight frequencies shows a clear voltage minimum at 7.59 MHz. (**Aii**) While trapping the mock sexual assault sample, the applied frequency was adjusted by 0.01 MHz each step. A local minimum in average voltage was observed at 7.61 MHz, so that frequency was applied for the remainder of the trapping. (**Bi–iv**) Visual monitoring of the trap site over 30 s. At the conclusion of the trapping event, a large aggregate of sperm cells was captured. (**Ci–iv**) Color adjusted images from trapping site. Image processing allows for sperm aggregate to be clearly identified. (**D**) Images of beads (**i**) and beads with sperm cells (**ii**) being trapped. The shadow above and behind the aggregate is clearly visible due to the angle of lighting above the microchip.

4. Discussion

The incorporation of this feedback process into the acoustic differential extraction prototype is a significant step towards a more automated, robust system that can handle any type of sexual assault sample. The proof-of-concept data shown in Figures 4–6 demonstrate that not only is there a predictable shift in voltage output in response to altering the applied frequency, but also that the optimal trapping frequency will correlate with the lowest voltage output of the piezo. More generally, it is also clear that increasing the viscosity of the solution will cause a shift in the optimal trapping frequency, which can be identified solely by monitoring changes in the minimum output voltage. This shift occurs in controlled samples of glycerol and water, as well as more realistic biofluids like serum and epithelial cell lysate. Figure 6 specifically shows that an initial test with a 'scanning solution' is necessary to determine the optimal frequency for each specific chip, but it also shows that the optimal frequency can change based on each specific sample. Most importantly, it shows that by monitoring the output voltage from the piezo, that shift in optimal frequency can be accounted for during the sample trapping, keeping the piezo at its resonant frequency regardless of changes in the liquid environment. This important demonstration has several implications. First, it indicates that the scanning process covering a range of frequencies can be conducted much more rapidly. This is enabled by the ability to measure piezo output voltage during trapping on the millisecond timescale, as opposed to the optical/visual image capture system requiring several seconds with the trapping of fluorescent beads and measure their aggregation. Second, the automated determination of the optimal trapping frequency can be conducted during a trapping of an unknown sample. There is no longer the need to acquire images from each trapping event and quantitate the bead aggregate size at multiple frequencies; in contrast, the local minimum in voltage output can be determined before the scan is even complete. This indicates that if, for example, the optimal trapping frequency is defined early in the frequency scan, a subsequent rise in voltage output solidifies the minimum (optimal frequency) as the previous scan step. At that point the system can abandon the scan and adopt the defined optimal frequency for the remainder of sample trapping. Third, and most important, these findings illustrate that acoustic trapping can be adjusted in real-time to prevent loss of sperm cells during a test. The initial frequency scan provides a starting point for sample trapping, but the ability to scan frequencies during trapping means that no matter what sample is being analyzed, the optimal trapping frequency will be found and implemented every time. In its future automated format, the rapidity of the frequency scan (~100 ms per frequency) combined with the flow rate of the system (45 µL/min) will result in less than 1 nL of sample traversing the trap site during each tested frequency, and the potential for sample loss during the scan is negligible.

In its current form, this feedback system requires manual manipulation and data analysis during acoustic trapping. This approach does effectively trap sperm cells from high E-cell samples, which was formerly problematic [11], but does not yet fully address the overarching problem of the sexual assault backlog. Namely, this technique must be faster and more automated than the current differential extraction protocols. The clear extension of this work is to incorporate automated feedback into the ADE system, by writing a program that will detect the minimum voltage output of the piezo and adjust the optimal trapping frequency independent of any operator. This will provide a form of 'cruise control' as the system will be able to adapt to any sample type in real time. Furthermore, the current manual feedback system is limited in terms of speed by the ability of the user to process information and enters in new frequencies at which to trap. When the automated cruise control system is applied, computer processing will conduct an entire scan in hundreds of milliseconds, meaning that near-constant analysis can take place through the sample trapping, ensuring that the optimal trapping frequency is always applied. As a more general future application, it may be feasible to use this technology for the express purpose of sensing material properties in real time. For example, a monitoring rate of cellular lysis via rapid voltage measurements could give precise information about the kinetics of certain reactions, or indicate the degree of cellular rupture.

5. Conclusions

Expanding upon excellent work from the Nilsson group, we have applied real time electronic feedback to our novel acoustic differential extraction prototype. By rapidly measuring the voltage output of a piezoelectric transducer and adjusting the applied frequency of sound during acoustic trapping, subtle variations in sample composition can be accounted for to prevent the loss of sperm cells. The novel aspect of this work stems from using different electronic measurements (i.e., voltage instead of impedance), as well as applying the described principles and methods to a range of mock samples and standards. This exciting development means that a broader range of samples will be able to be tested with ADE, and once the feedback system is made automated, no user input will be required.

Author Contributions: Conceptualization, C.P.C. and J.P.L.; methodology, C.P.C., V.F., and N.S.S.; software, L.S., V.F., and C.P.C.; investigation, C.P.C. and V.F.; writing – original draft preparation, C.P.C., J.P.L., M.S.W., and V.F., project administration, J.P.L. and N.S.S.

Funding: This research was supported by the U.S. Department of Justice, award number 2013-NE-BS-K027.

Acknowledgments: The authors wish to acknowledge the Palm Beach County Sheriff's Office and the Mesa Police Department for their guidance and knowledge regarding sexual assault sample processing.

Conflicts of Interest: The authors declare no conflict of interest. The funders had no role in the design of the study; in the collection, analyses, or interpretation of data; in the writing of the manuscript, or in the decision to publish the results.

References

1. Nelson, M.; Chase, R.; DePalma, L. *Making Sense of DNA Backlogs, 2012–Myths vs Reality*; US Department of Justice, Office of Justice Programs, National Institute of Justice: Washington, DC, USA, 2013; pp. 1–20.
2. Strom, K.; Ropero-Miller, J.; Jones, S.; Sikes, N.; Pope, M.; Horstmann, N. The 2007 Survey of Law Enforcement Forensic Evidence Processing. Final Report to the NIJ from Grant 2007F_07165, October 2009. Available online: https://www.ncjrs.gov/pdffiles1/nij/grants/228415.pdf (accessed on 4 September 2019).
3. Ritter, N. The road ahead: Unanalyzed Evidence in Sexual Assault Cases. National Institute of Justice Special Report, May 2011. Available online: https://ncjrs.gov/pdffiles1/nij/233279.pdf (accessed on 4 September 2019).
4. Hurst, L.; Lothridge, K. 2007 DNA Evidence and Offender Analysis Measurement: DNA Backlogs, Capacity and Funding. Final Report to the NIJ from Grant 2006-MU-BX-K002, January 2010. Available online: https://www.ncjrs.gov/pdffiles1/nij/grants/230328.pdf (accessed on 4 September 2019).
5. LaPorte, G.; Waltke, H.; Heurich, C.; Chase, R. *DNA Analysis, Capacity Enhancement, and other Forensic Activities*; National Institute of Justice Report Forensic Science: Washington, DC, USA, 2018; pp. 1–20.
6. Gill, P.; Jefferys, A.; Werrett, D. Forensic application of DNA 'fingerprints'. *Nature* **1985**, *318*, 577–579. [CrossRef] [PubMed]
7. Vuichard, S.; Borer, U.; Bottinelli, M.; Cossu, C.; Malik, N.; Meier, V.; Gehrig, C.; Sulzer, A.; Morerod, M.; Castella, V. Differential DNA extraction of challenging simulated sexual-assault samples: A Swiss collaborative study. *Investig. Genet.* **2011**, *2*, 11. [CrossRef] [PubMed]
8. Norris, J.; Evander, M.; Horsman-Hall, K.; Nilsson, J.; Laurell, T.; Landers, J. Acoustic differential extraction for forensic analysis of sexual assault evidence. *Anal. Chem.* **2009**, *81*, 6089–6095. [CrossRef] [PubMed]
9. Xu, K.; Clark, C.; Poe, B.; Lounsbury, J.; Nilsson, J.; Laurell, T.; Landers, J. Isolation of a low number of sperm cells from female DNA in a glass-PDMS-glass microchip via bead-assisted acoustic differential extraction. *Anal. Chem.* **2019**, *91*, 2186–2191. [CrossRef] [PubMed]
10. Clark, C.; Xu, K.; Scott, O.; Hickey, J.; Tsuei, A.; Jackson, K.; Landers, J. Acoustic trapping of sperm cells from mock sexual assault samples. *FSI Genet.* **2019**, *41*, 42–49. [CrossRef]
11. Clark, C.; Xu, K.; Scott, O.; Hickey, J.; Plean, B.; Edwards, C.; Sikorsky, J.; Crouse, C.; Woolf, M.; Landers, J. External evaluation of an acoustic differential extraction prototype in forensic laboratories. *FSI Genet.* in press.
12. Evander, M.; Nilsson, J. Acoustofluidics 20: Applications in acoustic trapping. *Lab Chip* **2012**, *12*, 4667–4676. [CrossRef]
13. Bellastella, G.; Cooper, T.; Battaglia, M.; Strose, A.; Torres, I.; Hellenkemper, B.; Soler, C.; Sinisi, A. Dimensions of human ejaculated spermatozoa in Papanicolaou-stained seminal and swim-up smears obtained from the Integrated Semen Analysis System (ISAS). *Asian J. Androl.* **2010**, *12*, 871–879. [CrossRef]
14. Bruus, H. Acoustofluidics 2: Perturbation theory and ultrasound resonance modes. *Lab Chip* **2012**, *12*, 20. [CrossRef]
15. Bao, M.; Onshage, A.; Iwansson, K. *Handbook of Sensors and Actuators*; Elsevier Science: Amsterdam, The Netherlands, 1996.
16. Newton, J.; Schofield, D.; Vlahopoulou, J.; Zhou, Y. Detecting cell lysis using viscosity monitoring in *E. coli* fermentation to prevent product loss. *Am. Inst. Chem. Eng.* **2016**, 1069–1076.
17. Dhillon, G.; Brar, S.; Kaur, S.; Verma, M. Rheological studies during submerged citric acid fermentation by *Aspergillus niger* in stirred fermentor using apple pomace ultrafiltration sludge. *Food Bioprocess. Technol.* **2013**, *6*, 1240–1250. [CrossRef]
18. Perley, C.; Swartz, J.; Cooney, C. Measurement of cell mass concentration with a continuous-flow viscometer. *Biotechnol. Bioeng.* **1979**, *21*, 519–523. [CrossRef]
19. Shimmons, B.; Svrcek, W.; Zajic, J. Cell concentration control by viscosity. *Biotechnol. Bioeng.* **1976**, *18*, 1793–1805. [CrossRef]
20. Atanov, Y.; Berdenikov, A. Relation between fluid viscosity and compressibility. *J. Eng. Phys.* **1983**, *43*, 878–879. [CrossRef]
21. Elert, G. Viscosity. The Physics Hypertextbook. 1998. Available online: https://physics.info/viscosity/ (accessed on 5 June 2019).
22. Doinikov, A. Acoustic radiation forces: Classical theory and recent advances. *Recent Res. Devel Acoust.* **2003**, *1*, 39–67.

23. Sepehrirahnama, S.; Chau, F.; Lim, K. Effects of viscosity and acoustic streaming on the interparticle radiation force between rigid spheres in a standing wave. *Phys. Rev.* **2016**, *93*, 023307. [CrossRef] [PubMed]

24. Hammarström, B.; Evander, M.; Wahlstrom, J.; Nilsson, J. Frequency tracking in acoustic trapping for improved performance stability and system surveillance. *Lab Chip* **2014**, *14*, 1005–1013. [CrossRef]

25. Dual, J.; Hahn, P.; Leibacher, I.; Moller, D.; Schwarz, T. Acoustofluidics 6: Experimental characterization of ultrasonic particle manipulation devices. *Lab Chip* **2012**, *12*, 852–862. [CrossRef]

26. Hawkes, J.; Coakley, W. A continuous flow ultrasonic cell-filtering method. *Enzyme Microb. Technol.* **1996**, *19*, 57–62. [CrossRef]

27. Kwiatkowski, C.; Marston, P. Resonator frequency shift due to ultrasonically induced microparticle migration in an aqueous suspension: Observations and model for the maximum frequency shift. *J. Acoust. Soc. Am.* **1998**, *103*, 3290–3300. [CrossRef]

28. McLucas, J. Wideband peak detector operates over wide input-frequency range. *EDN Des. Ideas.* **2007**. Available online: https://www.edn.com/design/analog/4324882/Wideband-peak-detector-operates-over-wide-input-frequency-range (accessed on 28 April 2019).

29. Uchino, K. Piezoelectric ceramics for transducers. In *Ultrasonic Transducers*; Woodhead Publishing Series in Electronic and Optical Materials: Sawston, UK; Cambridge, UK, 2012; pp. 70–116.

30. Millikan, R.; Grover, M. *Elements of electricity*; American Technical Society: Chicago, IL, USA, 1917; p. 54.

31. Aroom, K.; Harting, M.; Cox, C.; Radharkrishnan, R.; Smith, C.; Gill, B. Bioimpedance analysis: A guide to simple design and implementation. *J. Surg. Res.* **2009**, *153*, 23–30. [CrossRef] [PubMed]

32. Laurell, T.; Petersson, F.; Nilsson, A. Chip integrated strategies for acoustic separation and manipulation of cells and particles. *Chem. Soc. Rev.* **2006**, *36*, 492–506. [CrossRef] [PubMed]

33. Oncescu, V.; O'Dell, D.; Erickson, D. Smartphone based health accessory for colorimetric detection of biomarkers in sweat and saliva. *Lab Chip* **2013**, *13*, 3232–3238. [CrossRef] [PubMed]

34. Kim, H.; Awofeso, O.; Choi, S.; Jung, Y.; Bae, E. Colorimetric analysis of saliva–alcohol test strips by smartphone-based instruments using machine-learning algorithms. *Appl. Opt.* **2017**, *56*, 84–92. [CrossRef]

35. Soldat, D.J.; Barak, P.; Lepore, B.J. Microscale colorimetric analysis using a desktop scanner and automated digital image analysis. *J. Chem. Educ.* **2009**, *86*, 617. [CrossRef]

36. Cabaret, F.; Bonnot, S.; Fradette, L.; Tanguy, P.A. Mixing time analysis using colorimetric methods and image processing. *Ind. Eng. Chem. Res.* **2007**, *46*, 5032–5042. [CrossRef]

37. Capitan-Vallvey, L.F.; Lopez-Ruiz, N.; Martinez-Olmos, A.; Erenas, M.M.; Palma, A.J. Recent developments in computer vision-based analytical chemistry: A tutorial review. *Anal. Chim. Acta* **2015**, *899*, 23–56. [CrossRef]

38. Krauss, S.T.; Holt, V.C.; Landers, J.P. Simple reagent storage in polyester-paper hybrid microdevices for colorimetric detection. *Sens. Actuators B Chem.* **2017**, *246*, 740–747. [CrossRef]

39. Barthel, K.U. 3D-data representation with ImageJ, In Proceedings of the ImageJ Conference, Luxembourg, 18–19 May 2006.

40. Rasband, W.S. ImageJ, US National Institutes of Health, Bethesda, Maryland, USA, 2011. Available online: http://imagej.nih.gov/ij/ (accessed on 6 October 2019).

41. Schneider, C.A.; Rasband, W.S.; Eliceiri, K.W. NIH Image to ImageJ: 25 years of image analysis. *Nat. Methods* **2012**, *9*, 671. [CrossRef]

42. Rueden, C.T.; Schindelin, J.; Hiner, M.C.; DeZonia, B.E.; Walter, A.E.; Arena, E.T.; Eliceiri, K.W. ImageJ2: ImageJ for the next generation of scientific image data. *BMC Bioinform.* **2017**, *18*, 529. [CrossRef] [PubMed]

Article

Microfluidic In-Flow Decantation Technique Using Stepped Pillar Arrays and Hydraulic Resistance Tuners

Gangadhar Eluru, Pavan Nagendra and Sai Siva Gorthi *

Optics and Microfluidics Instrumentation Lab, Department of Instrumentation and Applied Physics, Indian Institute of Science, Bangalore 560012, India
* Correspondence: saisiva@iisc.ac.in; Tel.: +91-80-2293-3529

Received: 18 May 2019; Accepted: 10 July 2019; Published: 15 July 2019

Abstract: Separating the particles from the liquid component of sample solutions is important for several microfluidic-based sample preparations and/or sample handling techniques, such as plasma separation from whole blood, sheath-free flow focusing, particle enrichment etc. This paper presents a microfluidic in-flow decantation technique that provides the separation of particles from particle-free fluid while in-flow. The design involves the expansion of sample fluid channel in lateral and depth directions, thereby producing a particle-free layer towards the walls of the channel, followed by gradual extraction of this particle-free fluid through a series of tiny openings located towards one-end of the depth-direction. The latter part of this design is quite crucial in the functionality of this decantation technique and is based on the principle called wee-extraction. The design, theory, and simulations were presented to explain the principle-of-operation. To demonstrate the proof-of-principle, the experimental characterization was performed on beads, platelets, and blood samples at various hematocrits (2.5%–45%). The experiments revealed clog-free separation of particle-free fluid for at least an hour of operation of the device and demonstrated purities close to 100% and yields as high as 14%. The avenues to improve the yield are discussed along with several potential applications.

Keywords: in-flow decantation; self-sheath generation; microfluidics; sheath-free flow focusing; plasma separation; particle enrichment

1. Introduction

Microfluidics, in its immense potential to offer miniaturization, cost-effectiveness, precision, automation, and the use of ultra-small quantities of samples, is rapidly expanding into areas of health-care, water treatment, soil testing, biomedical research, chemical and biological sciences. In several applications involved in these areas, separation of suspended particles from the sample fluid is a must as a pre-preparatory step for further investigation/research. As most of the research in microfluidics is oriented towards health-care, the foregoing discussion presents the importance and latest developments of separation process in this area. In health-care, the majority of diagnostic tests are performed on body fluids such as blood, as the biomarkers associated with most of the body's medical conditions can be found in them.

Blood is an important body fluid, that is responsible for the delivery of nutrients, oxygen to the cells and takes away metabolic waste from the cells. Its main constituents are plasma ($\approx$55% by volume) and blood cells ($\approx$45% by volume). Approximately 92% by volume of plasma is water and contains dissipated proteins, glucose, mineral ions, carbon dioxide etc. The blood cells are majorly red blood cells (RBCs), white blood cells, and platelets. Many markers associated with changes in health reflect as either a change in the composition of plasma or blood cells. Blood-based tests can be broadly categorized into plasma/serum-based tests and blood cell-based tests.

The majority of plasma/serum-based tests require the separation/removal of blood cells from the whole blood as the cellular debris interferes with most of plasma/serum-based tests. The standard laboratory technique of cell separation from whole blood is based on centrifugation, which is very cumbersome, time consuming (about 30 min) and requires skilled personnel to operate. Microfluidics based devices for plasma separation can be majorly classified into active and passive. The devices based on active techniques use some form of forces such as dielectrophoretic, electrohydrodynamic, electro-osmotic, centrifugal, and acoustic to achieve the desired cellular separation [1–7]. The major disadvantages associated with these techniques are the inability to provide high quality plasma at higher hematocrits, complexity in fabrication and integration, and the typical dependence on bulky external power supplies.

Passive techniques take leverage on one or a combination of the parameters, such as the use of filters, device geometry, inertial or Dean's effects, gravity, and biophysical effects which include the Fahraeus effect and the Zweifach-Fung bifurcation law to achieve the desired separation. The techniques of plasma separation based on filtration suffer with major problems of clogging due to the very small pore size of filters [8]. Plasma separation designs based on geometrical and biophysical effects [9–13] do not cause clogging problems in general, however, they provide high purity plasma only at lower hematocrits and/or smaller flow rates, thereby limiting their use for rapid diagnosis of majority of biochemical tests [14]. Inertial and Dean's effects were utilized to separate plasma by pumping the blood at higher flow rates and were demonstrated to produce high purity plasma only at lower hematocrits [15]. Several other passive variant techniques that demonstrated plasma separation have been extensively reviewed by Han et al. and Siddhartha et al. [16,17]. However, the production of high purity plasma at larger hematocrits (from whole blood) at optimum flow rates (without causing hemolysis) in higher yields, in a simple to fabricated device, is still a challenge.

Most of the blood-cell based tests using microfluidics to analyze blood cells in flow on the flow cytometric principles of imaging or scattering. One of the crucial requirements of these microfluidic devices is flow focusing of the cells into a plane or a line for interrogation widely known as flow focusing. Most of the flow focusing techniques employ external sheath fluid to focus these cells. The efforts to generate sheath-free flow focusing have led to the development of both passive and active techniques. Due to the conventional drawbacks associated with active techniques, such as complexity in fabrication and integration, bio-cellular incompatibility and the requirement of bulky and expensive power supplies, passive techniques remained as an attractive choice for microfluidic based diagnostics.

Most of the passive techniques operate at high flow rates by utilizing either Dean's effects or inertial effects and are not suitable for imaging based applications [18–20]. One of the passive techniques that operate at low flow rates uses pillars to achieve the separation of particle-free fluid from the particles and reuse them to accomplish sheath-free flow focusing [21]. The proposed pillar-based design works well for rigid particles of larger size than the spacing between the pillars, but is not suitable for separating deformable particles, such as red blood cells from the fluid. To use such a device for blood cell (particle size ranges from 2 µm to 30 µm) separation and focusing requires fabricating the pillars with a spacing smaller than 2 µm while keeping the height of the pillar (and channel) large enough to accommodate the largest particle (30 µm), which is highly difficult to achieve using regular techniques of photolithography and soft lithography. This pillar design limits its applicability to only rigid particles and could not answer the problem of clogging due to the uncontrolled fluid flow through the pillar gaps. This leaves the problem of providing a clog-free passive separation design that is simple to fabricate, that can work for a wide range of particle ranges and flow rates, as open ended and demand solutions.

This paper presents a technique that provides solutions to the aforementioned problems in a single device. The presented design can separate the particles from the fluid for a wide range of flow rates, particle sizes and concentrations in very high purity and yield, in a single polydimethylsiloxane

(PDMS) layer device, through the use of the principle termed as wee-extraction. The device consists of stepped pillar arrays (different from the regular pillar arrays due to their size and placement) and hydraulic resistance tuners in a unique combination to control the fluid flow through the pillar gaps in the stepped pillar array. This unique combination offers the proposed technique needed that has advantages over the other techniques to accomplish decantation in high purity and yield. The applicability of this technique for low flow rates as well as high flow rates and for wide variety of particle sizes without clogging was demonstrated. The high purity (100%) and yield (14%) of the particle-free fluid separated, and the applicability, efficient working of the technique even at higher particle concentrations (45%) demonstrates the potential of the proposed technique and its practical utility. The principle of operation of this technique was demonstrated using beads (5 μm), platelets, and blood of various particle concentrations (hematocrits). This novel proposed technique in combination with any simple technique to fabricate a 3D flow focusing device that uses external-sheath fluid [22] can be used to develop a sheath-free flow cytometer that meets the needs of blood-cell based tests. Another important feature of this design is simplicity in fabrication, as it involves fabricating a master mold only once and the subsequent device fabrication can be done in a single step without the requirement of complicated alignment procedures, although the design involves multiple heights. The following sections present the theory and principle of operation, simulation and experimental results along with their consequent implications.

2. Theory, Simulations and Experiments

2.1. Theory and Principle of Operation

The device design and its operation can be well understood by artificially dividing the whole design into three sections namely Section I, Section II and Section III as shown in Figure 1a. Section I consists of the sample input and channel of depth D_1 for the sample inflow. Section II is further subdivided into central channel C and side channels S1 and S2, each having depths D_2 and D_3, respectively. The meeting point of Section I with Section II has a transition of depth from D_1 to D_2 and width W_1 to W_2. The side channels S1 and S2 are identical and are placed symmetrically on either side of central channel C. These side channels consist of series of pillars with spacing between them as W_{ps}. This configuration of pillar array with height (D_3) different from that of channel C (D_2) and placed towards one end of the channel (in the height direction) is termed the stepped pillar array. A schematic better illustrating the stepped pillar arrays is shown Figure 1b. The central channel depth D_2 and width W_2 are chosen larger than the sample channel depth D_1 and width W_1, to facilitate the redistribution of sample particles present inside the sample solution along the depth and width directions. For a dilute suspension of particles, as the sample passes from the sample channel to the central channel, the average separation of particles and the average distance of the particles from the top, bottom and side walls increases due to the expansion of the channel in the depth and width directions. Schematics illustrating the same are shown in Figure 2a,b. This increment in the distance of the particles from the walls leaves the room for a large quantity of particle-free fluid to be closer to the walls. This is the first essential accomplishment of the proposed design. The separation of this particle-free fluid, stepped pillar array design along with hydraulic resistance tuners is proposed.

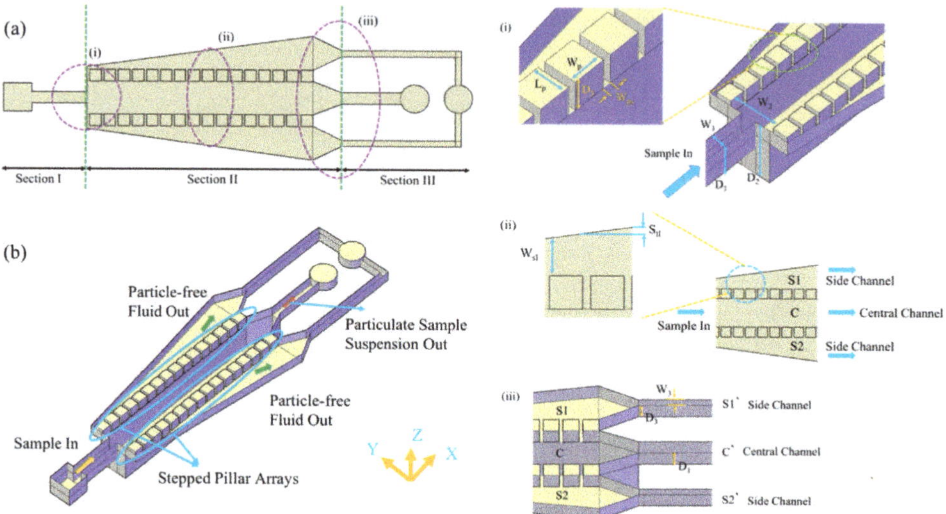

Figure 1. Schematics representing the proposed decantation device. (**a**) Schematic of the proposed microfluidic device. (**b**) An isometric view of the device schematic to better visualize the features. (**i**), (**ii**), and (**iii**) represent the representative schematic images at the appropriate locations.

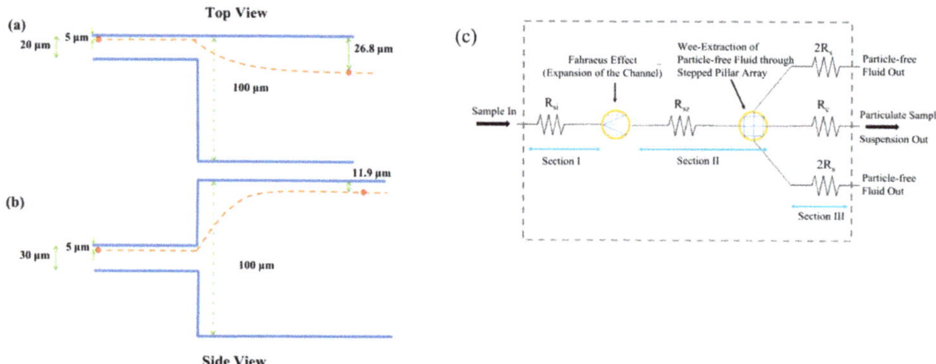

Figure 2. Schematics illustrating the movement of particle due to the expansion of the channel (Fahraeus effect) and a hydraulic resistance model of the device. (**a,b**) represent the schematics illustrating the movement of the particle due to the expansion of the channel. Motion of the particle which was originally moving at 5 μm away from the top and side walls prior to expansion of the channel is shown in (**a,b**), as seen in side and top views, respectively. The numbers shown in the figure were obtained from simulation results. (**c**) Schematic representing the hydraulic resistance model of the device. R_{si}, R_{sz}, R_c and $2R_s$ are the hydraulic resistances of the appropriate sections/channels of the device shown in Figure 1. New symbols were used to represent the phenomena happening on the particles due to Fahraeus effect and wee-extraction.

The stepped pillar array consists of pillars of size L_p, W_p and D_3 as shown in Figure 1. The width of the pillars (W_p), the spacing between them (W_{ps}), and the number of pillars (N) in each side channel decide the overall size of the device. The side channels lateral spacing W_{sl} was continuously increasing from the start of the pillars until the end in steps of S_{tl} per period. The purpose of using the pillars with constant spacing W_{ps} between them and a variable lateral spacing W_{sl} is to extract a small and controllable quantity of fluid through the pillar gaps continuously. If there is a zero step increment S_{tl}, it is expected to have no further accumulation of fluid through the side channels, due to the pressure of the already existing fluid in that region, except towards the end of Section II. Hence, to have a larger quantity of fluid to accumulate in the side channels, there is a need to have some nonzero S_{tl}.

The spacing between the pillars W_{ps} and the height of the pillar D_3 define the cross-section for the fluid flow through the pillar gaps, hence providing a measure for fluid extraction through them. The smaller the pillar gap cross-section, the better the control is over the quantity of fluid drawn through the pillar gaps, for a given Section III design. Designing this volume of extraction through each of these gaps smaller than the volume of the critical particle to be separated, ensures the particles are completely separated out of the fluid. The volume of the fluid drawn through pillar gaps (V_{pg}) depends on the cross-sectional area of the pillar gap ($W_{ps} \times D_3$), the average velocity of flow across the pillar gaps (v), and the average velocity of flow near the pillars in the direction of overall fluid flow (u) and is given by $V_{pg} = W_{ps} \times D_3 \times v \times (W_{ps}/u)$. Smaller V_{pg} can be obtained by making the terms $W_{ps}^2 \times D_3$ and v/u smaller. The ratio v/u is coupled to the cross-section of pillar gaps and Section III of the design. To obtain a smaller V_{pg}, in principle, W_{ps} or D_3 or v/u or any combination of them can be chosen arbitrarily small. Regarding the fabrication difficulties, W_{ps} and D_3 can be made small enough to conveniently fabricate and v/u can be manipulated by controlling the Section III design. This ensures a smaller quantity of fluid withdrawal through the pillar gaps. This process of extracting smaller quantities of fluids through the pillar gaps is defined as the principle of wee-extraction.

Section III consists of a central channel C' and the side channels S1' and S2' with hydraulic resistance R_c and R_s, respectively. These side channels S1' and S2' are the hydraulic resistance tuners of the proposed device, which can be changed by changing the length or cross-sectional parameters of these side channels. R_s is the hydraulic resistance of the two side channels S1' and S2' combined. As these two side channels are identical, the hydraulic resistance of each side channel can be $2 \times R_s$. The main function of this Section III design is to control the ratio of the fluid that gets drawn through the side channels with respect to the central channel, which in turn is controlled by the ratio of hydraulic resistances of the central and side channels C', S1', and S2'. The flow rate ratio of side channels S1' and S2' (combined total flow rate = Q_s) and that of central channel C' (Q_c) is given by $Q_s/Q_c = R_c/R_s$. The side channel flow rate Q_s in terms of the sample flow rate Q is given by $Q_s = (Q \times R_c)/(R_c + R_s)$. The above analysis is approximate and can be used to get only a qualitative understanding of the separation process. In the above analysis, it is presumed that the Section II device parameters do not have any significant effect on the flow rates Q_s and Q_c, except for the distribution of flow inside Section II. This was proven to be true through both simulations and experiments as presented in subsequent sections.

The summary of the whole functionality of the device is depicted in the hydraulic resistance model of the device as shown in Figure 2c. The hydraulic resistances of various sections and channels of the devices are modelled as resistors and the effect of phenomena, such as Fahraeus effect and wee-extraction through stepped pillar array, are represented with symbols. The transition from Section I of the design to Section II of the design leading to a depth and width increment has resulted in bringing more quantity of particle-free fluid towards the walls of the channel (Fahraeus effect). Sections II and III of the design together bring out the wee-extraction of the fluid through the stepped pillar array from the suspension. This latter process of wee-extraction first involves extracting the particle-free fluid into the side channel that was made available due to the Fahraeus effect, and subsequent extraction of the particle-free fluid that is present in the suspension, but was not made available as particle-free layers due to Fahraeus effect. This extraction of particle-free fluid through the stepped pillar array happens in Section II, but being controlled by the hydraulic resistances of Section III.

For scenarios that involve lower concentration of sample particles, the former process (Fahraeus effect) creates larger quantities of particle-free fluid near the walls. A suitable choice of parameters for the latter process can facilitate the extraction of this fluid into the side channels. This use of the Fahraeus effect helps in having a much greater yield of separation of particle-free fluid, when used with dilute suspension of particles, which in turn can be used for sheath-free flow focusing of particles. However, as the particle concentration increases, the Fahraeus effect may not be able to generate large quantities of particle-free fluid near the walls. This leaves the technique to solely depend upon the latter process to separate the particle-free fluid from the suspension. This can be accomplished by choosing the wee-extraction volume across the pillar gaps to be smaller than the critical particle volume to be separated. This process of extracting the particle-free fluid (wee-extraction) facilitates a possibility of extracting the particle-free fluid in very high purity and in modest yields. This technique offers a scope for further enhancing the yield through cascading as discussed in subsequent sections.

2.2. Device Fabrication

The design of the device is such that the three different depths required can be obtained into a single layer of PDMS, and hence is very simple to fabricate. The three heights on the Master were prepared by the usual process of multi-step variable height optical lithography using three SU-8 photoresists 2005, 2015 and 2100 as illustrated in Figure 3a. The first layer was fabricated by spin coating SU-8 2005 photoresist and exposing it to ultraviolet (UV) light using a mask design that contains the stepped pillar arrays (Mask-1 Supplementary Information Figure S1). This layer is developed using the developer solution and contains the stepped pillar arrays and the hydraulic resistance tuners. The second layer was fabricated on top of the first layer by spin coating SU-8 2015 and exposing it to UV light using the mask design-2 (Supplementary Information Figure S2). The second layer ensures the inlet channel is of desired height to intake the particles without clogging. The third layer is further fabricated by spin coating SU-8 2100 and exposing it to UV light using the mask design-3 (Supplementary Information Figure S3). This ensures the central channel height is much larger than inlet channel to facilitate the Fahraeus effect. The resulting depths D_1, D_2, and D_3 were measured using the Dektak surface profiler and were found to be 20 μm, 74 μm, and 4.5 μm respectively. PDMS devices were further fabricated from the Master using the standard process of soft lithography. The micrograph of the fabricated device that was taken towards the end of Section II that depicts the stepped pillar arrays, central and side channels is shown in Figure 3b.

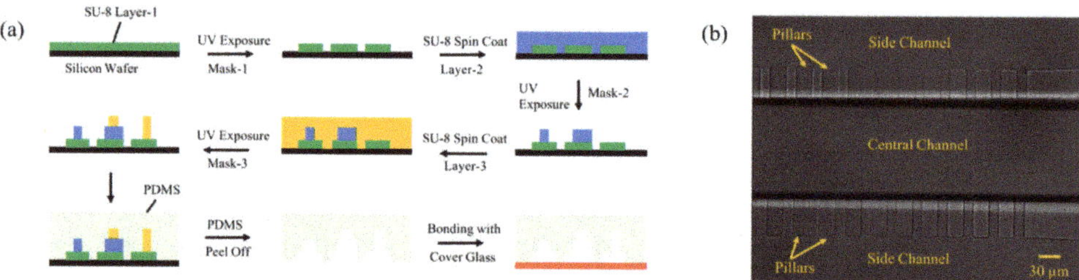

Figure 3. Schematics illustrating the multi-step variable height lithography and a micrograph of the fabricated device. (**a**) Schematic represents the various steps involved in achieving the 3 different heights into the Master mold and then transferring them into the PDMS device in a single shot. (**b**) Micrograph of the PDMS device that was fabricated and shows the image of the device towards the end of Section II.

2.3. Sample Preparation

Fresh venous blood was collected from healthy subjects in vacutainers containing EDTA (Ethylenediaminetetraacetic acid) anticoagulant. The blood was diluted with phosphate buffer saline (PBS) (135 mMNaCl, 2.7 mM KCL, 10 mM Na_2HPO_4, 2 mM KH_2PO_4 and pH adjusted to 7.4) accordingly as per the needed hematocrit. For experiments with platelets, the undiluted blood was centrifuged at 3000 rpm for 20 min and the resulting supernatant was diluted by a factor of 2 by adding PBS. Further, 5 μm polystyrene bead solution was prepared by diluting the 10 μL of the raw suspension obtained from Sigma-Aldrich in 1ml of water.

2.4. Simulations

As the exact design that was used for the experiments is complex and computationally intensive, the simulations were performed on a simpler model that captured the essence and provides insights into the behavior of particle separation. This design consists of Section I of the model discussed above, but of smaller length. The Section II consists of a smaller number of pillars than the actual device ranging from 50 to 200, but similar other parameters. The parameters used for the simulations are D_1 = 20 µm, D_2 = 100 µm, D_3 = 5 µm, W_{ps} = 10 µm, W_p = 20 µm, L_p = 40 µm. The Section III design has kept the hydraulic resistance ratios of the side channel (S′) to the central channel (C′), abbreviated as *SCRR* (side channel to central channel resistance ratio), fixed while performing the simulations with respect to variations of other parameters. The schematic of the design that was used for simulations is shown in Supplementary Information Figure S4. The fluid dynamic simulations were performed on COMSOL multi physics software version 5.2 using the inbuilt laminar flow module. The fluid for flow inside the design was chosen to be water and the mesh type was chosen to be tetrahedral and extremely fine. The wide varieties of simulations were performed by changing relevant parameters and the results have been presented in the results and discussion sections.

2.5. Experimental Procedure

Regarding the practical applicability of the proposed technique for biological applications, experimental characterization was performed on the fabricated devices by pumping blood at different hematocrit (*hct*) levels. The blood cells in flow were imaged using a Nikon microscope, 10× micro objective and Pike camera as shown in Figure 4. The number of blood cells escaping through the side channels were counted to quantify the purity of the side fluid collected. The purity of the side fluid is defined as,

$$\text{Purity} = \left[\frac{\left(N_{sampl} - N_s \right)}{N_{sampl}} \right] \times 100\%,$$

where N_{sampl} is the total number of particles expected to enter the device per second and N_s is the total number of particles entering the side channels S1 and S2 per second. N_{sampl} is computed based on hematocrit, flow rate (Q) and is given by $N_{sampl} = Q \times hct$, whereas N_s is computed based on the images of the cells entering S1 and S2. The images that were acquired have been post processed using morphological operations tool box available in MATLAB and a custom written code to obtain the number of cells entering S1 and S2 (Supplemenatry Information).

For all the experiments, the fluid from the side channels and fluid from the central channel were collected separately into vials using polyethylene tubing. The quantification of the volume of the fluid extracted in side channels was performed by measuring the length changes of the fluid inside the polyethylene tubing attached to side channels and multiplying with the inner area of the cross-section of the tubing. The measurement has a precision of less than 5% in determining the volume of fluid collected in side channels. This is because every measurement of the length change inside the polyethylene tubing for volume estimation was larger than 2 cm and the error in measuring the length change was less than 1 mm.

The above experimental procedure was carried out by varying a number of parameters to understand their effects. The parameters include the side channel to the central channel resistance ratio (*SCRR*), the hematocrit value (*hct*), the flow rate (Q) and time. Another study was carried out to characterize the purity of sheath collected by varying the particle size. Platelets and 5 µm beads were considered for these experiments in place of RBCs. Both these particle suspensions were then pumped into the microfluidic device at 100 µlh^{-1} and imaged for estimating the purity of the particle-free fluid collected.

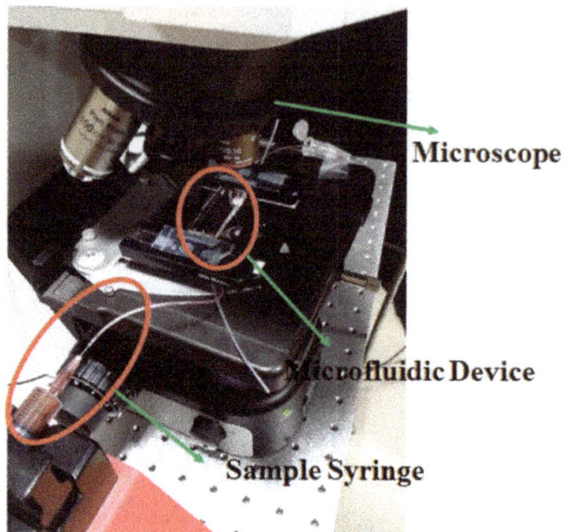

Figure 4. Experimental set-up that was used for the validation of proposed microfluidic technique.

3. Results and Discussion

3.1. Simulations

From the theory, it is expected that the successful development of a device should decant particles below a critical particle volume, understand the flow across all the pillar gaps and know the parameters that control the volume of the net fluid being drawn through the side channels S1′ and S2′. The simulations were performed to understand the effect of various parameters of Sections II and III on the net fluid being drawn through the side channels S1′ and S2′ and the effect of Section II parameters on flow uniformity across the pillar gaps. The results of these simulations along with the respective conclusions are presented below.

3.1.1. Effect of Section II Parameters [Height of Pillars (D_3), Spacing between the Pillars (W_{ps}), Number of Pillars (N), Slope of the Side Channel (S_{slp}), and Flow Rate (Q)] and Section III Parameters [Side Channel to Central Resistance Ratio ($SCRR$)] on Side Channel to Central Channel Flow Rate Ratio (SCF)

The simulations were performed on a device model that was presented in a simulations section with parameters $D_1 = 20$ μm, $D_2 = 80$ μm, $N = 50$, and with all other parameters as specified in Figures 5 and 6a accordingly.

The side channel slope S_{slp}, the side channel to the central channel hydraulic resistance ratio $SCRR$, and the side channel to the central channel flow rate ratio SCF are defined as $S_{slp} = S_{tl}/(W_p + W_{ps})$, $SCRR = R_s/R_c$, and $SCF = Q_s/Q_c$, respectively. The parameters S_{tl} is the change in distance of the side wall from the pillar array plane for a length change of a pillar array period ($W_p + W_{ps}$) along the pillar array direction as shown in Figure 1. R_c is the combined hydraulic resistance of the central channels C and C′, and R_s is the hydraulic resistance of the all the side channels S1, S2, S1′, and S2′. However, as the side channels S1 and S2 did not contribute significantly to the net hydraulic resistance R_s due to their larger width, the effective side channel resistance was computed based on the side channels S1′ and S2′ in this paper. Similarly, the net hydraulic resistance of the central channel R_c was effectively due to C′ because of its small width and depth compared to C. SCF is computed based on the total flow rate (Q_s) of the side channels S1′ and S2′, and the flow rate (Q_c) through the central channel C′.

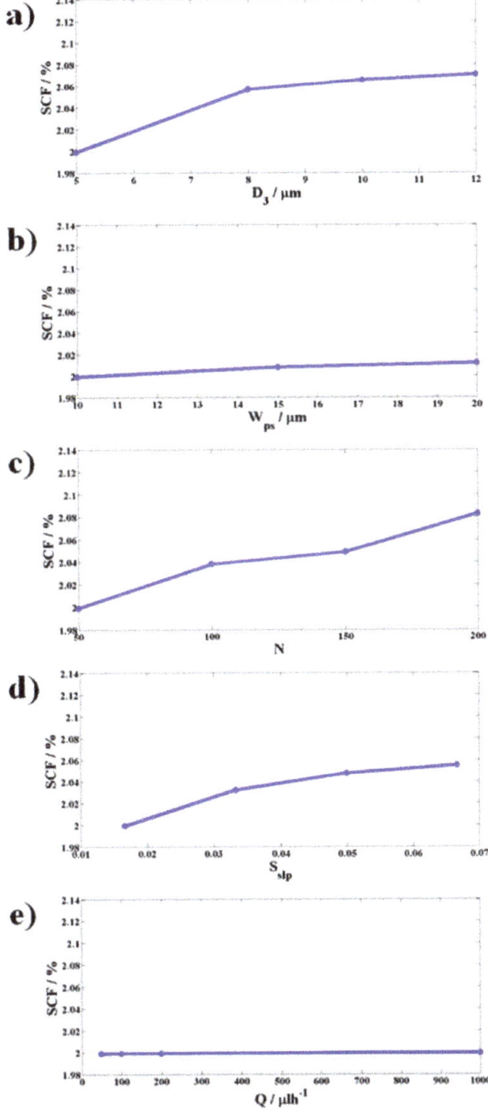

Figure 5. Each panel reports the trend of *SCF*. The simulation results showing the effect of parameters of Section II, namely (**a**) height of the pillar (D_3), (**b**) spacing between the pillars (W_{ps}), (**c**) number of pillars (N), (**d**) slope of the side channel S (S_{slp}), and (**e**) the sample flow rate (Q) on the side to central channel flow rate ratio (*SCF*), respectively. The results indicate there is no significant effect of the parameters of Section II and the flow rate on *SCF*.

Figure 5 shows that there is no significant change in *SCF* with respect to changes in Section II parameters D_3, W_{ps}, N, S_{slp} and Q whereas Figure 6a shows a clear variation of *SCF* with respect to the Section III parameter *SCRR*. This concludes that Section III alone had influence on the net fluid flow through the side channels S1′ and S2′. The reason for this can be understood from the perspective that the fluid flow rate through a channel depends to a great deal on the hydraulic resistance offered by the channel instead of the details of the various geometrical obstructions in the path. The latter observation was further demonstrated experimentally by measuring *SCF* with respect to variations in *SCRR* and has been shown in Figure 6b. The device used for the experiments was chosen to have all the parameters similar to that of the simulated device, except that the number of pillars were taken to be 1000 for the reasons discussed in the next section.

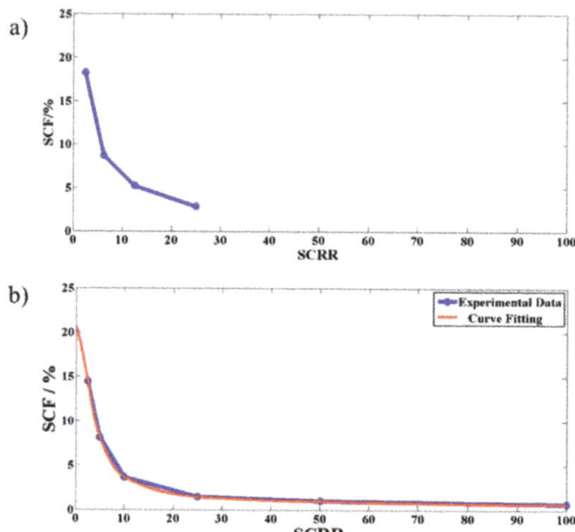

Figure 6. The effect of the side to the central channel hydraulic resistance ratio (*SCRR*) on the side to the central channel flow rate ratio (*SCF*). (**a**) The simulation results demonstrating the effect of *SCRR* on *SCF*. (**b**) The experimental results showing the variation of *SCF* with respect to *SCRR*.

3.1.2. Effect of Section II Parameters [Number of Pillars (N) and Side Channel Slope (S_{slp})] on the Uniformity of the Flow Across the Pillar Gaps

The simulations were performed on the model described in the simulations section with $D_1 = 20$ µm, $D_2 = 80$ µm, $D_3 = 5$ µm, $SCRR = 50$, and pillar spacing cross-Section 10 µm (width) × 5 µm (depth) while varying the number of pillars (N) or the slope of the side channel (S_{slp}). *SCRR* was kept fixed at a specific value to keep the *SCF* constant throughout the simulations. The choice of *SCRR* to 50 during the simulations has no practical relevance and can be kept at any number to understand the flow behavior. By keeping the number of pillars fixed to 50, the slope of the side channel wall (S_{slp}) was varied from 0.015 to 0.065 by varying the S_{tl} from 0.5 µm to 10 µm per pillar array period. The resulting flow rate variation between the pillars as a function of pillar position in the model is shown in Figure 7a.

One observation is that the flow rate distribution across each pillar gap was changing as a function of the position of the pillar. More specifically, this variation is more pronounced towards the end pillars. The flow rate rapidly shoots up towards the end pillars. This is an undesirable feature if this section of the design is expected to extract fluid volumes smaller than the volume of the critical particle across all pillar gaps. As S_{slp} is increased, a similar variation of the flow rate continued except for one major change. That is, the peak of this flow rate has fallen by redistributing this growing flow rate over larger number of pillars. This implies that the larger the slope of this wall, the more uniform the distribution of the flow rate across the pillars.

Figure 7b shows the variation of the flow rate across the pillar gaps as a function of the number of pillars, for S_{tl} as 0.5 µm per pillar array period. As the number of pillars has been increased from 50 to 200, the decrease in the peak flow rate and the redistribution of this growing flow rate across the pillars can be clearly seen. The number of pillars over which the variable flow rate is present is determined by finding the pillar at which the flow rate drops to 1% of the peak flow rate and counting the pillars backwards from the pillar at which peak flow rate was observed. These observations suggest that the choice of large values for S_{slp} and N seem to contribute to the increased uniformity of the flow rate across the pillar gaps.

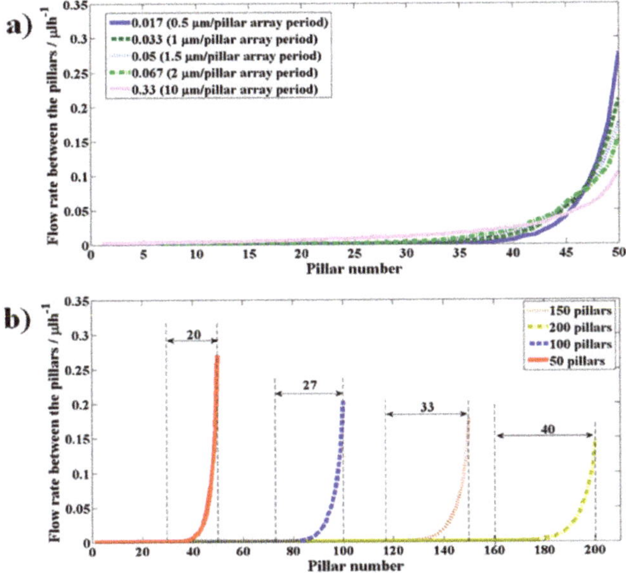

Figure 7. The variation of the flow rate across the pillar gaps, plotted as a function of the pillar number, when the slope of the side channel S_{slp} (**a**) and number of pillars (N) (**b**) are varied.

3.2. Experimental

Having found relevant parameters that affect the performance of the device from the simulations, the devices were fabricated with those optimum parameters. As the purity of the particle-free fluid was not obtained from our simulations, the experiments were conducted to further probe the performance of the device (purity of particle-free fluid) with respect to other relevant parameters, such as particle size, time, the flow rate, hematocrit and *SCRR*.

Variation of Purity of Plasma Generated as a Function of Time, Flow Rate (Q), Hematocrit, and *SCRR*

The proposed microfluidic device was fabricated with parameters $D_1 = 20$ µm, $D_2 = 74$ µm, $D_3 = 4.5$ µm, $W_{ps} = 12$ µm, $W_p = 18$ µm, $S_{tl} = 0.5$ µm and with *SCRR* = 2.5, 10, 25, 50, and 100. The experiments were performed on the fabricated device with respect to time, the flow rate, hematocrit, and *SCRR*, and the purity of plasma collected was evaluated using image processing as discussed in the section on the experimental procedure. Figure 8a shows the time independence of the purity of plasma generated and to a value of close to 100% at a hematocrit of 15% and *SCRR* = 25. This infers that the quality of plasma generated is very pure and is independent of time and can be used anytime during chip analysis.

As can be observed from Figure 8b, plasma purity remains constant at approximately 99% with a maximum variation of only up to 1% with respect to the variation in the flow rate to approximately three orders of magnitude ranging from 10 µlh^{-1} to 1500 µlh^{-1} at a hematocrit of 25% and *SCRR* = 50. This clearly demonstrates the flow rate independent performance of the proposed technique and can be used for a wide range of applications.

Figure 8c shows the variation of plasma purity as a function of hematocrit and *SCRR*. A smaller *SCRR* is expected to provide a larger side channel flow rate as it offers low resistance to the flow inside the channels, however, it may compromise purity at higher hematocrits. Larger *SCRR* is expected to provide a smaller side channel flow rate while offering very high purity even at higher hematocrits. Figure 8c shows experimental results that agree with the theory. The smaller *SCRR* of 2.5 offered 100% purity until 2.5% hematocrit and started dropping until 98.6% as the hematocrit reached 5%. However, larger *SCRR* of 100 offered 100% purity until 28% hematocrit and the purity dropped by only 1% even at a hematocrit of 45%. Figure 9a–d show the images of RBCs at various hematocrit values inside the central channel C and those that escaped into the side channels S1 and S2 at *SCRR* of 100.

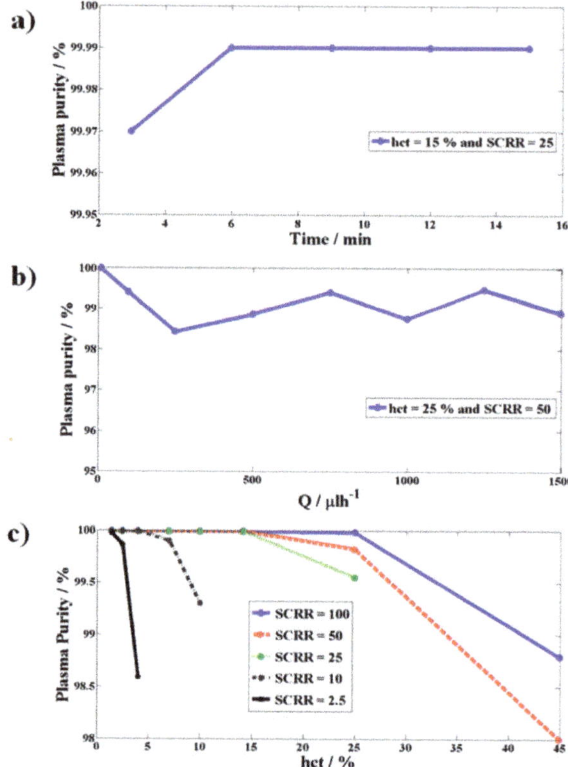

Figure 8. The variation of plasma purity as a function of (**a**) duration of the experiment (time), (**b**) flow rate (*Q*), (**c**) hematocrit (*hct*) and *SCRR*.

The yield of the device (*Y*) is the percentage fraction of the plasma that could be extracted into the side channels from the available plasma in the injected sample and is defined as,

$$Y = Q_s / [Q \times (1 - hct)] = SCF / [(1 + SCF) \times (1 - hct)],$$

as the true yield also depends on the quantity of the particle-free fluid available. The smaller *SCRR* of 2.5 (at *hct* = 2.5%) has contributed to a plasma yield of approximately 14% and the larger *SCRR* of 100 (at *hct* = 45%) contributed to a smaller plasma yield of approximately 4%. This lower yield at higher hematocrits is not an upper limit to this proposed device, and it can be further increased without compromising on purity, by multiplexing this design in stages and suitably adjusting the side channel resistances as discussed in the subsequent section. Further, 5 μm beads and platelets were pumped at 100 μlh⁻¹ into a microfluidic device with *SCRR* = 50, and the purity of the recovered particle-free fluid was close to 100%. Figure 9e,f show the representative images of the beads and platelets inside the microfluidic device close to the end of Section II, respectively. It can be clearly seen that no platelets and beads were observed in the side channels S1 and S2 and the whole of the particles were concentrated in the central channel.

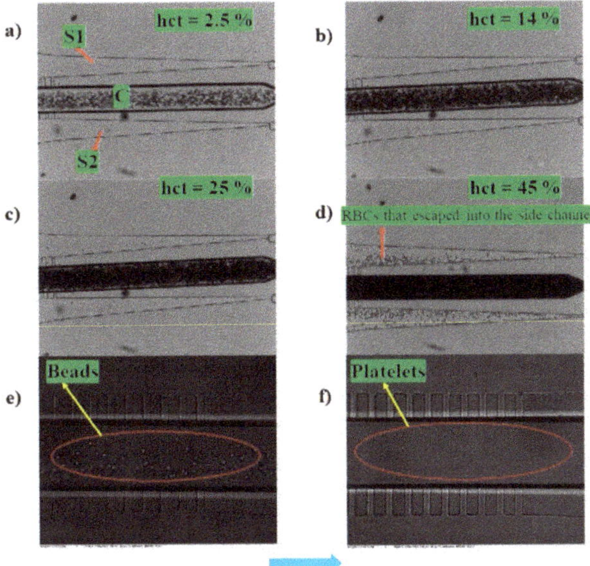

Figure 9. Experimental images illustrating the functioning of the proposed device for various particle concentrations and sizes. (**a–d**) represent the experimental images depicting the concentration of RBCs in central channel C and those that escaped into the side channels S1 and S2 for different values of hematocrit. (**a**) 2.5% hematocrit, (**b**) 14% hematocrit, (**c**) 25% hematocrit, and (**d**) 45% hematocrit. (**e,f**) represent the experimental images depicting the localization of particles (5 μm beads and platelets) to the central region C alone. The *SCRR* of the devices used in (**a–d**) is 100 and in (**e**,textbff) is 50.

3.3. Discussion on the Choice of Parameters for Wee-Extraction

Consider the case in which one value of *SCRR* gave a specific *SCF*. Hence, for a sample flow rate of Q, the side channel total flow rate is $Q_s = (Q \times SCF)/(1 + SCF)$. As the device has two side channels S1 and S2, the flow rate through each is half of Q_s. For N number of pillars on each side, the flow rate averaged through each pillar gap is $Q_{pg} = Q_s/(2 \times N)$. For a particle of critical volume V_p not to escape through the pillar gap, the volume being drawn through the pillar gaps should be smaller than V_p. As the particle passes closely adjacent to the pillars, it has two components of velocity u and v. u is along the direction of main sample flow and is responsible for the particles to retain in the central channel, whereas v is perpendicular to the direction of main sample flow and into the pillar gaps and is responsible for the particle escape into the side channels S1 and S2. Due to the motion of the particle along main sample flow, each particle spends time duration of approximately $t_{pg} = W_{ps}/u$. Within this duration, the volume of fluid that escapes into the pillar gaps is $W_{ps} \times D_3 \times v \times t_{pg} = Q_{pg} \times t_{pg}$. When this volume is smaller than V_p, the particles do not escape into the side channels, thereby holding a possibility to achieve 100% purity of the side channel fluid collected. The reduction in the expected purity can arise when this criterion that has been derived for the uniform flow across the pillars is not met.

From the devices used for the experiments, consideration is given to a device with $SCRR = 100$ for analysis. For this $SCRR$, the SCF is 2%. For $Q = 400$ μlh^{-1}, $W_{ps} = 12$ μm, and $N = 1000$, Q_{pg} will be 0.0039 μlh^{-1}. The simulations have been done to estimate u at a distance close enough to the pillars (2.5 μm away the center of the pillar and 2.5 μm away from the top of pillar) and was found to be

0.24 mms^{-1}. This gave rise to time duration of 50 ms, and the volume of suction as 54 fL per pillar gap. In the case of plasma extraction using whole blood, the quantity of fluid being extracted is smaller than the RBC particle size (100–120 fL), hence avoiding possibility of RBC escape through the pillar gaps.

However, certain considerations must be made in view of the observations from the simulations as shown in Figure 7. The flow will not be uniform across the pillars as was presumed in the above analysis. The flow rate across the pillars close to the end of the array in the direction of the flow can be high compared to the pillars in the beginning of the array. This may lead to the particles not meeting the desired criteria as the fluid volume escaping through the pillar gaps are smaller than the particle volume to be separated towards the end pillars thereby leading to a drop in purity of the fluid collected in the side channels. The fabricated device has 1000 pillars which is five times larger than the number of pillars used in the simulation (200), hence, it is expected to have a better uniformity of the flow rate across the pillar gaps than the simulated results show in Figure 7b. At higher hematocrits (45%) of operation of the device, the purity has slightly come down to 99%. This may be an indicator that the flow rate across the pillar gaps towards the end of the array is still larger than the volume of RBC. Increasing the number of pillars and the slope S_{slp} should help bring down this peak volume that is being extracted across the pillar gaps to smaller than the volume of RBC, thereby providing 100% purity even at much larger hematocrits (> 45%).

From Figure 5e, it can be observed that *SCF* remains unchanged as the sample flow rate is increased, indicating that the flow rate in the side channels and central channel proportionately increase. On an average, this indicates how proportionately u and v change, thereby leading to almost similar flow volumes across the pillar gaps that are independent of the sample flow rate. This also suggests how the purity can remain almost constant even for a wide variation of flow rates. This behavior can be observed experimentally as shown in Figure 8b. The above analysis is very general and can be applied to separating particles of any size, volume and concentration at any desired flow rate. Suitably choosing the large values for *SCRR*, number of pillars N (>1000), and the slope of the side channels S_{slp} (S_{tl} > 0.5 μm per period of the array), and smaller height D_3 and spacing between the pillars W_{ps} as per the above guidelines. This assists in designing a device that can separate particles smaller than platelets (<2 μm), without the need for complicated fabrication, and at dimensions that can be easily fabricated using the conventional techniques of microfabrication. The yield of this proposed device can be increased by multiple folds and a way of accomplishing this is presented in the following section.

3.4. Enhancing the Yield through Multiplexing

Consider one of the proposed designs that have been discussed so far as stage 1 and for example, it produces a yield of Y as shown in the resistance model of the device, Figure 10a. If this stage (stage 1) is added to the end of a similar such stage (stage 2), it adds additional resistance to the central channel of stage 2, thereby reducing the purity of the plasma collected. To ensure that the *SCRR* of the stage 2 of the newly formed combination device remains at its earlier value, an additional resistance X_1 needs to be added on either side of stage 2, as shown in Figure 10b. This process ensures the increment in the yield by a factor of [2 − *SCF*/(1 + *SCF*)], as the two stages contribute to the plasma generation. The inset in Figure 10 shows a representative schematic of the device with stages 1 and 2 as presented in Figure 10b. Similarly, for a k-stage device, the yield is expected to be:

$$\left[k + \frac{-SCF}{1 + SCF} + \left(\frac{-SCF}{1 + SCF} \right)^2 + \left(\frac{-SCF}{1 + SCF} \right)^3 + \cdots + \left(\frac{-SCF}{1 + SCF} \right)^{(k-1)} \right] \times Y,$$

when ensured that the purity is unchanged by adding extra resistances to each stage of the device (Figure 10c). As an example, the yield of the device with *SCRR* of 100 at hematocrit of 45% is 4% and this can be increased to approximately 20% by suitably cascading five such stages, considering the extra resistances that need to be incorporated during the design. In this analysis, it was presumed that the yield of each stage cannot change substantially with respect to the addition of a greater number of stages. However, it is noted that in cases when the yield of each stage changes substantially, which typically may happen at lower *hct* and higher *SCF*, the modified yield of each stage needs to be considered to arrive at an accurate value of the total yield.

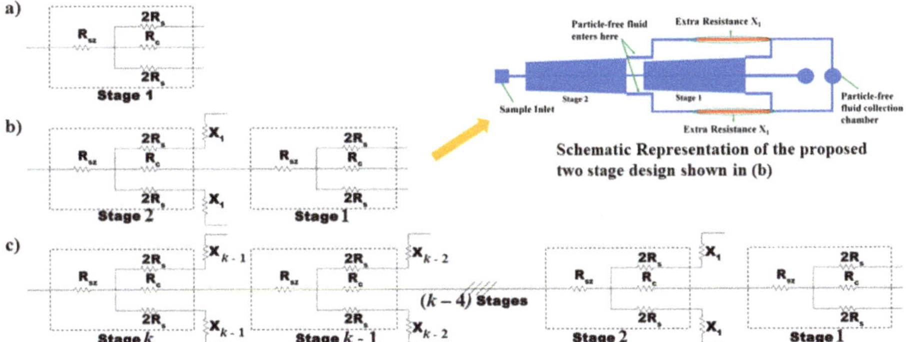

Figure 10. Schematic illustrating the strategies to enhance the yield. (**a**) Hydraulic resistance model of the device along with the side and central channels C', S1' and S2' and is depicted as Stage-I. (**b**) Hydraulic resistance model when Stage-II being added to Stage-I and the representative schematic is shown in the inset. (**c**) Hydraulic resistance model when *k* such stages are present in the device.

4. Conclusions

This paper presents a technique that can, in principle, separate the particles and the fluid in 100% purity and higher yields for a wide range of particle sizes, concentrations, and flow rates. The principle of operation was studied through the theory and simulations, while the proof-of-principle was demonstrated using 5 μm beads, platelets, and blood of ranging hematocrits (2.5%–45%). Lower hematocrit (2.5%) blood was separated in 100% purity and high yields (14%), whereas the high hematocrit blood (45%) was separated in purities ranging from 98% to 100% depending upon the flow rate of operation (10 μlh^{-1}–1500 μlh^{-1}). The yield can be increased in approximately *k*-folds without compromising purity by cascading the proposed design *k*-times appropriately. The purity can be enhanced further for any size and the concentration of particles by suitably choosing optimum values for the side channel to the central channel hydraulic resistance ratio (*SCRR*), the number of pillars (*N*), the slope (*S$_{slp}$*), the pillar spacing (*W$_{ps}$*), and the height of the pillars (*D$_3$*).

Unique features of this technique are the stepped pillar arrays and the hydraulic resistance tuners. Stepped pillar arrays play a unique and important role in extracting smaller quantities of fluid through the pillar gaps by providing the smaller cross-section for the fluid flow into the side channels. Another associated important feature of these stepped pillar arrays is that the pillar spacing cross-section can be much larger than the size of the particle to be separated from the particle-free fluid, thereby facilitating ease in fabrication even for separating submicron particles. This was very well demonstrated by choosing the pillar gap cross-section as 12 μm × 4.5 μm and the particles to be separated as platelets (2 μm–4 μm) with 100% purity. The hydraulic resistance tuners majorly control the net yield and play a major role in deciding the critical particle volume to be decanted in association with the stepped pillar arrays. The stepped pillar arrays along with hydraulic resistance tuners facilitate the flow-rate-independent the fluid flow through the pillar gaps, thereby ensuring the flow-rate-independent application of this technique. This was demonstrated for flow rates ranging from 10 μlh^{-1}–1500 μlh^{-1}.

This novel technique offers several potential applications towards plasma separation from whole blood for biochemical tests to achieve self-sheath flow by utilizing the particle-free fluid to surround the central particle-rich sample, and the enrichment of rare cells such as circulating tumor cells (CTCs) for cancer detection. As a summary, this paper offers the design, theory, and guidelines to achieve the decantation of the particles of any size and concentration at 100% purity and desired yield, while the in-flow at any flow rate of operation concurrently provides great ease of fabrication.

Supplementary Materials: The following are available online at http://www.mdpi.com/2072-666X/10/7/471/s1, Figure S1: Mask design for layer-1 that was used to fabricate the Master. Figure S2: Mask design for layer-2 that was used to fabricate the Master. Figure S3: Mask design for layer-3 that was used to fabricate the Master. Figure S4: Schematic of the device design that was used for performing simulations.

Author Contributions: Conceptualization, G.E. and S.S.G.; data curation, G.E. and P.N.; formal analysis, G.E. and P.N.; funding acquisition, S.S.G.; investigation, G.E. and P.N.; methodology, G.E. and P.N.; project administration, S.S.G.; supervision, S.S.G.; writing – original draft, G.E.; writing – review and editing, S.S.G.

Funding: The research was supported by the Infosys Grant, CIDR, Indian Institute of Science, Bangalore, India.

Acknowledgments: The authors would like to thank Rajesh Srinivasan and Abhishek Pathak for their generous help in providing the samples for experiments. The authors also would like to acknowledge Sanjeev Kumar's Research Group for generously letting us use the simulation facility, and the National Nano-fabrication facility in the Center for Nanoscience and Engineering, Indian Institute of Science for the fabrication and characterization facility.

Conflicts of Interest: The authors declare no conflict of interest.

References

1. Arifin, D.R.; Yeo, L.Y.; Friend, J.R. Microfluidic blood plasma separation via bulk electrohydrodynamic flows. *Biomicrofluidics* **2006**, *1*, 014103. [CrossRef]
2. Lenshof, A.; Ahmad-Tajudin, A.; Järås, K.; Swärd-Nilsson, A.-M.; Åberg, L.; Marko-Varga, G.; Malm, J.; Lilja, H.; Laurell, T. Acoustic Whole Blood Plasmapheresis Chip for Prostate Specific Antigen Microarray Diagnostics. *Anal. Chem.* **2009**, *81*, 6030–6037. [CrossRef]
3. Jiang, H.; Weng, X.; Chon, C.H.; Wu, X.; Li, D. A microfluidic chip for blood plasma separation using electro-osmotic flow control. *J. Micromech. Microeng.* **2011**, *21*, 085019. [CrossRef]
4. Shih, C.-H.; Lu, C.-H.; Yuan, W.-L.; Chiang, W.-L.; Lin, C.-H. Supernatant decanting on a centrifugal platform. *Biomicrofluidics* **2011**, *5*, 013414. [CrossRef]
5. Chen, C.-C.; Lin, P.-H.; Chung, C.-K. Microfluidic chip for plasma separation from undiluted human whole blood samples using low voltage contactless dielectrophoresis and capillary force. *Lab Chip* **2014**, *14*, 1996–2001. [CrossRef]
6. Chen, X.; Kuo, J. Blood separation and plasma preparation on a compact disk microfluidic chip. In Proceedings of the 9th IEEE International Conference on Nano/Micro Engineered and Molecular Systems (NEMS), Waikiki Beach, HI, USA, 13–16 April 2014; pp. 47–50.
7. Szydzik, C.; Khoshmanesh, K.; Mitchell, A.; Karnutsch, C. Microfluidic platform for separation and extraction of plasma from whole blood using dielectrophoresis. *Biomicrofluidics* **2015**, *9*, 064120. [CrossRef]
8. Shimizu, H.; Kumagai, M.; Mori, E.; Mawatari, K.; Kitamori, T. Whole blood analysis using microfluidic plasma separation and enzyme-linked immunosorbent assay devices. *Anal. Methods* **2016**, *8*, 7597–7602. [CrossRef]
9. Faivre, M.; Abkarian, M.; Bickraj, K.; Stone, H.A. Geometrical focusing of cells in a microfluidic device: An approach to separate blood plasma. *Biorheology* **2006**, *43*, 147–159.
10. VanDelinder, V.; Groisman, A. Separation of Plasma from Whole Human Blood in a Continuous Cross-Flow in a Molded Microfluidic Device. *Anal. Chem.* **2006**, *78*, 3765–3771. [CrossRef]
11. Rodríguez-Villarreal, A.I.; Arundell, M.; Carmona, M.; Samitier, J. High flow rate microfluidic device for blood plasma separation using a range of temperatures. *Lab Chip* **2010**, *10*, 211–219. [CrossRef]
12. Kersaudy-Kerhoas, M.; Dhariwal, R.; Desmulliez, M.P.Y.; Jouvet, L. Hydrodynamic blood plasma separation in microfluidic channels. *Microfluid. Nanofluid.* **2010**, *8*, 105. [CrossRef]
13. Tripathi, S.; Kumar, Y.V.B.; Agrawal, A.; Prabhakar, A.; Joshi, S.S. Microdevice for plasma separation from whole human blood using bio-physical and geometrical effects. *Sci. Rep.* **2016**, *6*, 26749. [CrossRef]
14. Yang, S.; Undar, A.; Zahn, J.D. A microfluidic device for continuous, real time blood plasma separation. *Lab Chip* **2006**, *6*, 871–880. [CrossRef]
15. Lee, M.G.; Shin, J.H.; Choi, S.; Park, J.-K. Enhanced blood plasma separation by modulation of inertial lift force. *Sens. Actuators B: Chem.* **2014**, *190*, 311–317. [CrossRef]

16. Hou, H.W.; Bhagat, A.A.S.; Lee, W.C.; Huang, S.; Han, J.; Lim, C.T. Microfluidic Devices for Blood Fractionation. *Micromachines* **2011**, *2*, 319–343. [CrossRef]
17. Tripathi, S.; Kumar, Y.V.B.V.; Prabhakar, A.; Joshi, S.S.; Agrawal, A. Passive blood plasma separation at the microscale: A review of design principles and microdevices. *J. Micromech. Microeng.* **2015**, *25*, 083001. [CrossRef]
18. Di Carlo, D. Inertial microfluidics. *Lab Chip* **2009**, *9*, 3038–3046. [CrossRef]
19. Xuan, X.; Zhu, J.; Church, C. Particle focusing in microfluidic devices. *Microfluid. Nanofluid.* **2010**, *9*, 1–16. [CrossRef]
20. Zhang, J.; Yan, S.; Yuan, D.; Alici, G.; Nguyen, N.-T.; Warkiani, M.E.; Li, W. Fundamentals and applications of inertial microfluidics: A review. *Lab Chip* **2015**, *16*, 10–34. [CrossRef]
21. Torino, S.; Iodice, M.; Rendina, I.; Coppola, G.; Schonbrun, E. Hydrodynamic self-focusing in a parallel microfluidic device through cross-filtration. *Biomicrofluidics* **2015**, *9*, 064107. [CrossRef]
22. Eluru, G.; Julius, L.A.N.; Gorthi, S.S. Single-layer microfluidic device to realize hydrodynamic 3D flow focusing. *Lab Chip* **2016**, *16*, 4133–4141. [CrossRef]

 *micromachines* 

Article

Hydrodynamic Microparticle Separation Mechanism Using Three-Dimensional Flow Profiles in Dual-Depth and Asymmetric Lattice-Shaped Microchannel Networks

Takuma Yanai [†], Takatomo Ouchi [†], Masumi Yamada * and Minoru Seki

Department of Applied Chemistry and Biotechnology, Graduate School of Engineering, Chiba University, 1-33 Yayoi-cho, Inage-ku, Chiba 263-8522, Japan; lyushi.15m@gmail.com (T.Y.); tktm888@gmail.com (T.O.); mseki@faculty.chiba-u.jp (M.S.)
* Correspondence: m-yamada@faculty.chiba-u.jp; Tel.: +81-43-290-3398
† These authors equally contributed to this work.

Received: 6 June 2019; Accepted: 21 June 2019; Published: 25 June 2019

Abstract: We herein propose a new hydrodynamic mechanism of particle separation using dual-depth, lattice-patterned asymmetric microchannel networks. This mechanism utilizes three-dimensional (3D) laminar flow profiles formed at intersections of lattice channels. Large particles, primarily flowing near the bottom surface, frequently enter the shallower channels (separation channels), whereas smaller particles flowing near the microchannel ceiling primarily flow along the deeper channels (main channels). Consequently, size-based continuous particle separation was achieved in the lateral direction in the lattice area. We confirmed that the depth of the main channel was a critical factor dominating the particle separation efficiencies, and the combination of 15-μm-deep separation channels and 40-μm-deep main channels demonstrated the good separation ability for 3–10-μm particles. We prepared several types of microchannels and successfully tuned the particle separation size. Furthermore, the input position of the particle suspension was controlled by adjusting the input flow rates and/or using a Y-shaped inlet connector that resulted in a significant improvement in the separation precision. The presented concept is a good example of a new type of microfluidic particle separation mechanism using 3D flows and may potentially be applicable to the sorting of various types of micrometer-sized objects, including living cells and synthetic microparticles.

Keywords: microfluidic device; particle separation; hydrodymanics; microchannel; cell sorting

---

1. Introduction

The need to separate micrometer-sized particles precisely, especially mammalian cells of specific phenotypes, is increasing with the recent progress in cell-based liquid biopsy technologies and stem cell engineering [1–3]. In the industrial production of synthetic microparticles, monodispersity at a particle size is a critical factor dominating the function and reliability of particle-based products, as represented by particle-based separation matrices. In the last decade, microfluidic systems have been recognized as a practical tool for precisely separating micrometer-sized cells and particles [4–6]. Several types of microfluidic cell separators are commercially available, most of which employ laminar flow systems or inertial forces of particle movement in microchannels. Representative examples of particle sorting mechanisms that do not necessitate the application of outer forces include deterministic lateral displacement (DLD) [7–9], pinched-flow fractionation (PFF) [10,11], hydrodynamic filtration [12–14], Dean-flow fractionation [15–17], hydrophoresis [18–20], inertial microfluidics [21,22], and multi-orifice fractionation [23,24]. Most of these techniques utilize precisely controlled flow profiles in a quasi-two-dimensional microchannel, i.e., laminar flow patterns in microchannels with a uniform depth, neglecting the flow rate distribution in the z (depth) direction. Meanwhile, recent studies have demonstrated that three dimensionally fabricated microchannels effectively function as new particle sorting/focusing devices [25,26]. Researchers have used secondary flows, Dean-flows, or microvortices formed in the microchannel cross section (x-z plane) [27–30]. From these examples, we expect that unprecedented but efficient microfluidic mechanisms for particle separation can be developed using three-dimensional (3D) flow profiles in microfluidic channels with non-uniform depths.

In our recent study, we proposed the concept of a purely hydrodynamic particle separation scheme using slanted, asymmetrically arranged, lattice-shaped microchannel networks [31]. The lattice structure was composed two types of perpendicularly crossing microchannels ("main channels" and "separation channels") with a uniform depth. To split a small amount of fluid flow from the main channel to the separation channel at each crossing point, the density of the separation channels was 30–100 times higher than that of the main channels. Using this microchannel configuration, large particles flow along the main channel, whereas small particles enter the separation channels, resulting in the size-dependent separation of particles in the lateral (x) direction. The lattice configuration of the microchannel was advantageous because it is robust against microchannel clogging. Additionally, the separation throughput can potentially be increased compared with single microchannels with a depth/width of several tens of micrometers. However, one concern remains: the number of main channels is limited, and only ~10 main channels can be placed in the lattice region with a width of ~10 mm. We pondered the outcome of placing multiple main channels more densely, instead of creating a significant difference in the densities of the two types of perpendicularly crossing channels. This was our primary motivation to test a new type of lattice microchannel-based particle separation technique.

Herein, we describe a new principle of size-based particle separation using asymmetric lattice channel networks composed of two channel types that are more densely arranged and have different depths. The schematic images exhibiting the particle separation behaviors are shown in Figure 1. To create an anisotropic flow distribution at each crossing point, we fabricated dual-depth lattices; the main channels, which were slanted to the lower right direction ($+ x$ and $+ y$ directions), were made deeper, whereas the separation channels, which were perpendicularly crossing the main channels, were shallower. From the precise observation of the particle behaviors in the channels, we noticed that complex 3D flows critically dominated the size-dependent difference in the particle behaviors, thereby achieving particle separation in the lateral direction. Particularly, it was interesting that the large particles frequently entered the shallower separation channels, whereas the small ones did not. These separation behaviors were completely different from those observed in our previous study using uniform-depth lattice channels [31], or the previously well-developed DLD techniques. In this study, we investigated several factors affecting the separation performances of particles and conducted several experiments to tune the separation size and improve the separation efficiencies.

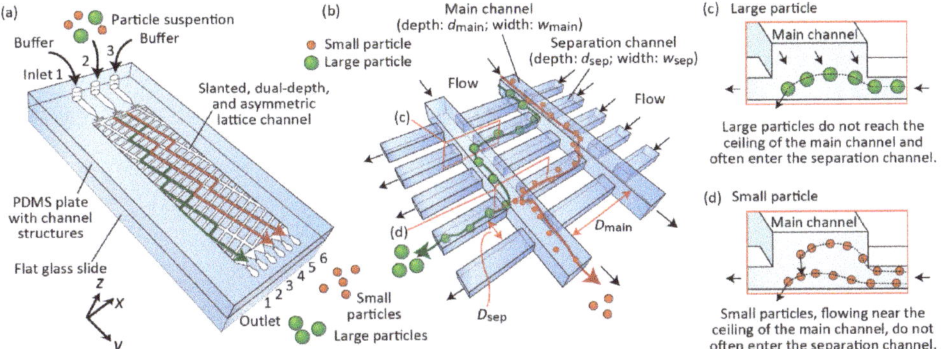

Figure 1. (**a**) Schematic image illustrating particle separation behaviors using dual depth, lattice-shaped channel networks. (**b**) Detailed behaviors of particles in the lattice region. (**c,d**) Illustrations showing the particle movement in the depth (z) direction from the shallower separation channels into the main channels. These images correspond to the cross sections shown in panel (**b**).

2. Materials and Methods

2.1. Separation Mechanism

The detailed separation mechanism is shown in Figure 1. We employed asymmetric lattice-channel networks with three inlets and six outlets. The lattice structure was composed of deep main channels and shallow separation channels (Figure 1b). We introduced a particle suspension from Inlet 2 and buffer solution from Inlets 1 and 3. Because the main channels are slightly slanted against the flow direction (y direction), and a small amount of the fluid flow is split from the main channel into the separation channel at every crossing point, particles alternately flow through the main channel and separation channels. At the moment when particles enter the main channel from the separation channel, larger particles with sizes comparable to the depth of the separation channel cannot reach the ceiling of the main channel, because of the hydrodynamic restriction effect in the widening region, as in the case of the PFF scheme (Figure 1c). Therefore, these large particles frequently enter the separation channel. Meanwhile, small particles can reach the ceiling of the main channel, and they are likely to flow along the main channel and do not often enter the main channel (Figure 1d). Consequently, the difference in the lateral positions (x position) of the particles is enhanced as the particles flow downstream, and these particles are separated based on size. In the following experiments, we examined if this concept explains the mechanism of particle separation using the presented microfluidic systems.

2.2. Fabrication and Design of Microfluidic Devices

Polydimethylsiloxane (PDMS)-glass microfluidic devices were fabricated using standard soft lithography and replica molding techniques [32]. Briefly, we first prepared SU-8 molds on Si wafers. The SU-8 spin-coating and ultraviolet light irradiation processes were repeated twice to obtain dual-depth structures. A PDMS prepolymer (Silpot 184, Dow Corning Toray, Tokyo, Japan) was poured onto the prepared mold, and subsequently cured at 85 °C for 30 min. After completing the crosslinking reaction of PDMS, the PDMS plate with channel structures was peeled off from the mold; subsequently, it was bonded against a flat glass slide (S1112, Matsunami Glass, Tokyo, Japan) after O_2 plasma-based surface activation. Finally, inlet/outlet silicone tubes were attached and subsequently glued to form the inlet/outlet ports.

We prepared six types of microfluidic devices (Microdevices A-F) with different microchannel geometries, as shown in Figure 2 and Table 1. Each inlet channel was branched into two channels to uniformly introduce fluid samples into the lattice region. Prefilter structures were placed in the inlet channels to avoid the introduction of large particulates into the lattice region. In the lattice region, two types of microchannels crossed perpendicularly. The deeper main channels (width of w_{main}, depth of d_{main}) were slanted against the flow direction (y-direction) with a slant angle of 15°, and they were periodically placed with the interchannel distance of D_{main}. The separation channels (width of w_{sep}, depth of d_{sep}) were shallower and narrower than the main channels, which were also placed at the interchannel distance of D_{sep}. Microdevices A–C were prepared to examine the effects of the main channel depth. Microdevices D–F were used to tune the critical size of the particle separation of Microdevice A; these channels exhibited a perfectly similar relationship, but with different sizes.

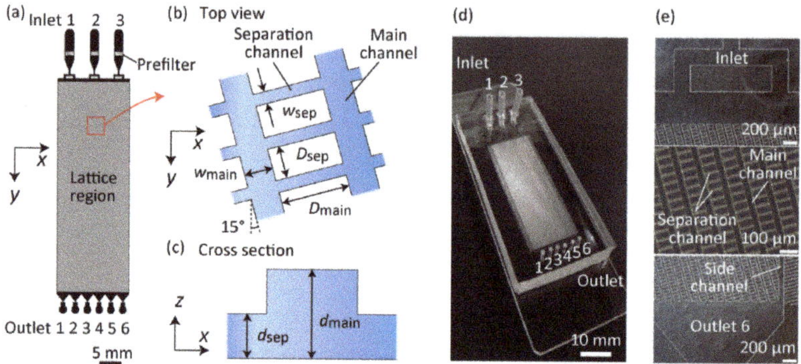

Figure 2. Microchannel design and photographs of the microfluidic devices. (**a**) Top view of Microdevices A–C, (**b**) enlarged image of the lattice region, and (**c**) cross-sectional view of the lattice region. w_{main}: width of the main channel; w_{sep}: width of the separation channel; d_{main}: depth of the main channel; d_{sep}: depth of the separation channel; D_{main}: interchannel distance of the neighboring main channels; D_{sep}: interchannel distance of the neighboring separation channels. (**d**) Photograph of the prepared Microdevice A and (**e**) microscopic images showing the inlet, lattice, and outlet regions of Microdevice A.

Table 1. Parameters of six types of fabricated microfluidic devices. The parameters correspond to those shown in Figure 2. The size ratio indicates the relative size ratios for Microdevices A, D, E, and F.

Microdevice	w_{main} (μm)	w_{sep} (μm)	d_{main} (μm)	d_{sep} (μm)	D_{main} (μm)	D_{sep} (μm)	Lattice Size (mm × mm)	Size Ratio
A	35	15	40	15	93	25	32 × 12	1
B	35	15	25	15	93	25	32 × 12	-
C	35	15	70	15	93	25	32 × 12	-
D	52	23	60	23	139	38	48 × 18	1.5
E	23	10	27	10	62	17	21.3 × 8	0.67
F	17	7	20	7	46	13	16 × 6	0.5

2.3. Particle Separation Experiments

We employed fluorescent/nonfluorescent standard polystyrene particles with different diameters. Microparticles with the average diameters of 2.1 μm (B0200; blue fluorescent), 3.0 μm (R0300; red fluorescent), 3.1 μm (G0300; green fluorescent), 4.8 μm (G0500; green fluorescent), 6.0 μm (4206A; nonfluorescent), 9.9 μm (G1000; green fluorescent), and 15 μm (4215A; nonfluorescent) were obtained from Thermo Fisher Scientific, MA, USA. These particles were suspended in an aqueous solution of 18% sucrose and 0.5% tween 20 at concentrations of 5×10^6–3×10^7 particles per milliliter. This solution prevents the precipitation of particles. The particle suspension was pumped from Inlet 2, whereas the same solution without particles (the "buffer") was introduced from Inlets 1 and 3 using syringe pumps (KDS200, KD Scientific, MA, USA). The behaviors of the particles were observed using a fluorescence microscope (IX71, Olympus, Tokyo, Japan) equipped with a charge-coupled device camera (DP80, Olympus, Tokyo, Japan). The absolute numbers of particles, separated and recovered from each outlet, were evaluated by measuring the output volumes and analyzing the particle concentrations using a hemocytometer. The recovery ratio of the particles was defined as the number of particles recovered from an outlet divided by the total particle number recovered from all six outlets. On average, ~200 particles were counted for each condition, and experiments were repeated at least thrice using individual microdevices. To investigate the flow behaviors of microparticles in the x-y planes at different depths (different z positions), a high-speed confocal microscope system (Confocal Scanning Micro PIV System, Seika Corp., Tokyo, Japan) was used.

3. Results and Discussion

3.1. Particle Separation Using Microdevice A

In the presented lattice-shaped microchannel networks, the flow rate distribution along the streamline (y-axis) is not necessarily uniform because of the asymmetrically placed channels with different depths. The values of the output volumes are key for evaluating the separation efficiency of the target particles. Hence, we first measured the volumetric flow rates distributed to each outlet using Microdevice A, by introducing distilled water from the inlets and measuring the output volumes by weighing. The result is shown in Supplementary Figure S1. As expected, the flow rates through the six outlets were not uniform and a distribution was shown; the flow rate to Outlet 1 was only ~10% of the input flow, whereas that to Outlet 6 was higher (~28%). This non-uniformity is attributable to the relatively wide and deep main channels slanted to the right direction (direction to Outlet 6). For the following particle separation experiments, the absolute number of particles recovered from each outlet was evaluated by multiplying the particle concentration and output volume.

Next, we observed the separation behaviors of fluorescent standard microparticles as a model. Figure 3 shows 4.8- and 9.9-µm green particles flowing through Outlets 1 and 3 of Microdevice A, when the input flow rates from Inlets 1, 2, and 3, denoted as Q_1, Q_2, and Q_3, respectively, were 20, 20, and 80 µL/min, respectively. Most of the large 9.9-µm particles flowed through Outlet 1, whereas small 4.8-µm particles were primarily distributed to Outlets 1, 2, and 3. This result clearly indicates that the presented dual-depth lattice-channel network can function as a size-selective sieving matrix for micrometer-sized particles. This separation result is similar to that reported in our previous study using uniform-depth lattice channels [31] or in the DLD scheme [7–9], in that the differences in the lateral positions (x positions) of particles are enlarged as the particles flow through the lattice region in the y direction. However, interestingly, the particle behaviors differ completely from those observed in our previous study because larger particles flow into the separation channels more frequently than the smaller particles, thus resulting in a large degree of lateral displacement distance for the larger particles.

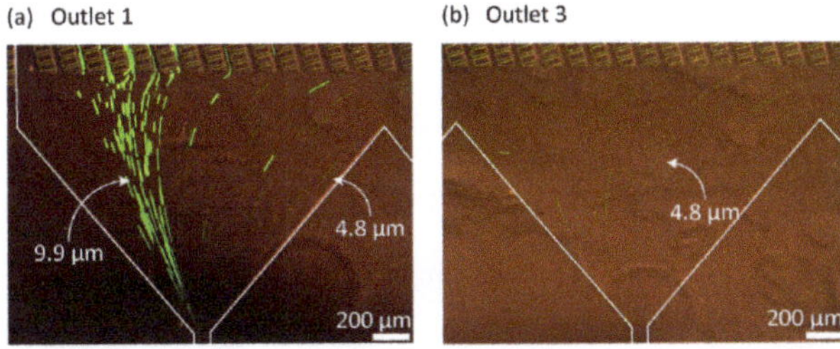

Figure 3. Behaviors of the fluorescent standard particles (4.8- and 9.9-µm green particles) flowing near (**a**) Outlet 1 and (**b**) Outlet 3 of Microdevice A.

Several key operation parameters may affect the separation performances of the particles using the presented concept. First, the ratio of the input flow rates was naturally regarded as a critical factor. We therefore investigated the effect of the input flow-rate ratio, while maintaining the total flow rate Q_{total} (= Q_1 + Q_2 + Q_3) at 120 µL/min. The result is shown in Figure 4. When the ratio of the input flow rates, Q_1, Q_2, and Q_3 was 1:1:1, the input position of particles was relatively broad (~3.8 mm; Figure 4a). The three types of different-sized particles introduced exhibited similar behaviors and were not clearly separated, even though the 9.9-μm particles were not recovered from Outlets 5 and 6 (Figure 4c). Meanwhile, when the relative ratio of the particle suspension (Q_2) was decreased, i.e., when the ratio was changed to 1:1:4 or 4:1:1, the input position was narrowed to ~1.6 mm (Figure 4b). Especially in the 1:1:4 condition, where the particles were introduced from the left-side region of the lattice, a good separation was obtained (Figures 3 and 4d). More than 90% of the 9.9-μm particles were recovered from Outlet 1, whereas 3.0- and 4.8-μm particles were dispersed and primarily recovered from Outlets 1 to 4. Although the separation efficiency was not sufficiently high in this condition, we clarified that a pinching effect of the particle input position had occurred. When the ratio was changed to 4:1:1, that is, the relative value of Q_1 was increased, the introduction position was shifted to the right side of the lattice. In this condition, most of the particles were recovered from Outlets 4 to 6, regardless of the particle size, and particle separation was not achieved (Figure 4e). This result clarified that the input position was a highly critical factor dominating the particle separation performances. A higher separation precision may be expected when the relative value of Q_2 is further lowered; however, we performed particle separation experiments at the input flow rate ratio of 1:1:4 in the following experiments, to ensure a relatively high throughput of particle separation.

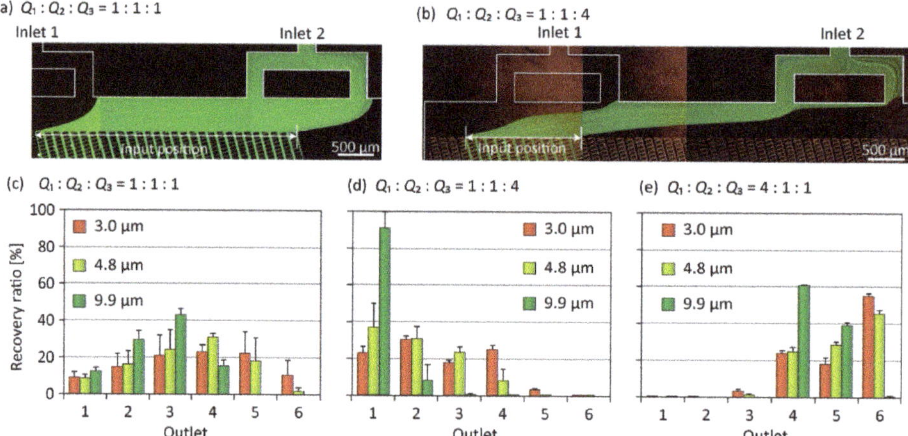

Figure 4. (**a**,**b**) Fluorescence micrographs exhibiting the behaviors of green fluorescent particles flowing into the lattice region of Microdevice A, when the ratios of the input flow rates were changed as indicated. (**c**–**e**) Recovery ratios of three types of particles when the ratio of the input flow rates was changed. The total flow rate Q_{total} was constant at 120 μL/min for these experiments. Each set of data represents the mean ± standard deviation (SD) from three individual experiments.

In most microfluidic schemes for particle separation using laminar flow systems, the particle separation performances are not significantly affected by the absolute value of the flow rate, provided that the particle inertia is negligible [33]. To examine if the particle separation performance of the presented method is affected by the absolute value of the flow rate, we performed particle sorting experiments under different flow-rate conditions using Microdevice A. The total flow rate Q_{total} was changed at 60 μL/min or 1200 μL/min, whereas the flow-rate ratio remained at 1:1:4. The result is shown in Supplementary Figure S2. The ratio of the 9.9-μm particles recovered from Outlet 1 slightly decreased when the total flow rate was increased to 1200 μL/min; the particle separation behaviors did not change significantly compared with those at 60 and 120 μL/min (Figure 4d). This result clearly indicated that the particle inertia did not dominate the separation mechanism of the presented method. Because a relatively high separation throughput was demonstrated, we expect that a high-throughput processing would be possible using the presented concept by further optimizing the operating conditions and/or microchannel geometries, including microchannel parallelization [34].

Additionally, we attempted to observe the flowing behaviors of mammalian cells to validate the applicability of the presented method for cell sorting/manipulation applications. We introduced a suspension of NIH-3T3 cells in phosphate buffered saline (PBS), whose nuclei were stained blue using Hoechst 33342 dye, from Inlet 2, and PBS without cells from Inlets 1 and 3. The input flow rates Q_1, Q_2, and Q_3 were 20, 20, and 80 µL/min, respectively. The results indicate that these cells, with an average diameter of ~12 µm, were mostly recovered from Outlet 1 (Supplementary Figure S3), as in the case of the 9.9-µm particles. Although we did not perform size-based fractionation of cells with specific sizes, the cell concentration was increased approximately twice after recovery from Outlet 1. This result demonstrates applicability to cell separation in addition to cell concentration and carrier medium exchange using the presented microfluidic device.

3.2. Observation of 3D Particle Behavior Using High-Speed Confocal Microscopy

To elucidate the separation mechanisms in detail and to strengthen our theory as depicted in Figure 1, we observed the behaviors of green fluorescent 3.1- and 9.9-µm particles in the lattice region using high-speed confocal microscopy. A buffer solution and the particle suspension were introduced into Microdevice A at Q_1, Q_2, and Q_3 of 2.0, 2.0, and 8.0 µL/min, respectively. Particles flowing near the ceiling of the main channel and those near the bottom surface were individually observed. The flows of the particles are shown in Figure 5 and Supplementary Video S1. Near the ceiling of the microchannel, we primarily observed 3.1-µm particles flowing along the main channel, but the number of large 9.9-µm particles was extremely small. It was assumed that the large particles could not reach the ceiling of the microchannel when they were flowing from the shallower separation channel into the deeper main channel, because of the hydrodynamic constriction effects for the large particles, as in the case of PFF [10]. By contrast, 9.9- and 3.1-µm particles, flowing near the bottom surface, frequently entered the separation channels (Figure 5b). Furthermore, we observed that 3.1-µm particles flowed in the upper and/or lower direction (+z or -z direction) more frequently than the 9.9-µm particles. Once the small particles reached the ceiling of the main channel, they flowed along the main channel. These observations clarified that the separation mechanism of this scheme utilized 3D flows in the planar lattice-shaped microchannel networks, as explained in Figure 1.

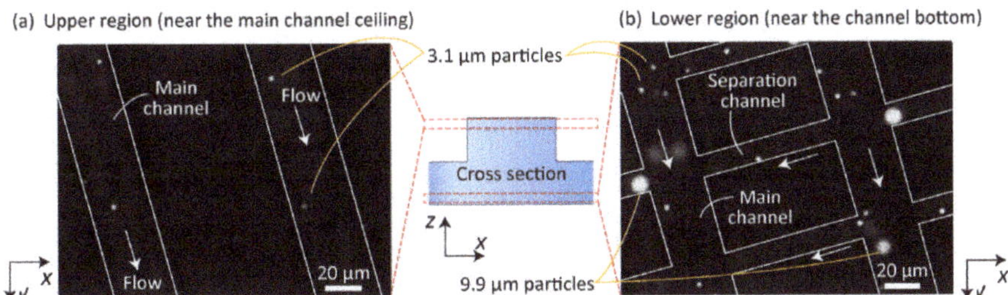

Figure 5. High-speed confocal microscopic images showing the flow behaviors of 3.1- and 9.9-µm particles (**a**) in the upper region (near the main-channel ceiling) and (**b**) in the lower region (near the channel bottom). Small circular dots indicate 3.1-µm particles, whereas the large circular objects indicate 9.9-µm particles. The microchannel is visualized by the white lines.

3.3. Effect of the Main Channel Depth on Particle Separation

In addition to the operation parameters, various geometric parameters of the microchannels affect the separation behaviors of the particles. Among various factors, we investigated the effect of the main channel depth d_{main} on particle separation performances. In addition to Microdevice A with d_{main} of 40 µm, we fabricated two microdevices, Microdevices B and C, with d_{main} values of 25 and 70 µm, respectively (Table 1). The results of particle separation are shown in Figure 6. In Microdevice B, the three types of particles introduced were primarily recovered from Outlet 2, and significant differences were not observed between these particles (Figure 6a). This result suggested that the particle migration in the upper (z) direction was suppressed, because the particles, once introduced from the separation channel into the main channel, could easily be reintroduced into the separation channel (Figure 6b). By contrast, when Microdevice C was used, the particles were almost randomly distributed to Outlets 1–4 (Figure 6c). It was likely that the particles, once they had reached near the ceiling of the main channel, flowed along the main channel and never flowed near the channel bottom, and were recovered from Outlets 3–4. Meanwhile, particles flowing near the bottom surface would be frequently introduced into the separation channels, regardless of the particle size (Figure 6d). Consequently, we confirmed that Microdevice A with d_{main} of 40 µm was the optimal device among the three types of devices for separating microparticles with this size range.

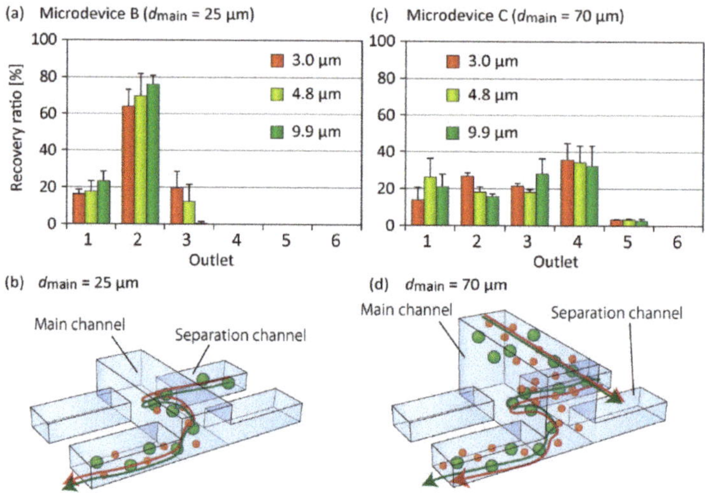

Figure 6. Recovery ratios of three types of microparticles and schematic images showing the particle behaviors in the dual-depth lattice channels. (**a,b**) Microdevice B with the main channel depth of 25 µm and (**c,d**) Microdevice C with the main channel depth of 70 µm. In (**a,c**), each set of data represents the mean ± SD from three individual experiments.

3.4. Tuning of Separation Size

The ability to tune the separation size is significantly important for a microfluidic particle separation scheme to widen its application range. It may be possible to alter one or more geometrical parameters of Microdevice A to tune the separation size; however, we prepared three additional types of microfluidic devices (Microdevices D, E, and F), all of which exhibited a perfectly similar relationship with Microdevice A. The relative sizes of Microdevices D, E, and F to Microdevice A were 1.5, 0.67, and 0.5, respectively. We performed particle separation experiments with the input flow rates Q_1, Q_2, and Q_3 of 20, 20, and 80 µL/min, respectively, for all the devices. To examine the controllability of the separation size, we used several types of fluorescent and/or nonfluorescent model particles with a diameter from 2.1 to 15 µm. The results are shown in Figure 7. For each microdevice, the largest particles were primarily recovered from Outlet 1, whereas the smallest particles were distributed to Outlets 1 to 4, as in the case of Microdevice A. For example, 4.8-µm particles were distributed to Outlets 1–4 of Microdevice D, but were mainly recovered from Outlet 1 of Microdevice F. These results clearly suggest that the usage of different-sized microfluidic devices in a similar relationship is a reasonable strategy to effectively tune the separation behaviors of the particles.

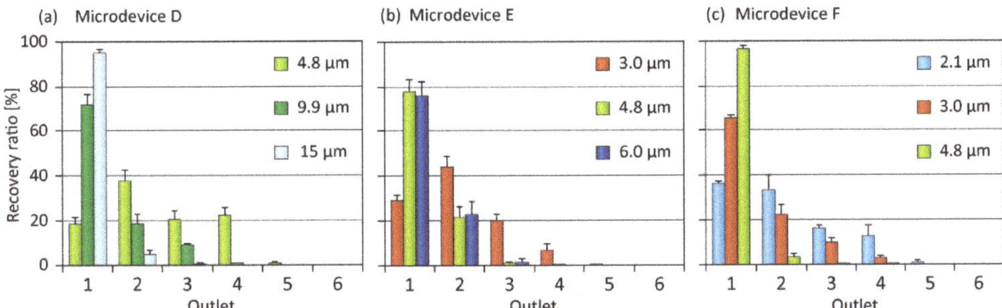

Figure 7. Separation results of model particles with different similitude ratios; (**a**) similitude ratio was 1.5 (microdevice D), (**b**) similitude ratio was 0.66 (microdevice E), and (**c**) similitude ratio was 0.5 (microdevice F). The Q_{total} and Q_1:Q_2:Q_3 were 120 µL/min and 1:1:4, respectively. Each dataset presents the mean ± SD from three individual demonstrations.

3.5. Control of the Vertical Position of Particles

From the results on particle separation and observations, we assumed that the difference in particle position in the z direction might be the primary reason for the insufficient separation efficiency. This could be especially severe for the smaller particles introduced; large particles were primarily collected from Outlet 1, but smaller particles were distributed to Outlets 1–4. To improve the precision of particle separation, we employed a Y-shaped connector that was inserted into Inlet 2 of Microdevice A (Figure 8a). The particle suspension was introduced from one of the two branch inlets, whose position was close to the outlets (+y direction), whereas the buffer without particles was introduced from another inlet, which was close to the inlets (−y direction). Using such a setup, almost all the particles were pushed against the ceiling of the microchannel in the lattice region. The flow rates Q_1, $Q_{2\_particles}$, Q_{2_buffer}, and Q_3 were 20, 2, 18, and 80 µL/min, respectively.

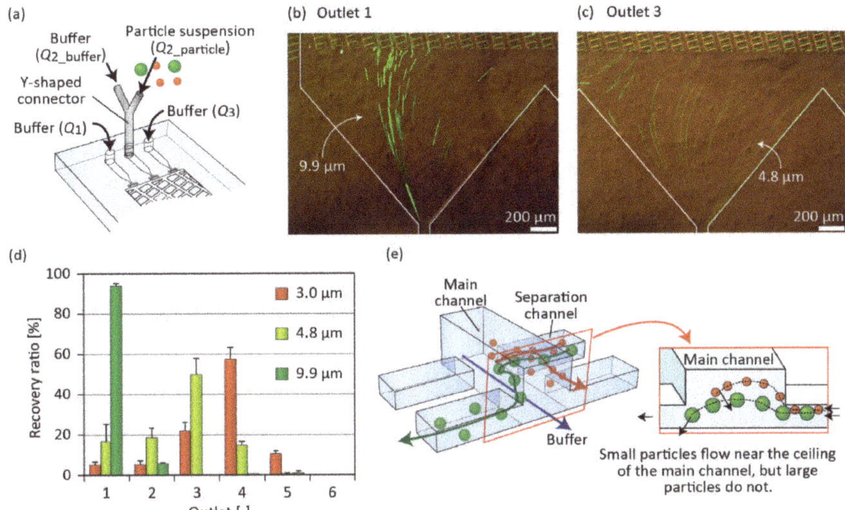

Figure 8. (**a**) Schematic image showing Microdevice A with an attached Y-shaped connector. (**b**,**c**) 4.8- and 9.9-µm particles flowing near Outlets 1 and 3. (**d**) Result of particle separation when Q_1, $Q_{2\_particles}$, Q_{2_buffer}, and Q_3 were 20, 2, 18, and 80 µL/min, respectively. Each set of data represents the mean ± SD from three individual experiments. (**e**) Schematic image showing the particle behaviors in the lattice channel, when the input position of the particles was restricted to the upper region.

Figure 8b,c show the 4.8- and 9.9-μm particles flowing near Outlets 1 and 3, respectively. Compared with the separation results without using the Y-shaped connector (Figure 3), the number of 4.8-μm particles flowing through Outlet 1 decreased significantly. The separation results of the three types of particles are shown in Figure 8d. The introduced 3.0-, 4.8-, and 9.9-μm particles were mainly recovered from Outlets 1, 3, and 4, respectively. The 3.0- and 4.8-μm particles were separated, indicating that the separation resolution had improved. This result indicated that the small particles were likely to selectively flow near the ceiling of the main channel, whereas the large particles, flowing near the bottom, frequently entered the separation channels (Figure 8e). The separation resolution of the 3.0- and 4.8-μm particles was not high in these experiments, possibly because these particles are relatively small compared to the critical separation size of the presented microfluidic device. A possible strategy to improve the separation resolution of these particles is to sequentially employ several types of microfluidic devices with different separation sizes. From these results, we confirmed that the introduction of particles from the limited area in the x-z plane of the lattice region was a highly effective strategy to improve the separation precision of dual-depth, asymmetric microchannel systems. Although the throughput of separated particles decreased compared with the result shown in Figure 3, the presented approach could offer useful insights into microfluidic particle/cell sorting using 3D flow profiles in stereoscopic microchannel structures.

4. Conclusions

We have successfully demonstrated a new concept of a microfluidic particle separation scheme using dual-depth, asymmetric microfluidic lattice structures. By precisely observing the separation behaviors of model particles, we revealed that the 3D laminar flow profile was utilized for the size-selective differences in particle behaviors. Additionally, the control of the input position in both x and z directions was significantly effective in improving the separation performances. The lattice configuration was advantageous because it was robust against microchannel clogging, as bypass flows were generated even when microchannels were clogged at several points. With the recent progress in 3D fabrication techniques, including 3D printers, microfluidic channels with complex 3D structures have recently been gaining increasing attention. The results obtained in this study would provide useful insights into the utilization of 3D flows in microchannels for the separation and manipulation of various types of micrometer-sized particles.

Supplementary Materials: The following are available online at http://www.mdpi.com/2072-666X/10/6/425/s1: Figure S1: The output volumes from the six outlets of Microdevice A; Figure S2: Results of particle separation using Microdevice A when the total flow rate was changed; Figure S3: Fluorescence micrograph showing the behaviors of mammalian cells in Microdevice A; Video S1: High-speed confocal microscopic movie showing the flowing behaviors of 3.1- and 9.9-μm particles in Microdevice A, which correspond to those shown in Figure 5.

Author Contributions: M.Y. conceived and designed the experiments. T.Y. and T.O. performed the experiments and analyzed the data. T.Y., T.O., and M.Y. wrote the manuscript. M.S. supervised the project. All authors discussed the results and contributed to the manuscript.

Funding: This research was financially supported in part by Grants-in-Aid for Scientific Research (19H02520, 17H03463, and 26286032) from the Ministry of Education, Culture, Sports, Science, and Technology, Japan.

Acknowledgments: The authors thank Wataru Seko for his technical support.

Conflicts of Interest: The authors declare no conflicts of interest.

References

1. El-Ali, J.; Sorger, P.K.; Jensen, K.F. Cells on chips. *Nature* **2006**, *442*, 403–411. [CrossRef] [PubMed]
2. Contreras-Naranjo, J.C.; Wu, H.J.; Ugaz, V.M. Microfluidics for exosome isolation and analysis: Enabling liquid biopsy for personalized medicine. *Lab Chip* **2017**, *17*, 3558–3577. [CrossRef] [PubMed]
3. Kulasinghe, A.; Wu, H.; Punyadeera, C.; Warkiani, M.E. The use of microfluidic technology for cancer applications and liquid biopsy. *Micromachines* **2018**, *9*, 397. [CrossRef] [PubMed]

4. Gossett, D.R.; Weaver, W.M.; Mach, A.J.; Hur, S.C.; Tse, H.T.; Lee, W.; Amini, H.; Di Carlo, D. Label-free cell separation and sorting in microfluidic systems. *Anal. Bioanal. Chem.* **2010**, *397*, 3249–3267. [CrossRef] [PubMed]

5. Shields, C.W., IV; Reyes, C.D.; Lopez, G.P. Microfluidic cell sorting: A review of the advances in the separation of cells from debulking to rare cell isolation. *Lab Chip* **2015**, *15*, 1230–1249. [CrossRef] [PubMed]

6. Dalili, A.; Samiei, E.; Hoorfar, M. A review of sorting, separation and isolation of cells and microbeads for biomedical applications: Microfluidic approaches. *Analyst* **2018**, *144*, 87–113. [CrossRef] [PubMed]

7. Huang, L.R.; Cox, E.C.; Austin, R.H.; Sturm, J.C. Continuous particle separation through deterministic lateral displacement. *Science* **2004**, *304*, 987–990. [CrossRef]

8. Ranjan, S.; Zeming, K.K.; Jureen, R.; Fisher, D.; Zhang, Y. DLD pillar shape design for efficient separation of spherical and non-spherical bioparticles. *Lab Chip* **2014**, *14*, 4250–4262. [CrossRef] [PubMed]

9. Tottori, N.; Nisisako, T.; Park, J.; Yanagida, Y.; Hatsuzawa, T. Separation of viable and nonviable mammalian cells using a deterministic lateral displacement microfluidic device. *Biomicrofluidics* **2016**, *10*, 014125. [CrossRef]

10. Yamada, M.; Nakashima, M.; Seki, M. Pinched flow fractionation: Continuous size separation of particles utilizing a laminar flow profile in a pinched microchannel. *Anal. Chem.* **2004**, *76*, 5465–5471. [CrossRef]

11. Lu, X.; Xuan, X. Elasto-inertial pinched flow fractionation for continuous shape-based particle separation. *Anal. Chem.* **2015**, *87*, 11523–11530. [CrossRef] [PubMed]

12. Yamada, M.; Seki, M. Hydrodynamic filtration for on-chip particle concentration and classification utilizing microfluidics. *Lab Chip* **2005**, *5*, 1233–1239. [CrossRef] [PubMed]

13. Mizuno, M.; Yamada, M.; Mitamura, R.; Ike, K.; Toyama, K.; Seki, M. Magnetophoresis-integrated hydrodynamic filtration system for size- and surface marker-based two-dimensional cell sorting. *Anal. Chem.* **2013**, *85*, 7666–7673. [CrossRef] [PubMed]

14. Jung, H.; Chun, M.S.; Chang, M.S. Sorting of human mesenchymal stem cells by applying optimally designed microfluidic chip filtration. *Analyst* **2015**, *140*, 1265–1274. [CrossRef] [PubMed]

15. Bhagat, A.A.; Kuntaegowdanahalli, S.S.; Papautsky, I. Continuous particle separation in spiral microchannels using Dean flows and differential migration. *Lab Chip* **2008**, *8*, 1906–1914. [CrossRef] [PubMed]

16. Warkiani, M.E.; Khoo, B.L.; Wu, L.; Tay, A.K.; Bhagat, A.A.; Han, J.; Lim, C.T. Ultra-fast, label-free isolation of circulating tumor cells from blood using spiral microfluidics. *Nat. Protoc.* **2016**, *11*, 134–148. [CrossRef]

17. Petchakup, C.; Tay, H.M.; Li, K.H.H.; Hou, H.W. Integrated inertial-impedance cytometry for rapid label-free leukocyte isolation and profiling of neutrophil extracellular traps (NETs). *Lab Chip* **2019**, *19*, 1736–1746. [CrossRef]

18. Choi, S.; Song, S.; Choi, C.; Park, J.K. Hydrophoretic sorting of micrometer and submicrometer particles using anisotropic microfluidic obstacles. *Anal. Chem.* **2009**, *81*, 50–55. [CrossRef]

19. Song, S.; Choi, S. Design rules for size-based cell sorting and sheathless cell focusing by hydrophoresis. *J. Chromatogr. A* **2013**, *1302*, 191–196. [CrossRef]

20. Kim, B.; Lee, J.K.; Choi, S. Continuous sorting and washing of cancer cells from blood cells by hydrophoresis. *Biochip. J.* **2015**, *10*, 81–87. [CrossRef]

21. Di Carlo, D.; Edd, J.F.; Irimia, D.; Tompkins, R.G.; Toner, M. Equilibrium separation and filtration of particles using differential inertial focusing. *Anal. Chem.* **2008**, *80*, 2204–2211. [CrossRef] [PubMed]

22. Zhou, Y.; Ma, Z.; Tayebi, M.; Ai, Y. Submicron particle focusing and exosome sorting by wavy microchannel structures within viscoelastic fluids. *Anal. Chem.* **2019**, *91*, 4577–4584. [CrossRef] [PubMed]

23. Park, J.S.; Song, S.H.; Jung, H.I. Continuous focusing of microparticles using inertial lift force and vorticity via multi-orifice microfluidic channels. *Lab Chip* **2009**, *9*, 939–948. [CrossRef] [PubMed]

24. Fan, L.L.; He, X.K.; Han, Y.; Du, L.; Zhao, L.; Zhe, J. Continuous size-based separation of microparticles in a microchannel with symmetric sharp corner structures. *Biomicrofluidics* **2014**, *8*, 024108. [CrossRef] [PubMed]

25. Fan, Y.J.; Wu, Y.C.; Chen, Y.; Kung, Y.C.; Wu, T.H.; Huang, K.W.; Sheen, H.J.; Chiou, P.Y. Three dimensional microfluidics with embedded microball lenses for parallel and high throughput multicolor fluorescence detection. *Biomicrofluidics* **2013**, *7*, 044212. [CrossRef] [PubMed]

26. Chen, Y.; Wu, T.H.; Kung, Y.C.; Teitell, M.A.; Chiou, P.Y. 3D pulsed laser-triggered high-speed microfluidic fluorescence-activated cell sorter. *Analyst* **2013**, *138*, 7308–7315. [CrossRef] [PubMed]

27. Li, M.; Munoz, H.E.; Schmidt, A.; Guo, B.; Lei, C.; Goda, K.; Di Carlo, D. Inertial focusing of ellipsoidal Euglena gracilis cells in a stepped microchannel. *Lab Chip* **2016**, *16*, 4458–4465. [CrossRef] [PubMed]

28. Wang, S.; Thomas, A.; Lee, E.; Yang, S.; Cheng, X.; Liu, Y. Highly efficient and selective isolation of rare tumor cells using a microfluidic chip with wavy-herringbone micro-patterned surfaces. *Analyst* **2016**, *141*, 2228–2237. [CrossRef] [PubMed]

29. Rafeie, M.; Zhang, J.; Asadnia, M.; Li, W.H.; Warkiani, M.E. Multiplexing slanted spiral microchannels for ultra-fast blood plasma separation. *Lab Chip* **2016**, *16*, 2791–2802. [CrossRef]

30. Mukherjee, P.; Wang, X.; Zhou, J.; Papautsky, I. Single stream inertial focusing in low aspect-ratio triangular microchannels. *Lab Chip* **2019**, *19*, 147–157. [CrossRef]

31. Yamada, M.; Seko, W.; Yanai, T.; Ninomiya, K.; Seki, M. Slanted, asymmetric microfluidic lattices as size-selective sieves for continuous particle/cell sorting. *Lab Chip* **2017**, *17*, 304–314. [CrossRef] [PubMed]

32. McDonald, J.C.; Whitesides, G.M. Poly(dimethylsiloxane) as a material for fabricating microfluidic devices. *Acc. Chem. Res.* **2002**, *35*, 491–499. [CrossRef] [PubMed]

33. Li, X.; Chen, W.Q.; Liu, G.Y.; Lu, W.; Fu, J.P. Continuous-flow microfluidic blood cell sorting for unprocessed whole blood using surfacemicromachined microfiltration membranes. *Lab Chip* **2014**, *14*, 2565–2575. [CrossRef] [PubMed]

34. Ozawa, R.; Iwadate, H.; Toyoda, H.; Yamada, M.; Seki, M. A numbering-up strategy of hydrodynamic microfluidic filters for continuous-flow high-throughput cell sorting. *Lab Chip* **2019**, *19*, 1828–1837. [CrossRef] [PubMed]

Article

Two-dimensional Simulation of Motion of Red Blood Cells with Deterministic Lateral Displacement Devices

Yanying Jiao, Yongqing He * and Feng Jiao

School of Chemical Engineering, Kunming University of Science and Technology, Kunming 650500, China; 18388195203@163.com (Y.J.); jiaofeng0526@163.com (F.J.)
* Correspondence: yqhe@kmust.edu.cn

Received: 20 May 2019; Accepted: 11 June 2019; Published: 12 June 2019

Abstract: Deterministic lateral displacement (DLD) technology has great potential for the separation, enrichment, and sorting of red blood cells (RBCs). This paper presents a numerical simulation of the motion of RBCs using DLD devices with different pillar shapes and gap configurations. We studied the effect of the pillar shape, row shift, and pillar diameter on the performance of RBC separation. The numerical results show that the RBCs enter "displacement mode" under conditions of low row-shift ($\Delta\lambda < 1.4$ μm) and "zigzag mode" with large row shift ($\Delta\lambda > 1.5$ μm). RBCs can pass the pillar array when the size of the pillar ($d > 6$ μm) is larger than the cell size. We show that these conclusions can be helpful for the design of a reliable DLD microfluidic device for the separation of RBCs.

Keywords: red blood cells; deterministic lateral displacement; trajectories; row shift

1. Introduction

Human blood consists of plasma and mainly red and white blood cells and platelets [1]. Through continuous circulation, blood provides oxygen and nutrients, removes metabolic waste from tissue, regulates body pH and temperature, in addition to other biological functions. RBCs are the most critical component, and have a direct impact on hemodynamics and hemorheology. RBCs are highly deformable due to their biconcave, seedless, and highly flexible membrane [2]. Researchers has recognized the deformability of RBCs as an inherent indicator of diseases such as diabetes and malaria. The separation of deformable RBCs through Lab-on-a-chip techniques, which is based on the formers' mechanical properties, is becoming an essential process for medical research and clinical disease diagnosis [3–5]. Further understanding of the dynamic behavior of the RBCs in a microchannel, and exploring new methods by which to efficiently separate RBCs from plasma, are urgent to the development of biomedical engineering.

In medical research such as liquid biopsies and cytopathology, researchers have begun to focus on the direct isolation of target cells from whole blood. In 2018, a new method for continuously focusing and separating biological particles directly with shear-induced diffusion from whole blood was successfully demonstrated; the method effectively combines the inherent complexity of blood with the migration from the flow inertia while causing little pollution [6]. They also demonstrated a new multi-flow micro-fluidic (MFM)system for unlabeled circulating tumor cells (CTCs) from the peripheral blood of patients, and proposed a promising alternative method by which to realize CTC capture. Their results show that the method will offer the possibility of achieving chemical individuality treatment [7].

Moreover, several techniques for cell separation have been proposed, such as label-free discrimination and fractionation of cell populations [8], fluorescence-activated cell sorting (FACS), magnetic-activated cell sorting (MACS), and centrifugation separation. Fluorescence-activated cell sorting (FACS) is an active sorting method in which tested cells are stained with a specific fluorescent dye to fluoresce the cells with a complementary fluorophore-conjugated antibody. The detected fluorescence data can be used to characterize the immune situation, size, and cell type, providing rich data support for analyses of gene expression. Magnetically-activated cell sorting (MACS) is a passive separation technique. Magnetic particles are introduced to label cells to bind specific proteins on cells and separate the sample from non-magnetic cells under the action of an external magnetic field [8,9]. Although FACS and MACS can provide rich data for high-throughput screening, these two methods are not widely available due to the limitations of cost of their labeled antibody magnetic nanoparticles and sheath fluids [10]. The macroscopic separation method consists of centrifugation, which uses different concentrations of blood cell components and different centrifugal forces at different settling rates. However, because of its reactivity to environmental changes, centrifugation may alter the immune properties [11]. The phase separation method based on the hematocrit effect in a microchannel with multiple outlets has also been adopted to separate RBCs. Yin et al. [2] investigated the separation process of multiple RBC flows through a symmetric microvascular bifurcation model with different cell deformability, aggregation, and hematocrit. Their symmetric bifurcation model is relatively simple, given the limitations of the 2-D model.

With the development of microfabrication technology, the post arrays of the pillars in a microchannel have been used for the selective delivery on the proteins [12] and drugs [13], and to separate RBCs [3], known as "deterministic lateral displacement" (DLD) [14,15]. The method has the advantages of requiring lower analytical reagent dosages, shorter detection periods, and yielding higher levels of precision, and this technology opens up the possibility of experimental microchip diagnostics replacing traditional blood tests. By collecting sample fractions to quantify the separation efficiency at each outlet, Xavier et al. [16] found that the purities can reach 88.5% and 98.3% for large and small beads, respectively. Their experimental results demonstrated that the DLD has outstanding potential for high purity separation.

Zhang et al. [4] numerically studied the influences of the pillar shapes on the dynamic behavior of the RBCs in the DLD devices, including cylindrical, diamond, and triangle. Sharp obstacles can significantly enhance the deformability of cells; their sharper edges can cause more significant distortion of RBCs. The cells around triangles bend more strongly than around diamonds, which may be related to DLD-based deformable sorting. Li et al. [17] considered a cylindrical microfluidic bifurcation channel, where three-dimensional fluid dynamics codes were used to simulate the trajectories of RBCs. Since healthy RBCs have the characteristic of high deformability, and the distortions induced by bifurcation are considerable, the healthy ones cannot uniformly distribute in bifurcated microchannels like rigid RBCs. That means that the deformability of RBCs has a significant influence on blood separation, and that healthy RBCs have an extremely high separation efficiency.

Fu et al. [18] combined the lattice-Boltzmann method (LBM), the immersion boundary method (IBM), and the discrete element method (DEM) to calculate the separation of particles of different sizes and shapes (sphere, triangle, diamond, and pentagon) through square and round pillars. Kruger et al. [19] investigated the shapes of RBCs between pillars using the capillary number Ca to indicate the deformability. Ca-dependent trajectories have been observed, and the direct collision of RBCs with the pillars may increase separation. Hou et al. [20] examined a continuous filtration method for the separation of the infected red blood cells (iRBCs) in a microfluidic device. iRBCs behave like white blood cells and move toward the side walls due to cell-cell interactions.

In this paper, we simulate the motions of RBCs in DLD devices by considering their deformability. Firstly, we changed the pillar shapes (circular and triangular) and their arrangements to evaluate the trajectories, surface stresses, and velocities of the rigid RBCs (regarding them as microspheres). We then introduced capillary number Ca to represent the deformability of RBCs, and discussed the variation of Ca under the configuration of the triangular pillars. Our work on the DLD device provides underlying understanding for the efficient separation of the RBCs in the future.

2. Simulation Methods and Models

We adopted the COMSOL® Multiphysics software as the computational tool, which is relatively well-suited to the calculation of microspheres manipulation [21]. We optimized the simulation code of [22], in which the separation trajectories in a DLD device are accurately predicted using the finite element method. Furthermore, we studied the velocity field, surface stress, trajectories, and velocity of red blood cell movement. During the simulation, it was shown that when the height of the microspheres/cells is consistent with the thickness of the device, the geometric nature of the microspheres/cells can be simplified to two-dimensions [20]. Also, compared to the 3-D model, the 2-D model can greatly reduce the number of calculations. Additionally, the friction between the microspheres and the device wall can be neglected.

2.1. The DLD Model

Figure 1 shows a typical schematic of a DLD device. The microfluidic channel has one inlet on the left and two outlets on the right; the separated particles/cells will flow out of one of the outlets. To enhance the analysis, we chose two different pillar shapes (circle and the triangle) as the calculation cases. The detailed dimensions of the pillar posts are shown in Figure 1.

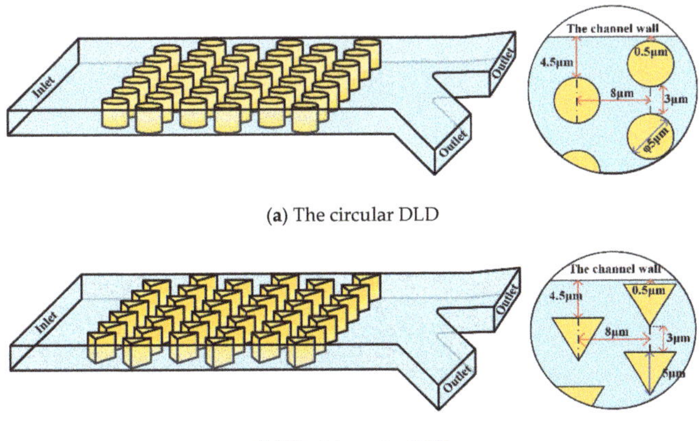

(a) The circular DLD

(b) The triangular DLD

Figure 1. Schematic of a deterministic lateral displacement array device. The microfluidic channel has an inlet on the left and two outlets on the right. The circular and triangular obstacles in the channel are made of polydimethylsiloxane (PDMS). The obstacles are periodically arranged, and each row is horizontal in terms of one row and offset. A liquid solution carrying rigid microspheres of different sizes flows through the passage from the inlet. The pillar shape within the DLD device is (a) circular and (b) triangular.

Table 1 lists the channel length, width, the pillar diameter, the height, and the obstacle spacing of the DLD device. Table 2 lists the physical properties of the fluid and the RBCs. Two working fluids were tested.

Table 1. The geometric parameters of the DLD array.

The Setting Parameters	The Size (μm)
L (the channel length)	109
W (the channel width)	54
d (the round obstacle diameter)	5
h (the equilateral triangle obstacle)	5
S (the obstacle spacing)	3

Table 2. The physical properties of the RBCs and working fluids.

The Working Fluid	The Density (kg/m^3)	The Dynamic Viscosity (Pa·s)	The Young's Modulus (Pa)	The Poisson's Ratio
Water	1000	0.001	2.16e9	0.414
RBCs	1090	-	2.5e2	0.3
PDMS	970	0.001	3e9	0.49

2.1.1. Governing Equations

In this two-dimensional simulation, we need to calculate an incompressible liquid flow by solving the Navier-Stokes equations [23,24] and the continuity equation:

$$\rho \frac{\partial u_{fluid}}{\partial t} = \nabla \left[-pI + \mu \left(\nabla u_{fluid} + \left(\nabla u_{fluid} \right)^T \right) \right] - 12 \frac{\mu u_{fluid}}{d_z^2} + F \tag{1}$$

$$\rho \nabla \times u_{fluid} = 0 \tag{2}$$

where $\rho \frac{\partial u_{fluid}}{\partial t}$ represents the unsteady inertia force (N/m^3), $12 \frac{\mu u_{fluid}}{d_z^2}$ represents the nonlinear inertia force (N/m^3), and F represents the volume force. ρ represents the fluid density (kg/m^3), p is the pressure (Pa), $\nabla()$ represents the gradient operator, I is the unit diagonal matrix, μ_f is the dynamic fluid viscosity (Pa·s), and d_z represents the channel height (mm). When gravity or other volumetric forces are not considered, $F = 0$. u_{fluid} represents the fluid velocity field (m/s).

In a microfluidic system, the flow rate is small and the Reynolds number ($R_e << 100$) is expressed as

$$R_e = \frac{lU\rho}{\mu}, \tag{3}$$

where l is the characteristic length of the rectangular channel ($l = 109$ μm), and U is the average velocity.

2.1.2. Boundary Conditions for Fluid-solid Interaction (FSI)

In a fluid-solid coupling boundary setting, the solid boundary is affected by the viscous force and flow pressure. Therefore, the velocity of the fluid can be used to derive the displacement rate of change of the solid microspheres.

$$u_{fluid} = u_w \tag{4}$$

$$u_w = \frac{\partial u_{solid}}{\partial t} \tag{5}$$

$$\sigma \cdot n = \Gamma \cdot n \tag{6}$$

$$\Gamma = \left[-pI + \mu \left(\nabla u_{fluid} + \left(\nabla u_{fluid} \right)^T \right) \right] \tag{7}$$

where Γ is the force on the solid boundary, which includes the fluid pressure and viscous resistance, and n denotes the outward normal bound. At the fluid-solid coupling boundary, the fluid velocity, u_{fluid}, is equal to the rate of change of the solid displacement, u_w. u_{solid} represents the solid displacement field (m/s), u_w represents the displacement rate of change of the solid microspheres and satisfies the no-slip condition at the channel wall, and σ is the Cauchy stress.

2.1.3. Initial Conditions

The fluid flow in the channel is fully developed laminar flow, and is driven by the pressure difference. To ensure that the nonlinear solver has the best possible convergence at the beginning, we set the inlet velocity to be non-constant. According to the parabolic characteristics of the laminar flow velocity distribution, we used the following formula to calculate the normal inlet velocity:

$$U = 6u_0(H - Y)Y/H^2 \tag{8}$$

where u_0 is the average flow velocity at the inlet (m/s), H is the channel height (mm), and Y is the value of the y-coordinate of the center of the microsphere/RBCs.

At the exit, a fixed boundary condition, also known as the Dirichlet boundary condition, was used to determine the value of the pressure.

$$\left[-pI + \mu\left(\nabla u_{fluid} + \left(\nabla u_{fluid}\right)^T\right)\right]n = -\hat{p}n \tag{9}$$

2.2. Red Blood Cell Model

Due to their non-expandable nature, RBCs can preserve their area and arc length in two dimensions. The deformability of cells depends on the elasticity of the cell membrane, the viscosity of the cytoplasm, and the applied flow rate [3,25]. Therefore, for the simulation of the cell screening process, it was necessary to consider the deformability of RBCs. Young's modulus, which is the modulus of elasticity detected by atomic force microscopy, is a physical quantity used to characterize the surface properties of cells in the biological field. It is determined only by the physical properties of the material itself; the larger Young's modulus, the greater the rigidity of the cells, and the more difficult it is to deform them.

2.2.1. Linear Elastic Material

If the rigid microspheres underwent small levels of deformation and were subjected to a low load, their displacement and deformation satisfy the linear elastic momentum conservation (see the Equation (11)) and the governing equilibrium equation [23,24]. In this simulation, the Young's modulus $E = (0.25 \pm 0.08)$ kPa, the Poisson's ratio $v = 0.3$ for RBCs.

$$\rho\frac{\partial u^2_{solid}}{\partial t^2} - \nabla \cdot \sigma = F_v \tag{10}$$

$$\sigma = J^{-1}FSF^T \tag{11}$$

$$F = (I + \nabla u_{solid}) \tag{12}$$

$$J = \det(F) \tag{13}$$

$$\varepsilon = \frac{1}{2}\left[(\nabla u_{solid})^T + \nabla u_{solid} + (\nabla u_{solid})^T \nabla u_{solid}\right] \tag{14}$$

$$S - S_0 = C : (\varepsilon - \varepsilon_0 - \varepsilon_{inet}) \tag{15}$$

Equation (10) represents Newton's equation of motion; Equation (11) is the linear elastic stress-strain law, and Equation (14) represents the strain-displacement (compatibility) equation. F_v is the boundary force and J as the stiffness matrix. ε is the infinitesimal strain tensor (we also use ε to present the row shift fraction in the discussion section); finally, C is the stiffness matrix.

2.2.2. The Capillary Number

To consider the factors that affect cell deformation, such as cell membrane, cytoplasm, and the flow rate of the surrounding fluid, we introduced a dimensionless number, capillary number Ca, by combining the characteristics of these aspects [3].

The bending stiffness k_b of a healthy RBC is 1×10^{-19} and the effective radius R_{eff} is 2.5 μm [1,3]. The maximum flow velocity U_{max} between the pillars is from 10 μm/s to 10 mm/s, and the post-gap $G = 1$ μm. For healthy RBCs, the capillary number Ca range is [0.0375, 375], while the Ca range of pathological RBCs will increase nearly 10 times due to their stiffness, which means that the value is [0.0038, 37.5] [26]. Therefore, the deformation of RBCs can be judged by the capillary number *Ca*; the larger the capillary number *Ca*, the more easily the RBCs are deformed.

2.3. The Mesh and Independence Verification

After creating a mesh of the domain and discretizing the equation, we use the Any Lagrangian-Eulerian (ALE) method to describe the interface between the fluid and the RBCs. The geometry of the domain can be defined by moving the mesh, and the deformation and movement of the boundary are the same. COMSOL can calculate the new grid coordinates of the channel region based on the moving boundary and mesh smoothing of the structure, and solve the Navier-Stokes equations in the moving coordinate system in the mesh and fluid domain. The microsphere meshes in the domain move and deform freely with the simulated motion, and are automatically divided.

Also, we need to conduct grid independence verification to select the appropriate mesh. The most basic method to change the mesh size; we will choose this mesh scale when the difference between the results of two consecutive mesh scales reaches a point at which it can be ignored. Table 3 gives the corresponding degrees of freedom for solving the various equations at different grid scales.

Table 3. The grid division scales.

The Cell Type	Maximum Unit Size (μm)	Minimum Unit Size (μm)	Maximum Unit Growth Rate	Curvature Factor	Narrow Area Resolution	Solving the Degree of Freedom
Conventional	2.67	0.119	1.15	0.30	1	11,161
Refine	2.08	0.0594	1.13	0.30	1	11,621
Refined	1.66	0.0238	1.10	0.25	1	16,132
Ultra-fine	0.772	0.00891	1.08	0.25	1	20,140
Extremely detailed	0.398	0.00119	1.05	0.20	1	48,624

Figure 2a plots the rate of change of displacement in the X direction for circular microspheres with $t = 0$ to $t = 0.3$ seconds at different grid scales. The grid is very close and the error is small in terms of regularity, refinement, ultra-fineness and extreme refinement. To consider the computational efficiency and the solver convergence, the ultra-thinning ratio is used for subsequent simulations. According to the COMSOL mesh quality detection method, the meshing is consistent with the rule that the mesh quality is close to 1, meaning that the mesh quality is good.

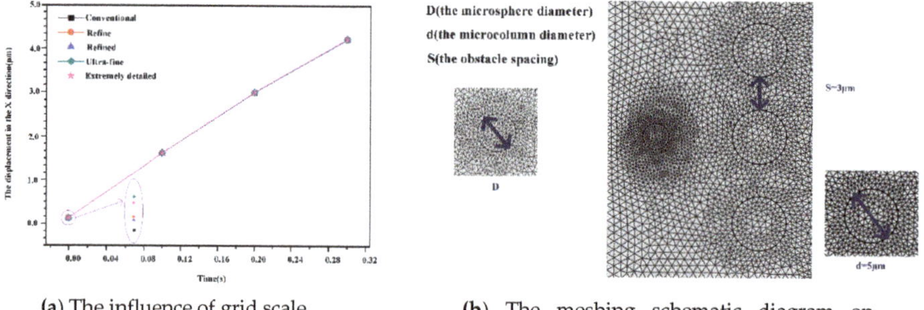

(a) The influence of grid scale

(b) The meshing schematic diagram on velocity of the microspheres

Figure 2. Grid-independent verification (**a**) describes the variation of the velocity of the microspheres along the x-direction at five different grid scales, and (**b**) plots the ultra-fine grids used for the simulation of this number of times.

Figure 2b is a meshing diagram and shows enlarges the microsphere area. The mesh is divided into free triangle meshes, and the ultra-fine meshing method is selected. Inside the obstacle, the moving mesh will change with the deformation of the obstacle; in the outer boundary of the watershed, the deformation in all directions is set to zero. Therefore, the initial mesh at $t = 0$ s is not evenly divided, but it can be seen that the mesh is evenly distributed around the microspheres. The mesh is looser on the fluid domain, while the fluid-solid boundary has a denser and smaller cell grid.

3. Results and Discussion

To simplify the simulation, we consider the RBCs to be microscopically deformed microspheres by ignoring their deformability in order to study the deflection effect of DLD devices on different sized microspheres. Three kinds of microspheres with different diameters were used for numerical simulations; the trajectories of the microspheres flowing through the DLD microchannel were analyzed.

3.1. The Trajectories of Microspheres

In a DLD device, the microspheres will separate according to the arrangement of the pillars under the laminar flow at low Reynolds number.

From the Figure 3, there are two main modes in which the rigid microspheres move between the gaps of pillar arrays. It is interesting to note that the position of the pillars will change the trajectories of the microspheres, leading their trajectories into a "zigzag mode"; however, the trajectories return to the original state after three rows of obstacles. When the radius is greater than the width of the channel, the microspheres adopt "displacement mode", and may not be able to enter channel 1 of the fluid in the gap, thus behaving differently in the pillar array [27].

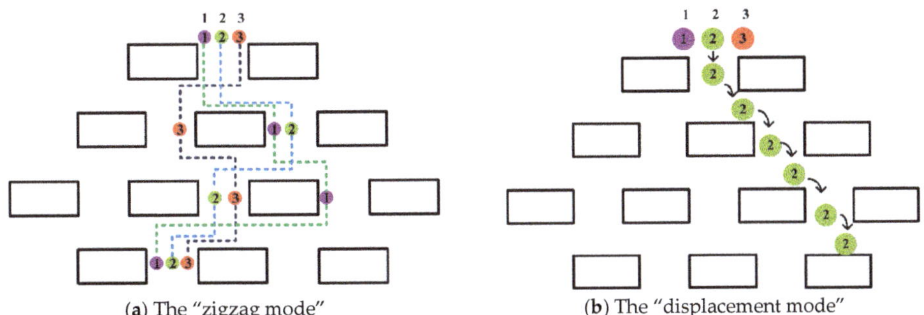

(a) The "zigzag mode" (b) The "displacement mode"

Figure 3. A schematic diagram of a human-defined barrier matrix with three fluid channels. (**a**) The three microspheres in the channel (shown in purple, green, and red, respectively) are incompatible. After three rows of obstacles, the trajectories of the microspheres return to the original channel position, following a zigzag pattern. (**b**) The microspheres with a radius which is greater than the width of the channel follow a displacement pattern, passing through the streamline at the center of the microspheres. The arrow marks the trajectories of the microsphere.

Compared with the longitudinal rectangular microchannels described in reference [27], we used horizontal microchannels in the numerical simulation to analyze the trajectories of the microspheres. We studied the factors affecting the separation of microspheres in the DLD device, the row shifts fraction ε, and the diameter of the microspheres D. The row-shift fraction ε is expressed as

$$\varepsilon = \frac{\Delta\lambda}{\lambda},\tag{16}$$

As shown in Figure 4, $\Delta\lambda$ is row shift and λ is the center-to-center distance.

There is a critical diameter D_c between the "displacement mode" and the "zigzag mode". Based on the gap size and microsphere diameters, the critical diameter of the microspheres is [28],

$$D_c = 1.4G\varepsilon^{0.48}, \tag{17}$$

where G is the post-gap, indicating the distance between the two pillars.

The result shows the occurrence of "displacement mode" at low row shifts ($\Delta\lambda \leq 2.5$ μm) and "zigzag mode" for larger row shifts ($\Delta\lambda \geq 3$ μm). This means that the separation trajectories of the microspheres is dependent on the row shifts in the presently-studied DLD device.

To investigate the effect of the complex pillar array on the separation of microspheres, we used different arrangements of pillars, i.e., by varying the row shift, the post-gap and the center-to-center distance between the two pillars. In Figure 4a, the flow pattern of microspheres 1.4 μm and 3 μm in diameter demonstrates "displacement mode" when $\Delta\lambda = 1.0$ μm, $D_c = 1.95$ μm, $G = 4$ μm, and $\varepsilon = 0.1111$. In Figure 4b, the trajectories of microspheres with the same diameter adopt "zigzag mode" when $\Delta\lambda = 4.0$ μm, $D_c = 3.79$ μm, $G=4$ μm, and $\varepsilon = 0.4444$. This indicates that the trajectories of the microspheres are related to the row shift, which is consistent with the conclusions found in the literature.

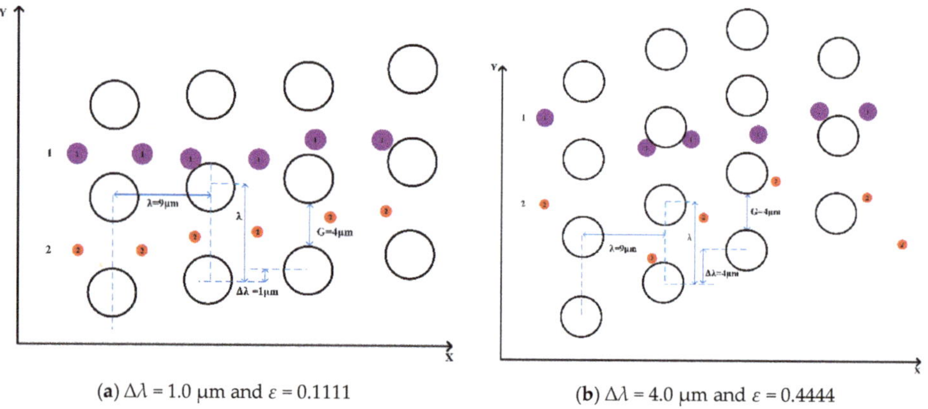

(**a**) $\Delta\lambda = 1.0$ μm and $\varepsilon = 0.1111$ (**b**) $\Delta\lambda = 4.0$ μm and $\varepsilon = 0.4444$

Figure 4. Two types of microspheres with diameters of 1.4 μm and 3.0 μm, marked in red and purple. (**a**) The trajectories of the microspheres are "displacement mode" in a circular pillar; the different row shift fraction is $\varepsilon = 0.1111$ and $\Delta\lambda = 1.0$ μm. (**b**) The trajectories of the microspheres are in "zigzag mode" in a circular pillar; the row shift fraction is $\varepsilon = 0.4444$ and $\Delta\lambda = 4.0$ μm.

Compared to [20], we found that the diameter of the microspheres affects their trajectories; we also selected different diameters of the microspheres in our simulation to verify this result. Three different diameters of microspheres were released from the same position in order to analyze the trajectories when staggered pillars were used. From top to bottom, the microspheres have diameters of 1.0 μm, 0.4 μm and 0.2 μm with the $\Delta\lambda = 1.5$ μm, $D_c = 1.88$ μm, $G = 3$ μm, and $\varepsilon = 0.1875$ respectively, and the trajectories of the three microspheres is in "zigzag mode". When $\Delta\lambda = 3.5$ μm, $D_c = 2.82$ μm, $G = 3$ μm, and $\varepsilon = 0.4375$, it was found that microspheres with diameters of 0.1 μm, 0.2 μm, 0.4 μm, 1 μm and 2 μm have different displacement rates, but that the trajectories are the same in "zigzag mode". Therefore, in this arrangement of pillars in Figure 5, the diameter of the microspheres does not change their trajectories; therefore, we suspect that this indicates a new law between "zigzag mode" and "displacement mode" regarding row shifts in these pillar arrangements.

To explore the critical row shifts between the "displacement mode" and the "zigzag mode" in this pillar arrangement, the same diameters, i.e., 1.0 μm, 0.4 μm and 0.2 μm, of the microspheres are in $\Delta\lambda = 1.0$ μm, $D_c = 1,55$ μm, $G = 3$ μm, and $\varepsilon = 0.125$ and another in $\Delta\lambda = 1.4$, $D_c = 1.82$ μm, $G = 3$ μm, and $\varepsilon = 0.175$ again in Figure 6. The results show that they all lead to "displacement mode", and that in the arrangement of the pillars, the "displacement mode" shifts at the low row shifts ($\Delta\lambda \leq 1.4$ μm), and the "zigzag mode" shifts at the larger row shifts ($\Delta\lambda \geq 1.5$ μm). With the trajectories of microspheres changing from "zigzag mode" to "displacement mode", the critical diameters become smaller, and the row shift fraction becomes larger.

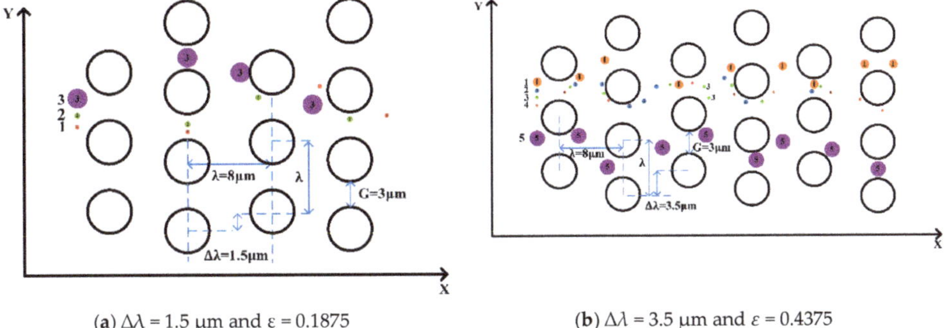

(a) $\Delta\lambda = 1.5$ μm and $\varepsilon = 0.1875$ (b) $\Delta\lambda = 3.5$ μm and $\varepsilon = 0.4375$

Figure 5. The trajectories of microspheres in a circular pillar are in "zigzag mode". (a) The row shift is $\Delta\lambda = 1.5$ μm and $\varepsilon = 0.1875$. We indicate three microspheres in the channel (in purple, green, and red) of different sizes. (b) The row shift is $\Delta\lambda = 3.5$ μm and $\varepsilon = 0.4375$. Five fluid channels and five differently sized microspheres (shown in orange, green, blue, red, and purple, respectively) are labeled.

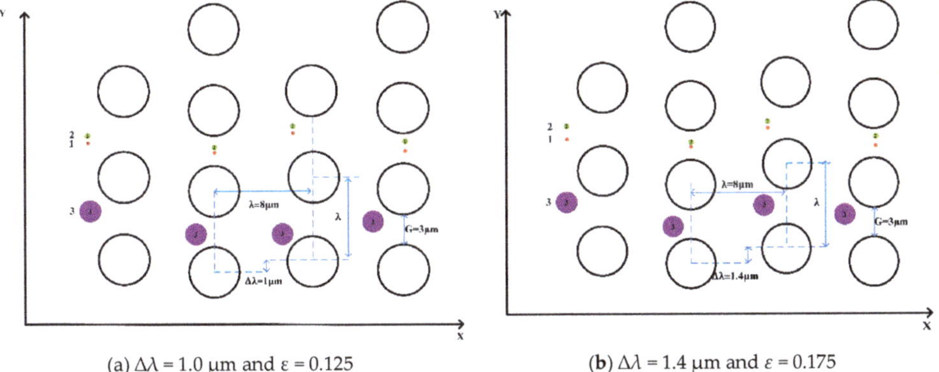

(a) $\Delta\lambda = 1.0$ μm and $\varepsilon = 0.125$ (b) $\Delta\lambda = 1.4$ μm and $\varepsilon = 0.175$

Figure 6. The trajectories of microspheres in a circular pillar array are in "displacement mode". Marked with two fluid channels, the three microspheres in the channel (shown here in purple and red, respectively) are different in size. (a) The row shift is $\Delta\lambda = 1.0$ μm and $\varepsilon = 0.125$. (b) The row shift is $\Delta\lambda = 1.4$ μm and $\varepsilon = 0.175$.

The microspheres showed periodic variations in this DLD device. We mainly studied microsphere separation in "zigzag mode" using staggered pillars. In this section, we considered RBCs as being rigid, and simulated the microspheres with the same parameters as those of the RBCs. We believe that this arrangement can effectively predict the trajectory of RBCs during separation. We analyzed the velocity and surface stress of different microsphere shapes (circular, elliptical, triangular and diamond) and the effect of the shape of the pillars.

3.2. The Surface Stress on the Microsphere

The deformability of RBCs is mainly due to stress on the surface. Next, we will discuss the stress induced by the flow. RBCs can be damaged or even rupture under complex loading forces, including viscous resistance and the pressure of the fluid. Therefore, it is necessary to analyze the surface stress on the microspheres in order to determine whether the microspheres will remain intact in DLD devices with a pillar structure [29].

The Young's modulus causes elastic deformation by affecting the level of stress on the microspheres. The surface stress of the microspheres is scalar, calculated from a stress tensor, and can be used to determine whether the microspheres remain intact when subjected to the loading force. In Figure 7a, when $t = 0$ s, at 0.126 s, 0.156 s, 0.171 s, and 0.19 s, the position of the microspheres of different shapes is shown in different colors. The maximum and minimum stress points are also presented (Figure 7b,c). When the microspheres are far from the pillars in the DLD device, the stress distribution is almost constant and uniform. The surface stress increases with increasing pressure. Circular and triangular microspheres are always symmetrical when they are in movement, so the maximum surface and minimum surface stress values are close to each other. Because of their different long and short axes, and the fact that microspheres collide with the pillars in the DLD device, the surface stress of elliptical and rhomboid microspheres will change abruptly. Under the low elasticity of the microspheres, surface stress is insufficient to cause apparent deformation.

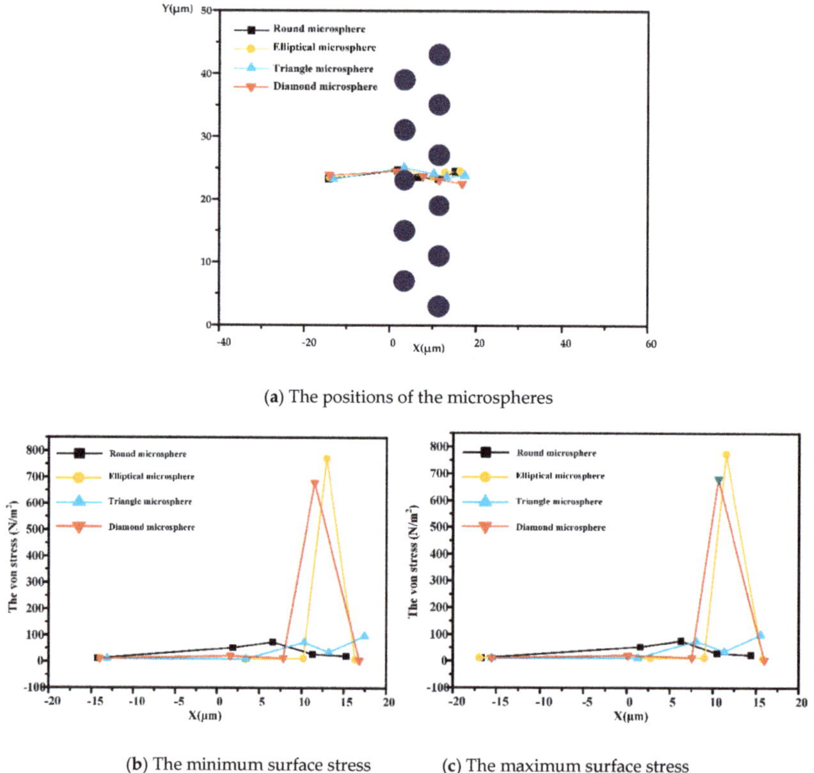

(**a**) The positions of the microspheres

(**b**) The minimum surface stress   (**c**) The maximum surface stress

Figure 7. The surface stress on round, elliptical, triangular and rhomboid microspheres (at t =0 s, 0126 s, 0.156 s, 0.171 s, and 0.19 s). (**a**) The microspheres are in different positions at five moments. The minimum and maximum stresses (von stress) are shown in (**b**) and (**c**), respectively.

3.3. The Velocity of the Microspheres

According to the five moments shown in Figure 7 above, we found that the velocity of the microspheres is only related to the viscous resistance at the entrance region. Since the viscous resistance is negligible, we consider the initial velocity of the microspheres to be close to the inlet velocity of the fluids ($u_0 = 12$ μm/s).

Next, we observed the velocity of the liquid in the DLD device. High velocities were measured in the gaps between the pillars. From the velocity of the microspheres, it was observed that the initial velocity of the microspheres was close to the inlet flow velocity, and in the case of a low level of velocity resistance, the shear stress of the fluid on the microspheres was negligible. This confirms that the microspheres have excellent follow-up performance in DLD devices. The result shows that the pillar structure can increase the velocity of the microspheres.

From Figure 8, we can see when the microspheres pass theough the DLD device, their velocity increases, with the maximum velocity occuring in the gaps between the pillar posts. When flowing between the pillars, the velocity was lower; this is caused by the immense surface stress required to produce negative resistance. When the microspheres approach the outlet, the velocity will gradually converge, eventually becoming zero.

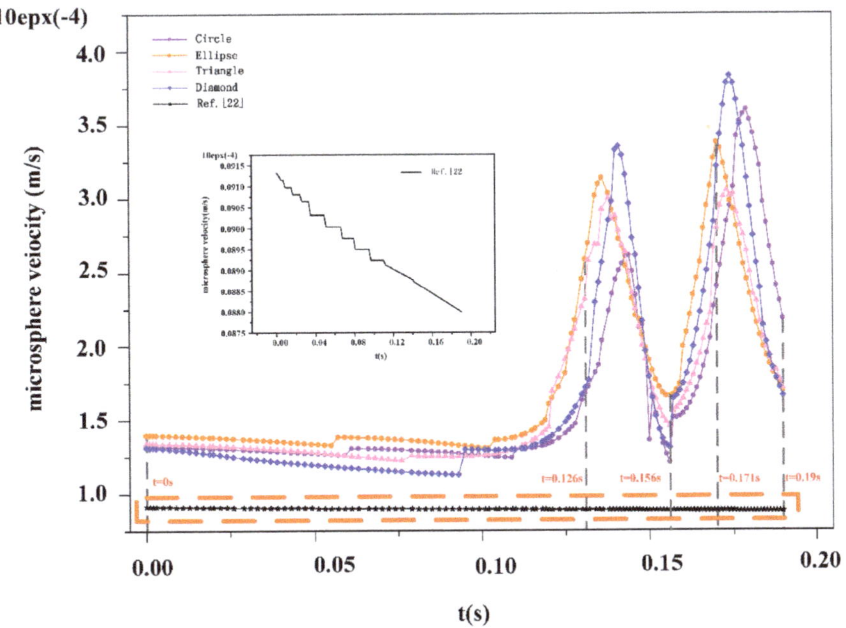

Figure 8. The velocity of the microspheres changes over time. Four kinds of microspheres with circular, elliptical, triangular, and diamond shapes were compared with the data of the reference [22], and a series of time points were selected to analyze their velocity: $t = 0$ s, 0126 s, 0.156 s, 0.171 s, and 0.19 s.

3.4. The Effect of Pillar Shapes

We investigated the effect of circular and triangular pillar structures on the flow characteristics in DLD devices. Figure 9 shows that there are some zero-velocity areas between the two pillars in these two shapes of pillar posts. We found that the fluid velocity in the circular pillar posts was much greater than that of the triangular ones. We thus optimize the shape of the pillars and analyzed the effect of velocity on the motion and stress levels of the microspheres.

The paraments in these DLD devices are all in $\Delta\lambda = 3.5$ μm, $D_c = 2.82$ μm, $G = 3$ μm, and $\varepsilon = 0.4375$ in Figure 10. The liquid gap between the microspheres and the pillars are clear, while the boundary is slightly weaker. We can see that the microspheres tend to move closer to the pillars in the triangular DLD device, and that the stress level is uniform. This indicates that triangular pillars can increase the levels of interaction with the microspheres, significantly affecting their trajectories.

Figure 9. The flow velocity distribution within different shapes of the pillars. The row shift fraction is the same as $\varepsilon = 0.1111$. All velocity in the flow direction (x-axis) is represented by the maximum velocity values (m/s) observed in the DLD array, where red indicates a high flow rate region and dark blue a boundary with a no-slip condition, i.e., a zero-flow velocity region.

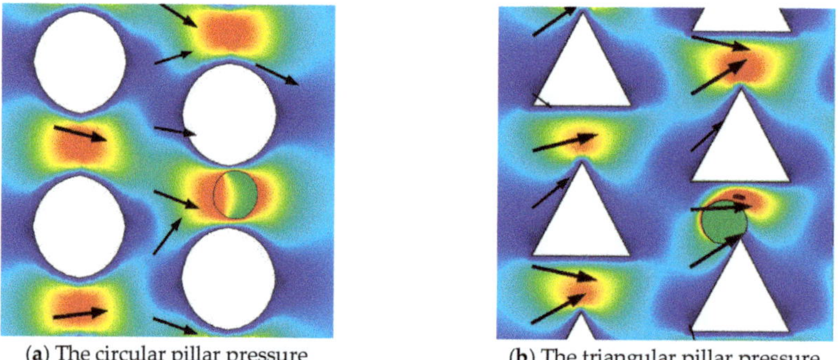

(**a**) The circular pillar pressure   (**b**) The triangular pillar pressure

Figure 10. The movement of microspheres identical in size in two types of DLD devices.

According to the setting of Young's modulus and Poisson's ratio, it was found from the simulation that the deformation of the microspheres was almost consistent with that of RBCs. To better observe RBC deformation in the DLD device with triangular pillars, we decided to combine the RBC model with the triangular pillars to attain the trajectory of the RBCs in the DLD device. And the Figure 11 shows the schematic diagram of the RBCs in 2-D and 3-D.

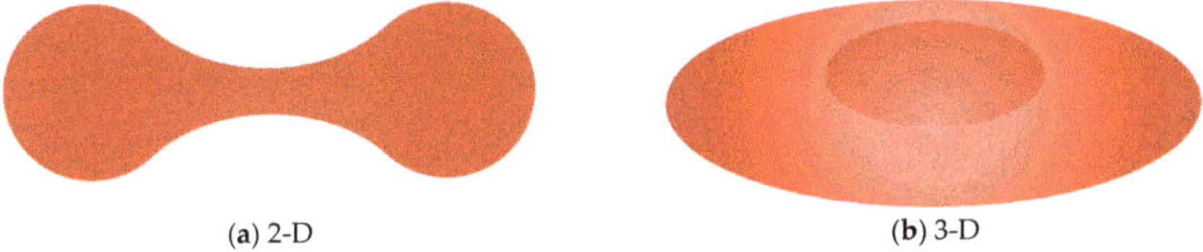

(**a**) 2-D   (**b**) 3-D

Figure 11. The RBCs model (2-D and 3-D).

3.5. The Deformation of Red Blood Cells

We analyzed the deformability of RBCs; asphericity needs to be introduced her eto describe the deviation from the spheric geometry [28],

$$\delta = \frac{(\lambda_1 - \lambda_2)^2}{(\lambda_1 + \lambda_2)^2},\tag{18}$$

where λ_1 and λ_2 are the square roots of two non-zero eigenvalues of the radius of the rotation tensor. The δ from 0 to 1 reflects the extent to which perfect RBCs or microspheres changed in terms of elongation. In the 2-D model, δ equalled 0.29.

A DLD device with triangular pillars and with a gap of $s = 3$ μm and height of $D = 5$ μm was selected to analyze the level of deformation of RBCs. It was found that there was no clear flow layer between the RBC model and the pillars, indicating that the DLD device has considerable sensitivity. In Figure 12a, the RBCs stayed near the top of the pillar for a long time, and could not continue to move. To explore the relationship between the trajectories of RBCs and the pillar scale, we compared the size of the pillars with a gap of $s = 10$ μm and a height of $D = 15$ μm.

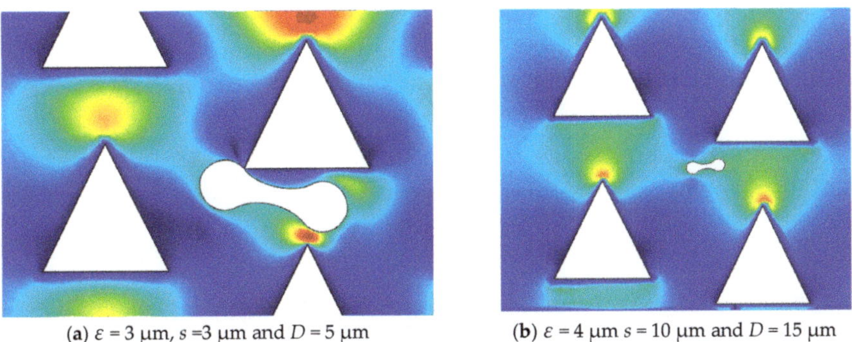

(**a**) $\varepsilon = 3$ μm, $s = 3$ μm and $D = 5$ μm   (**b**) $\varepsilon = 4$ μm $s = 10$ μm and $D = 15$ μm

Figure 12. Snapshot of RBCs in the triangular arrays with $s = 3$ μm and $D = 5$ μm and $s = 10$ μm and $D = 15$ μm.

In Figure 12b, it can see that the RBCs are able to pass through the DLD device of the pillar structure when $D > 6$ μm. Simultaneously, by increasing the diameter and the gap of the pillars, the flow velocity of the pillars decreases from the top to the bottom. The velocity of the RBCs near the pillars is less affected by the flow field, and the deformation is negligible.

To characterize the separation model of RBCs, the function of row shift fraction ε and capillary number Ca was used to verify the trajectories of the microspheres. It was found that the boundary between the displacement model and the "zigzag model" was very close to a simple exponential function [19].

$$\varepsilon(Ca) = 2.9 + 5.4e^{-1.72C_a} \tag{19}$$

According to Equation 20, $\varepsilon(Ca) = 3$ μm, $Ca = 2.1$ within the "zigzag mode" and the trajectory of RBCs is "zigzag mode" in the row shift $\varepsilon = 3$ μm. $\varepsilon(Ca) = 4$ μm and $Ca = 0.9$ in "displacement mode" and when the trajectory of RBCs is in "displacement mode" in the row shift fraction $\varepsilon = 4$ μm. Following Ref. [19], we know that the softer and more rigid gradient of the RBC yield Ca ranges from 0.4 to 1.2.

4. Conclusions

We studied RBC separation with a DLD device using the finite element simulation. This research will provide an essential reference for further research of RBC separation using DLD devices in the future. The following conclusions were obtained:

(a)   The trajectories of the RBCs are related to the row shifts. We found that the RBCs enter "displacement mode" under conditions of low row-shift ($\Delta\lambda < 1.4$ μm) and "zigzag mode" with large row-shift ($\Delta\lambda > 1.5$ μm).

(b)   The velocity and the surface stress of the microspheres are related to the shape of the pillars. The triangular pillars will produce high velocities and the uniform stress on the microspheres, which will enhance separation.

(c)   By considering the deformation of RBCs, the row shift fraction ε and the capillary number Ca are used to determine the RBC separation mode.

Author Contributions: Conceptualization, Y.H.; formal analysis, Y.J.; investigation, Y.J.; methodology, Y.H.; software, F.J.; supervision, Y.H.; writing–original draft, Y.J.; writing–review and editing, Y.H. & F.J.

Funding: This work was funded by the National Natural Science Foundation of China (grant number 11502102).

Conflicts of Interest: The authors declare no conflict of interest.

References

1.   Popel, A.S.; Johnson, P.C. Microcirculation and Hemorheology. *Annu. Rev. Fluid Mech.* **2005**, *37*, 43–69. [CrossRef] [PubMed]

2.   Yin, X.; Thomas, T.; Zhang, J. Multiple red blood cell flows through microvascular bifurcations: cell free layer, cell trajectory, and hematocrit separation. *Microvasc. Res.* **2013**, *89*, 47–56. [CrossRef] [PubMed]

3.   Kabacaoğlu, G.; Biros, G. Sorting same-size red blood cells in deep deterministic lateral displacement devices. *J. Fluid Mech.* **2018**, *859*, 433–475. [CrossRef]

4.   Zhang, Z.; Chien, W.; Henry, E.; Fedosov, D.; Gompper, G. Sharp-edged geometric obstacles in microfluidics promote deformability-based sorting of cells. *Phys. Rev. Fluids* **2019**, *4*, 024201. [CrossRef]

5.   Di Carlo, D. A mechanical biomarker of cell state in medicine. *Jala-J. Lab. Autom* **2012**, *17*, 32–42. [CrossRef] [PubMed]

6.   Zhou, J.; Tu, C.; Liang, Y.; Huang, B.; Fang, Y.; Liang, X.; Papautsky, I.; Ye, X. Isolation of cells from whole blood using shear-induced diffusion. *Sci.Rep-UK* **2018**, *8*, 9411. [CrossRef] [PubMed]

7.   Zhou, J.; Kulasinghe, A.; Bogseth, A.; O'Byrne, K.; Punyadeera, C.; Papautsky, I. Isolation of circulating tumor cells in non-small-cell-lung-cancer patients using a multi-flow microfluidic channel. *Microsyst. Nanoeng.* **2019**, *5*, 8. [CrossRef] [PubMed]

8.   Gossett, D.R.; Weaver, W.M.; Mach, A.J.; Hur, S.C.; Tse, H.T.; Lee, W.; Amini, H.; Di Carlo, D. Label-free cell separation and sorting in microfluidic systems. *Anal. Bioanal. Chem.* **2010**, *397*, 3249–3267. [CrossRef] [PubMed]

9.   Miltenyi, S.; Muller, W.; Weichel, W.; Radbruch, A. High gradient magnetic separation with MACS. *Cytometry* **1990**, *11*, 231–238. [CrossRef] [PubMed]

10. Després, D.; Flohr, T.; Uppenkamp, M.; Baldus, M.; Hoffmann, M.; Huber, C.; Derigs, H.G. CD34+ cell enrichment for autologous peripheral blood stem cell transplantation by use of the Clini MACs device. *J. Hematother Stem Cell Res.* **2000**, *9*, 557–564. [CrossRef] [PubMed]

11. Chiju, W.; Laijun, L.; Irving, G. Pre-pro-B cell growth-stimulating factor (PPBSF) upregulates IL-7Ralpha chain expression and enables pro-B cells to respond to monomeric IL-7. *J. Interferon Cytokine Res.* **2002**, *22*, 823–832. [CrossRef]

12. Adiga, S.; Curtiss, L.A.; Elam, J.; Pellin, M.; Shih, C.-C.; Shih, C.-M.; Lin, S.-J.; Su, Y.-Y.; Gittard, S.; Zhang, J.; et al. Nanoporous materials for biomedical devices. *JOM* **2008**, *60*, 26–32. [CrossRef]

13. Adiga, S.; Jin, C.; Curtiss, L.A.; Monteiro-Riviere, N.; Narayan, J. Nanoporous membranes for medical and biological applications. *Wires. Nanome. Nanobi.* **2009**, *1*, 568–581. [CrossRef] [PubMed]

14. Inglis, D.W.; Davis, J.A.; Austin, R.H.; Sturm, J.C. Critical particle size for fractionation by deterministic lateral displacement. *Lab Chip* **2006**, *6*, 655–658. [CrossRef] [PubMed]

15. Majid Ebrahimi, W.; Bhagat, A.A.S.; Bee Luan, K.; Jongyoon, H.; Chwee Teck, L.; Qing, G.H.; Anthony Gordon, F. Isoporous micro/nanoengineered membranes. *Acs Nano* **2013**, *7*, 1882–1904. [CrossRef]

16. Xavier, M.; Holm, S.H.; Beech, J.P.; Spencer, D.; Tegenfeldt, J.O.; Oreffo, R.O.C.; Morgan, H. Label-free enrichment of primary human skeletal progenitor cells using deterministic lateral displacement. *Lab Chip* **2019**, *19*, 513–523. [CrossRef] [PubMed]

17. Li, X.; Popel, A.S.; Karniadakis, G.E. Blood-plasma separation in Y-shaped bifurcating microfluidic channels: a dissipative particle dynamics simulation study. *Phys. Biol.* **2012**, *9*, 026010. [CrossRef]

18. Fu, X.; Yao, Z.; Zhang, X. Numerical investigation of polygonal particle separation in microfluidic channels. *Microfluid. Nanofluid.* **2016**, *20*, 106. [CrossRef]

19. Krüger, T.; Holmes, D.; Coveney, P. Deformability-based red blood cell separation in deterministic lateral displacement devices—A simulation study. *Biomicrofluidics* **2014**, *8*, 054114. [CrossRef]

20. Hou, H.W.; Bhagat, A.A.S.; Han, J.; Lim, C.T. Deformability based cell margination—A simple microfluidic design for malarial infected red blood cell filtration. In Proceedings of the 6th World Congress of Biomechanics, Singapore, 1–6 August 2010. [CrossRef]

21. Li, Q.; Ito, K.; Wu, Z.; Lowry, C.S.; Loheide Ii, S.P. COMSOL multiphysics: A novel approach to ground water modeling. *Ground Water* **2009**, *47*, 480–487. [CrossRef]

22. Xu, X.; Li, Z.; Nehorai, A. Finite element simulations of hydrodynamic trapping in microfluidic particle-trap array systems. *Biomicrofluidics* **2013**, *7*, 054108. [CrossRef] [PubMed]

23. Erturk, E.; Dursun, B. Numerical solutions of 2-D steady incompressible flow in a driven skewed cavity. *ZAMM-Z. Angew. Math. Mech.* **2007**, *87*, 377–392. [CrossRef]

24. Witold, G.; Kowalewski, T.A. *A Guide to Using Finite Element Software. Applied Mechanics of Solids*, 1st ed.; Francis, F., Ed.; CRC Press LLC. Palgrave Macmillan: Florence, Italy, 2009; Volume 10, pp. 85–92.

25. Jay, A.W.; Canham, P.B. Viscoelastic properties of the human red blood cell membrane. II. Area and volume of individual red cells entering a micropipette. *Biophys. J.* **1977**, *17*, 169–178. [CrossRef]

26. Suresh, S.; Spatz, J.; Mills, J.P.; Micoulet, A.; Dao, M.; Lim, C.T.; Beil, M.; Seufferlein, T. Connections between single-cell biomechanics and human disease states: gastrointestinal cancer and malaria. *Acta Biomater.* **2005**, *23*, S3–S15. [CrossRef]

27. Huang, L.; Cox, E.; Austin, R.; Sturm, J. Continuous particle separation through deterministic lateral displacement. *Science* **2004**, *304*, 987–990. [CrossRef]

28. Zhang, Z.; Henry, E.; Gompper, G.; Fedosov, D.A. Behavior of rigid and deformable particles in deterministic lateral displacement devices with different post shapes. *J. Chem. Phys.* **2015**, *143*, 243145. [CrossRef]

29. Sang, S.; Feng, Q.; Jian, A.; Li, H.; Ji, J.; Duan, Q.; Zhang, W.; Wang, T. Portable microsystem integrates multifunctional dielectrophoresis manipulations and a surface stress biosensor to detect red blood cells for hemolytic anemia. *Sci. Rep.* **2016**, *6*, 33626. [CrossRef]

 *micromachines*

Article

Evaluation of Positive and Negative Methods for Isolation of Circulating Tumor Cells by Lateral Magnetophoresis

Haeli Kang, Jinho Kim, Hyungseok Cho and Ki-Ho Han *

Department of Nanoscience and Engineering, Center for Nano Manufacturing, Inje University, Gimhae 50834, Korea; rkdgofl9388@naver.com (H.K.); injemems@naver.com (J.K.); elshaddai88@naver.com (H.C.)
* Correspondence: mems@inje.ac.kr; Tel.: +82-55-320-3715

Received: 21 May 2019; Accepted: 6 June 2019; Published: 8 June 2019

Abstract: We developed an epithelial cell adhesion molecule (EpCAM)-based positive method and CD45/CD66b-based negative method for isolating circulating tumor cells (CTCs) by lateral magnetophoresis. The CTC recovery rate, white blood cell depletion rate, and purity of CTCs isolated using the positive and negative methods were analyzed using blood samples spiked with cancer cells with different expression levels of EpCAM. The aim was to assess the strengths and weaknesses of the positive and negative isolation methods for CTC-based diagnostics, prognostics, and therapeutics for cancer. The EpCAM-based positive method yielded CTCs of high purity, while the CD45/CD66b-based negative method yielded a large number of CTCs. In conclusion, the positive method shows promise for detecting somatic oncogenic mutations and the negative method shows promise for discovery of cellular and transcriptomic biomarkers of cancer.

Keywords: circulating tumor cell; positive isolation; negative isolation; lateral magnetophoresis

1. Introduction

Circulating tumor cells (CTCs) are an important biomarker for the diagnosis, prognosis, and treatment of cancer [1–3]. Despite their clinical importance, there is no standard method for isolating CTCs [4] due to their extreme rarity, typically 1–100 CTCs per 1 mL of peripheral blood [5,6]. Therefore, novel methods for isolating CTCs are required, e.g., the CellSearch system (Menarini Silicon Biosystems), the gold-standard CTC isolation system that was the first to be approved by the United States Food and Drug Administration. Most label-dependent positive isolation methods are dependent on the epithelial cell adhesion molecule (EpCAM; epithelial-specific surface marker) on CTCs [7,8]. Because EpCAM-based positive isolation methods are based on immunoaffinity, they have high selectivity and specificity for isolating CTCs from blood. However, their specificity decreases if CTCs acquire a mesenchymal-like phenotype (including decreased surface expression of EpCAM) following the epithelial–mesenchymal transition and is low for CTCs derived from non-epithelial origin cancers (e.g., melanoma) [9–11]. In addition, CTCs expressing mesenchymal markers are related to a poor prognosis [12,13]. Consequently, current EpCAM-based positive isolation methods are inadequate due to the rarity and heterogeneity of CTCs and may miss important subtypes of CTCs [14–16].

Negative methods of isolating CTCs from blood involve the removal of white blood cells (WBCs) [17–20]. Negative methods can be used to harvest all types of CTCs, because they do not depend on the surface-marker profiles or the size of CTCs. In addition, because negative methods do not use antibodies specific to CTCs, they are suitable for downstream analyses, such as genetic assays, CTC culture, and xenografts [21,22]. Several negative methods for isolating CTCs have been developed, but they have lower specificity than positive methods. To date, no study has compared positive and negative microfluidic methods for isolating CTCs under identical conditions.

In this study, we introduce the microfluidic NegCTC-μChip, a negative method of isolating CTCs by lateral magnetophoresis. Anti-CD45/CD66b magnetic nanobeads are used to remove WBCs, which are typically expressed on the membrane of leukocytes and granulocytes, respectively. In addition, we used the previously developed PosCTC-μChip [23] (positive method) to isolate CTCs using anti-EpCAM magnetic nanobeads. We compared the CTC recovery rates, WBC depletion rates, and CTC purity of the PosCTC-μChip and the NegCTC-μChip, using blood samples spiked with cancer cells with different expression levels of EpCAM. Finally, we discussed the strengths and weaknesses of the two methods for CTC-based diagnostics, prognostics, and therapeutics for cancer.

2. Materials and Methods

2.1. Design and Working Principle

The PosCTC-μChip is a microfluidic device for isolating CTCs, based on an immunomagnetic approach that contains a lateral magnetophoretic microchannel where CTCs bound to anti-EpCAM magnetic nanobeads are positively isolated by lateral magnetophoresis (Figure 1a). The NegCTC-μChip consists of a free-bead capture microchannel (for removing free magnetic nanobeads) and a lateral magnetophoretic microchannel (for depleting WBCs bound to anti-CD45/CD66b magnetic nanobeads) (Figure 1b). The lateral magnetophoretic microchannels of both the PosCTC-μChip and the NegCTC-μChip include ferromagnetic wires, which are inlaid into the substrate at 5.7°, relative to the direction of flow. Large numbers of magnetic nanobeads are used in negative isolation methods, because each milliliter of blood contains 5×10^6 WBCs. This results in the presence of unbound residual magnetic nanobeads. As blood flows through the lateral magnetophoretic microchannel, the residual magnetic nanobeads bind to the ferromagnetic wires. Furthermore, DNA strands and serum factors adhere to the stacked nanobeads, and the agglomeration of WBCs and CTCs clogs the microchannel. This phenomenon decreases the recovery rate of CTCs and the depletion of WBCs, which in turn degrades the overall isolation performance. Therefore, residual magnetic nanobeads must be removed prior to isolation of CTCs by negative methods. Because positive methods require fewer magnetic nanobeads, no free-bead capture microchannel is needed. The free-bead capture microchannel is 1 mm in width, 50 μm in height, and 42.5 mm in length to promote capture of residual magnetic nanobeads and to provide stable fluid flow by acting as a fluid impedance, and consists of zones 1 and 2. Because the ferromagnetic wires in the free-bead capture microchannel are parallel to the direction of fluid flow, the nano-sized residual magnetic beads are captured, but micro-sized WBCs are not, thereby preventing microchannel blockage. The ferromagnetic wires in zone 2 are arranged so as to capture any residual magnetic nanobeads not captured in zone 1. The lateral magnetophoretic microchannel of the NegCTC-μChip is 2.8 mm in width and 100 μm in height, which is 5.6-fold larger than the cross-sectional area of the free-bead capture microchannel. Because the buffer solution is injected at a rate equal to that of the blood sample, the mean flow rate is 2.8-fold lower than that of the free-bead capture microchannel. This prolongs WBC exposure to the lateral magnetophoretic force, promoting depletion. To promote isolation of CTCs, the CTC and waste outlets of the NegCTC-μChip are 2.24 and 0.56 mm in width, respectively.

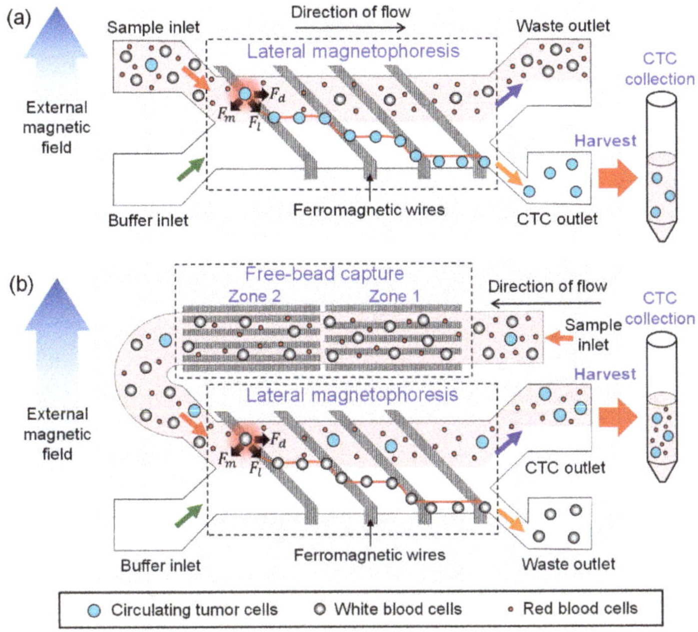

Figure 1. Schematic of (**a**) the PosCTC-µChip, which contains a lateral magnetophoretic microchannel where circulating tumor cells (CTCs) labeled with anti-EpCAM magnetic nanobeads are isolated by lateral magnetophoresis; and (**b**) the NegCTC-µChip, which has a free-bead capture microchannel for removing free magnetic nanobeads and a lateral magnetophoretic microchannel for depleting white blood cells (WBCs) tagged with anti-CD45/CD66b magnetic nanobeads.

When a uniform external magnetic field is laterally applied to the free-bead-capture and the lateral magnetophoretic microchannels, the external magnetic field near the wires is deformed, generating a high-gradient magnetic field over the microchannels. For the PosCTC-µChip, CTCs are labelled with magnetic nanobeads and behave as paramagnetic particles. When the magnetized CTCs pass over the wire, they simultaneously experience a magnetic force, F_m, and a hydrodynamic drag force, F_d, which generates a lateral magnetic force, F_l, on the CTCs (Figure 1a). Next, the CTCs flow into the CTC outlet and other blood cells flow into the waste outlet. In the case of the NegCTC-µChip, WBCs are tagged with magnetic nanobeads and behave as paramagnetic particles. As a blood sample passes through the free-bead capture microchannel, residual magnetic nanobeads, but not WBCs, are captured on the ferromagnetic wires. Because WBCs are 6−10 µm in diameter, they experience a high drag force and so pass through the free-bead capture microchannel into the lateral magnetophoretic microchannel. As magnetized WBCs pass over the ferromagnetic wires in the lateral magnetophoretic microchannel, they are forced laterally into the waste outlet, while other blood cells (including red blood cells (RBCs) and CTCs) flow into the CTC outlet. Because RBCs lack a nucleus, the sample can be used directly for monitoring CTC populations and for CTC-based genetic analyses.

2.2. Fabrication Process

Both the PosCTC-µChip and the NegCTC-µChip were produced by vacuum assembly of a disposable microchannel superstrate and a reusable substrate with inlaid ferromagnetic wires (Figures S1 and S2) [24]. The disposable microchannel superstrate was made by bonding a 12-µm-thick silicone-coated polymer film to a microstructured PDMS replica. An SU-8 mold on a glass master was first fabricated to create a PDMS replica with a microchannel and vacuum trench. An SU-8

3050 photoresist (MicroChem Corp., Westborough, MA, USA) was spun to create the SU-8 mold-pattern for the microchannels on the glass master, and an evaporated 1000 Å Cr layer was applied to promote adhesion between the SU-8 and the glass master. The SU-8 mold pattern for the lateral magnetophoretic microchannel of the PosCTC-μChip was 50 μm in thickness. The free-bead capture and lateral magnetophoretic microchannels of the NegCTC-μChip were 50 and 100 μm in thickness, respectively. Next, an acrylic square bar of 2×2 mm^2 was bonded to the glass master to define the vacuum trench (Figures S1a and S2a). An aluminum mold frame was used to pour the liquid PDMS. The SU-8 mold was completed by assembling the glass master and the aluminum mold frame. Liquid phase PDMS, prepared by mixing resin and curing agent at a 10:1 ratio (Sylgard 184; Dow Corning, Midland, MI, USA), was poured into the SU-8 mold and cured for 1 h at 75 °C in an oven (Figures S1b and S2b). After peeling the PDMS replica off the SU-8 mold, the inlet and outlet reservoirs and the vacuum hole of the PDMS replica were generated using a 1.5-mm-diameter punch. Next, the PDMS replica and the 12-μm-thick silicone-coated polymer film (5 g release force, Shanghai Guangtai Adhesive Products Co., Shanghai, China) were bonded by oxygen plasma treatment for 60 s at 6.8 W radiofrequency power (PDC-32G-2; Harrick Plasma, Ithaca, NY, USA) (Figures S1c and S2c). The reusable substrate was fabricated using a 0.7-mm-thick glass slide (Borofloat33 Pyrex; Schott, New York City, NY, USA). A Ti/Cu/Cr seed layer was first evaporated by electron beam onto the glass slide, and an SU-8 3050 photoresist was spun and patterned to create 40-μm-thick micromolds for ferromagnetic wires (Figures S1d and S2d). Permalloy ($Ni_{0.8}Fe_{0.2}$) was electroplated onto the glass slide (Figures S1e and S2e) and 40-μm-thick ferromagnetic wires were formed by chemical and mechanical polishing and inlaid in the reusable substrate (Figures S1f and S2f).

To set up the instrument, the reusable substrate was first placed at the center of two stacked neodymium–iron–boron (Nd–Fe–B) permanent magnets. After alignment of the disposable superstrate to the reusable substrate, an air vacuum pressure of −50 kPa was applied through the vacuum hole to induce bonding of the superstrate to the substrate (Figure 2). The vacuum contact also removed the air gap between the polymer thin film and the ferromagnetic wires, enabling the high-gradient magnetic field generated by the ferromagnetic wires to penetrate the micrometer-thick polymer film into the microchannel and manipulate magnetized CTCs (PosCTC-μChip) or WBCs (the NegCTC-μChip). Two syringe pumps were used to inject the blood sample and the buffer, and another syringe pump was used to suck the solution from the waste outlet to create a stable fluidic flow in the lateral magnetophoretic microchannel. The sample and the buffer flow rates of the PosCTC-μChip were both 2 mL/h and the suction flow rate was 3.2 mL/h, resulting in a flow rate to the CTC outlet of 0.8 mL/h. For the NegCTC-μChip, the sample and the buffer injection flow rates were 2.8 mL/h, to promote the capture of free magnetic nanobeads, but not WBCs, on the ferromagnetic wires and prolong the residence time of WBCs in the lateral magnetophoretic microchannel, promoting their depletion. A high flow rate hampers depletion of WBCs, due to the low capture efficiency of residual magnetic nanobeads in the free-bead capture microchannel and the reduced residence time of WBCs in the lateral magnetophoretic microchannel. A low flow rate results in capture of WBCs in both the free-bead capture and the lateral magnetophoretic microchannels, decreasing the CTC yield. CTCs and WBCs in the CTC outlet were enumerated under a fluorescence microscope and the CTC recovery rate, WBC depletion rate, and CTC purity were calculated.

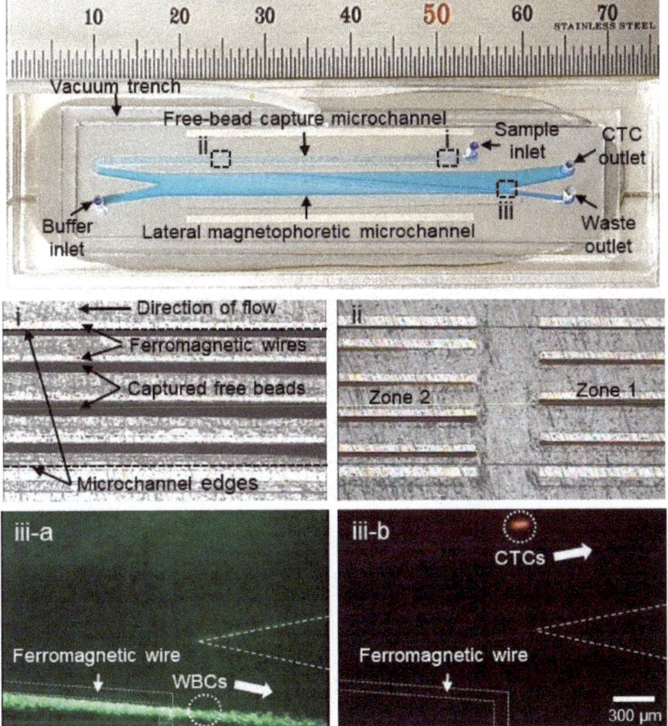

Figure 2. Photographs of the NegCTC-μChip, which consists of free-bead capture and lateral magnetophoretic microchannels. Enlarged views of (**i**) residual magnetic nanobeads captured on the ferromagnetic wires and (**ii**) zones 1 and 2 in the free-bead capture microchannel. In the lateral magnetophoretic microchannel, (**iii-a**) WBCs (green) bound to magnetic nanobeads are forced laterally into the waste outlet, and (**iii-b**) spiked cancer cells (red) flow into the CTC outlet.

2.3. Sample Preparation

Human peripheral blood was obtained from healthy donors, using a protocol approved by the institutional review board of Inje University (Inje 2017-05-013-003), collected in Vacutainer tubes (G-Tube™, Green Cross, Yongin, Korea) containing anticoagulant ethylenediaminetetraacetic acid, and processed within 6 h. MDA-MB-231 and MCF-7 cells were cultured in Dulbecco's Modified Eagle's Medium (DMEM, Invitrogen, Waltham, MA, USA), whereas PC-3 and SKBR-3 cells were grown in RPMI-1640 culture medium (Invitrogen) in a 95% humidified atmosphere and 5% CO_2 at 37 °C. Both media contained 10% (v/v) fetal bovine serum, 100 units/mL penicillin, and 1 mg/mL insulin. Prior to use in experiments, they were harvested from culture plates, using 0.25% trypsin at 37 °C for 3 min and transferred to 15 mL conical tubes. The harvested cancer cells were stained with a membrane-permeable red fluorescent nucleic acid dye (SYTO 64, Invitrogen) for 10 min at room temperature. The fluorescently stained cancer cells were washed in phosphate-buffered saline (PBS) with 0.2% bovine serum albumin (BSA) and suspended to a density of approximately 10^4 per mL. Non-tumor cells in 500 μL of healthy whole blood were stained with a membrane-permeable green fluorescent nucleic acid dye (SYTO 13, Invitrogen). Next, 100 fluorescent-stained CTCs were spiked into the 500-μL blood sample in a 1.5-mL microcentrifuge tube. The spiked blood sample was added to antibodies (anti-EpCAM antibody (Human EpCAM Positive Selection Kit, STEMCELL Technologies, Vancouver, BC, Canada) for the PosCTC-μChip and anti-CD45 and -CD66b antibodies (EasySep Human CD45 Depletion Kit and EasySep Human Whole Blood CD66b Positive Selection Kit,

STEMCELL Technologies) for the NegCTC-µChip) and magnetic nanobeads (about 50 nm in diameter) in sequence and incubated on ice for 60 and 90 min, respectively, according to the manufacturer's instructions (STEMCELL Technologies). During incubation, the antibodies bound to target cells and magnetic nanobeads, because they were bispecific to combine surface antigens expressed on target cells and dextran coated on magnetic nanobeads [25]. Finally, the sample was diluted with a fourfold volume (2 mL) of ice-cold PBS containing 0.2% BSA.

3. Results and Discussion

To evaluate the performance of the PosCTC-µChip and the NegCTC-µChip, MDA-MB-231, PC-3, SKBR-3, and MCF-7 cancer cells were used. The EpCAM expression levels of the cancer cell lines were quantified by flow cytometry (FACSCalibur, BD Biosciences, Franklin Lakes, NJ, USA), using a fluorescein isothiocyanate-labeled anti-EpCAM antibody (anti-human CD326, BioLegend) (Figures S3 and S4). The CTC recovery rate, WBC depletion rate, and CTC purity were calculated using the following equations:

$$\text{CTC recovery rate (\%)} = \frac{\text{The number of isolated cancer cell lines}}{\text{The number of spiked cancer cell lines}} \times 100 \tag{1}$$

$$\text{WBC depletion rate (Log)} = \log\left(\frac{\text{The initial number of WBCs}}{\text{The number of contaminated WBCs}}\right) \tag{2}$$

$$\text{CTC purity (\%)} = \frac{\text{The number of isolated cancer cell lines}}{\text{The number of isolated total nucleated cells (cancer cell lines and WBCs)}} \times 100 \tag{3}$$

CTCs and WBCs were enumerated by counting red-fluorescent and green-fluorescent cells, respectively. The number of isolated total nucleated cells is the sum of the number of isolated CTCs and WBCs. We confirmed the number of spiked CTCs by counting those in the CTC outlet and in the sample syringe, microfluidic device, and waste outlet.

The CTC recovery rate, WBC depletion rate, and CTC purity were evaluated using three measurement datasets. Using the PosCTC-µChip, the recovery rates of SKBR-3 and MCF-7 cells (high EpCAM expression) were 93.9 ± 1.0% (mean ± SD) and 98.4 ± 1.5%, respectively (Figure 3). In contrast, the recovery rates of MDA-MB-231 and PC-3 cells (low EpCAM expression) were 0% and 5.1 ± 1.7%, respectively. Thus, the EpCAM-based positive method had limited ability to isolate CTCs with low EpCAM expression. Using the NegCTC-µChip, the recovery rates of SKBR-3 and MCF-7 cells were 85.2 ± 4.2% and 80.7 ± 7.6%, respectively, and those of MDA-MB-231 and PC-3 cells were 91.0 ± 2.0% and 75.7 ± 9.3%, respectively. Therefore, the NegCTC-µChip enabled isolation of CTCs with a mean recovery rate of 83.1%, irrespective of their EpCAM expression level (Figure 3). CTC recovery rates of the negative isolation method showed greater variations than those of the positive isolation method, because spiked cancer cells could easily be damaged during the sample preparation procedure, causing non-specific binding with magnetic nanobeads.

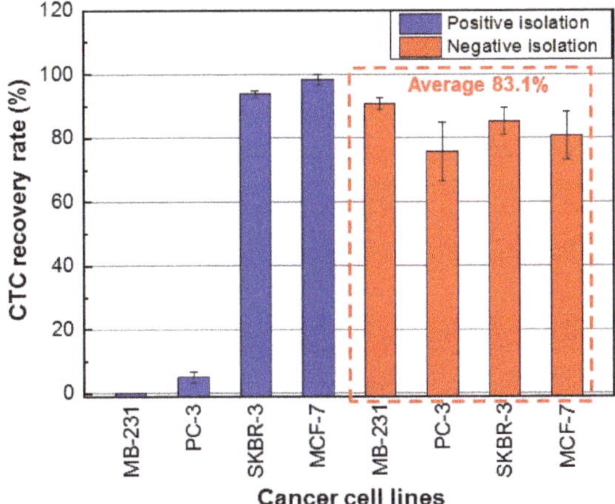

Figure 3. CTC recovery rates using the PosCTC-µChip and NegCTC-µChip from 500-µL blood samples spiked with CTCs with different expression levels of EpCAM. The data are means of triplicate determinations.

For the positive isolation method, the number of contaminating WBCs was 22-213 per 500 µL of blood (Figure 4). The mean number of contaminating WBCs was 92; thus, based on a value of 2.5×10^6 WBCs per 500 µL of blood, the average WBC depletion rate was 28,261-fold (4.5 log). For the negative isolation method, the number of contaminated WBCs was 605-1830 per 500 µL of blood (Figure 4). The mean number of contaminating WBCs was 1379 cells, for a mean WBC depletion rate of 1813-fold (3.3 log). Therefore, some WBCs were not labeled with anti-CD45/CD66b magnetic nanobeads, suggesting the presence of WBC subtypes with low CD45 and CD66b expression.

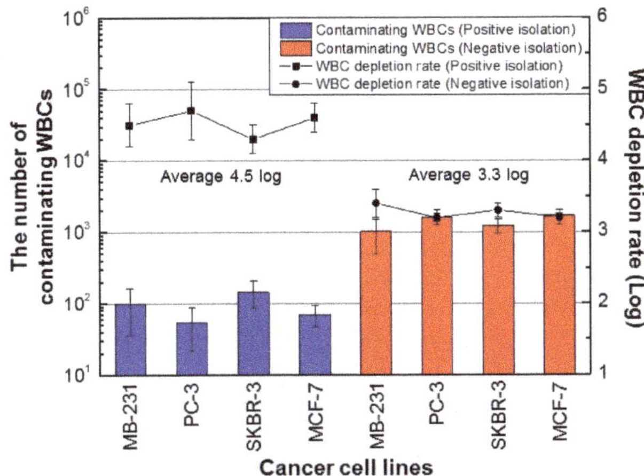

Figure 4. Number of contaminating WBCs using the PosCTC-µChip and the NegCTC-µChip from 500-µL blood samples spiked with CTCs. Data are means of triplicate determinations.

We calculated the purity of the CTCs isolated using the PosCTC-µChip and the NegCTC-µChip as the ratio of the number of CTCs to the sum of the number of CTCs and WBCs (Equation (3)). Using the positive method, the purities of SKBR-3 and MCF-7 cells (high EpCAM expression) were 38.9 ± 8.7% and 51.3 ± 1.6%, respectively (Figure 5). In contrast, the purities of MDA-MB-231 and PC-3 cells were 0% and 7.7 ± 3.9%, likely due to their low expression level of EpCAM. CTC purities of the positive isolation method showed large variations, because the number of contaminating WBCs showed a significant variation, despite being 10 times lower than that of the negative isolation method. Using the negative method, the purities of MDA-MB-231, PC-3, SKBR-3, and MCF-7 cells were 9.0 ± 3.8%, 4.6 ± 0.2%, 5.7 ± 1.7%, and 4.2 ± 0.4%, respectively. Despite the high recovery rate, the purity of CTCs isolated using the NegCTC-µChip was insufficient, due to the number of contaminating WBCs.

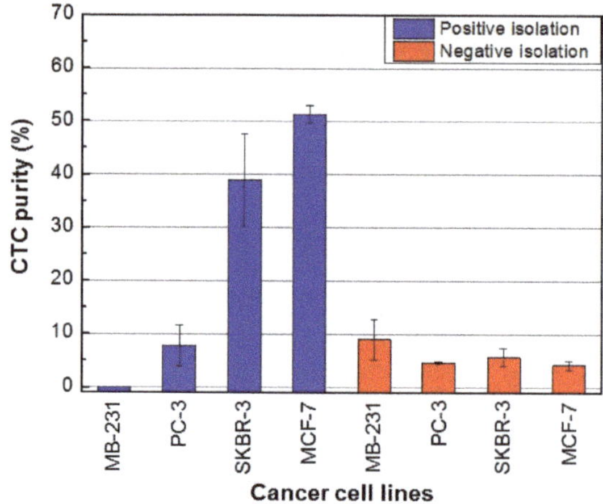

Figure 5. Purity of cancer cells isolated using the PosCTC-µChip and the NegCTC-µChip from 500-µL blood samples spiked with the CTCs. Data are means of triplicate determinations.

4. Conclusions

We compared an EpCAM-based positive isolation method and a CD45/CD66b-based negative isolation method, using MDA-MB-231, PC-3, SKBR-3, and MCF-7 cancer cells, which have different surface expression levels of EpCAM. The recovery rates of the positive isolation method increased with increasing EpCAM expression level. In contrast, the negative method enabled isolation of cancer cells from blood samples with an average recovery rate of 83.1%, irrespective of the expression level of EpCAM. Therefore, the negative isolation method yielded a larger quantity of CTCs from blood samples than the positive method, irrespective of their EpCAM expression level. WBCs were also extracted from the blood samples using the positive and negative isolation methods. In the case of the positive isolation method (mean, 92 WBCs per 500 µL of blood), this may have been due to non-specific binding of WBCs to anti-EpCAM magnetic nanobeads. In the case of the negative isolation method, the contaminating WBCs (mean 1379; 10-fold increase compared to the positive isolation method) were those not bound to anti-CD45/CD66b magnetic nanobeads, due to their abnormally low expression of CD45 and CD66b. Of note, the number of contaminating WBCs was similar in both the positive and negative isolation methods (Figure 4). The results of CTC recovery rate and WBC depletion rate are in line with studies reported by the Toner group [17], although they did not use the same cancer cells for positive and negative isolation.

Regarding the purity of isolated CTCs, because the positive method showed a relatively constant number of contaminating WBCs, CTC purity increased with increasing EpCAM expression level. In the case of the negative method, because the numbers of isolated CTCs and WBCs were similar, irrespective of the EpCAM expression level of the former, CTC purity was 4.2-9.0%. The purity of isolated CTCs was higher with the positive method than with the negative isolation method, with the exception of MDA-MB-231 (low EpCAM expression), due to the larger number of contaminating WBCs using the latter method. Our results indicate that the positive isolation method can be used to obtain highly pure CTCs, enabling detection of somatic oncogenic mutations using advanced genetic analysis techniques (such as next-generation sequencing) that require genomic DNA samples of >5%

purity [26,27]. The negative isolation method has a higher yield than the positive isolation method, making it suitable for discovery of cellular and transcriptomic biomarkers of cancer. Therefore, the isolation method suitable for the application in question should be selected.

Supplementary Materials: The following are available online at http://www.mdpi.com/2072-666X/10/6/386/s1, Figure S1: Fabrication process for the PosCTC-µChip, Figure S2: Fabrication process for the NegCTC-µChip, Figure S3: EpCAM expression level of cancer cell lines used.

Author Contributions: Conceptualization, K.-H.H.; Data curation, H.K.; Formal analysis, H.K.; Funding acquisition, K.-H.H.; Investigation, H.K. and J.K.; Methodology, H.K., J.K., H.C., and K.-H.H.; Project administration, K.-H.H.; Resources, H.K. and K.-H.H.; Supervision, K.-H.H.; Validation, H.K., J.K., H.C., and K.-H.H.; Visualization, H.K. and K.-H.H.; Writing—original draft, H.K. and K.-H.H.; Writing—review and editing, H.K., J.K., H.C., and K.-H.H.

Funding: This work (NRF-2016R1A2B4012077) was supported by the Midcareer Researcher Program through an NRF grant funded by the MEST.

Conflicts of Interest: The authors declare no conflict of interest.

References

1. Pantel, K.; Brakenhoff, R.H.; Brandt, B. Detection, clinical relevance and specific biological properties of disseminating tumour cells. *Nat. Rev. Cancer* **2008**, *8*, 329–340. [CrossRef] [PubMed]

2. Racila, E.; Euhus, D.; Weiss, A.J.; Rao, C.; McConnell, J.; Terstappen, L.W.M.M.; Uhr, J.W. Detection and characterization of carcinoma cells in the blood. *Proc. Natl. Acad. Sci. USA* **1998**, *95*, 4589–4594. [CrossRef] [PubMed]

3. Yu, M.; Stott, S.; Toner, M.; Maheswaran, S.; Haber, D.A. Circulating tumor cells: Approaches to isolation and characterization. *J. Cell Boil.* **2011**, *192*, 373–382. [CrossRef] [PubMed]

4. Cho, H.; Kim, J.; Song, H.; Sohn, K.Y.; Jeon, M.; Han, K.-H. Microfluidic technologies for circulating tumor cell isolation. *Analyst* **2018**, *143*, 2936–2970. [CrossRef] [PubMed]

5. Ross, A.A.; Cooper, B.W.; Lazarus, H.M.; Mackay, W.; Moss, T.J.; Ciobanu, N.; Tallman, M.S.; Kennedy, M.J.; Davidson, N.E.; Sweet, D. Detection and viability of tumor cells in peripheral blood stem cell collections from breast cancer patients using immunocytochemical and clonogenic assay techniques. *Blood* **1993**, *82*, 2605–2610. [PubMed]

6. Krishnamurthy, S.; Cristofanilli, M.; Singh, B.; Reuben, J.; Gao, H.; Cohen, E.N.; Andreopoulou, E.; Hall, C.S.; Lodhi, A.; Jackson, S.; et al. Detection of minimal residual disease in blood and bone marrow in early stage breast cancer. *Cancer* **2010**, *116*, 3330–3337. [CrossRef] [PubMed]

7. Allard, W.J. Tumor Cells Circulate in the Peripheral Blood of All Major Carcinomas but not in Healthy Subjects or Patients with Nonmalignant Diseases. *Clin. Cancer Res.* **2004**, *10*, 6897–6904. [CrossRef]

8. De Bono, J.S.; Scher, H.I.; Montgomery, R.B.; Parker, C.; Miller, M.C.; Tissing, H.; Doyle, G.V.; Terstappen, L.W.; Pienta, K.J.; Raghavan, D. Circulating Tumor Cells Predict Survival Benefit from Treatment in Metastatic Castration-Resistant Prostate Cancer. *Clin. Cancer Res.* **2008**, *14*, 6302–6309. [CrossRef]

9. Thiery, J.P. Epithelial–mesenchymal transitions in tumour progression. *Nat. Rev. Cancer* **2002**, *2*, 442–454. [CrossRef]

10. Kalluri, R.; Weinberg, R.A. The basics of epithelial-mesenchymal transition. *J. Clin. Investig.* **2009**, *119*, 1420–1428. [CrossRef]

11. Yu, M.; Bardia, A.; Wittner, B.S.; Stott, S.L.; Smas, M.E.; Ting, D.T.; Isakoff, S.J.; Ciciliano, J.C.; Wells, M.N.; Shah, A.M.; et al. Circulating Breast Tumor Cells Exhibit Dynamic Changes in Epithelial and Mesenchymal Composition. *Science* **2013**, *339*, 580–584. [CrossRef] [PubMed]

12. Aktas, B.; Tewes, M.; Fehm, T.; Hauch, S.; Kimmig, R.; Kasimir-Bauer, S. Stem cell and epithelial-mesenchymal transition markers are frequently overexpressed in circulating tumor cells of metastatic breast cancer patients. *Breast Cancer Res.* **2009**, *11*, R46. [CrossRef] [PubMed]

13. Yokobori, T.; Iinuma, H.; Shimamura, T.; Imoto, S.; Sugimachi, K.; Ishii, H.; Iwatsuki, M.; Ota, D.; Ohkuma, M.; Iwaya, T.; et al. Plastin3 Is a Novel Marker for Circulating Tumor Cells Undergoing the Epithelial-Mesenchymal Transition and Is Associated with Colorectal Cancer Prognosis. *Cancer Res.* **2013**, *73*, 2059–2069. [CrossRef] [PubMed]

14. Cristofanilli, M.; Budd, G.T.; Ellis, M.J.; Stopeck, A.; Matera, J.; Miller, M.C.; Reuben, J.M.; Doyle, G.V.; Allard, W.J.; Terstappen, L.W.; et al. Circulating Tumor Cells, Disease Progression, and Survival in Metastatic Breast Cancer. *N. Engl. J. Med.* **2004**, *351*, 781–791. [CrossRef] [PubMed]

15. Nagrath, S.; Sequist, L.V.; Maheswaran, S.; Bell, D.W.; Irimia, D.; Ulkus, L.; Smith, M.R.; Kwak, E.L.; Digumarthy, S.; Muzikansky, A.; et al. Isolation of rare circulating tumour cells in cancer patients by microchip technology. *Nature* **2007**, *450*, 1235–1239. [CrossRef] [PubMed]

16. Kim, S.; Han, S.-I.; Park, M.-J.; Jeon, C.-W.; Joo, Y.-D.; Choi, I.-H.; Han, K.-H. Circulating Tumor Cell Microseparator Based on Lateral Magnetophoresis and Immunomagnetic Nanobeads. *Anal. Chem.* **2013**, *85*, 2779–2786. [CrossRef] [PubMed]

17. Ozkumur, E.; Shah, A.M.; Ciciliano, J.C.; Emmink, B.L.; Miyamoto, D.T.; Brachtel, E.; Yu, M.; Chen, P.-I.; Morgan, B.; Trautwein, J.; et al. Inertial Focusing for Tumor Antigen–Dependent and –Independent Sorting of Rare Circulating Tumor Cells. *Sci. Transl. Med.* **2013**, *5*, 179ra47. [CrossRef] [PubMed]

18. Liu, Z.; Fusi, A.; Klopocki, E.; Schmittel, A.; Tinhofer, I.; Nonnenmacher, A.; Keilholz, U. Negative enrichment by immunomagnetic nanobeads for unbiased characterization of circulating tumor cells from peripheral blood of cancer patients. *J. Transl. Med.* **2011**, *9*, 70. [CrossRef]

19. Gourikutty, S.B.N.; Chang, C.-P.; Poenar, D.P. An integrated on-chip platform for negative enrichment of tumour cells. *J. Chromatogr. B* **2016**, *1028*, 153–164. [CrossRef]

20. Bu, J.; Kang, Y.-T.; Moon, B.-I.; Kim, Y.J.; Cho, Y.-H.; Chang, H.J. Dual-patterned immunofiltration (DIF) device for the rapid efficient negative selection of heterogeneous circulating tumor cells. *Lab Chip* **2016**, *16*, 4759–4769. [CrossRef]

21. Münz, M.; Murr, A.; Kvesic, M.; Rau, D.; Mangold, S.; Pflanz, S.; Lumsden, J.; Volkland, J.; Fagerberg, J.; Riethmüller, G.; et al. Side-by-side analysis of five clinically tested anti-EpCAM monoclonal antibodies. *Cancer Cell Int.* **2010**, *10*, 44. [CrossRef] [PubMed]

22. Chen, C.-L.; Chen, K.-C.; Pan, Y.-C.; Lee, T.-P.; Hsiung, L.-C.; Lin, C.-M.; Chen, C.-Y.; Lin, C.-H.; Chiang, B.-L.; Wo, A.M. Separation and detection of rare cells in a microfluidic disk via negative selection. *Lab Chip* **2011**, *11*, 474–483. [CrossRef] [PubMed]

23. Cho, H.; Kim, J.; Han, S.-I.; Han, K.-H. Analytical evaluation for somatic mutation detection in circulating tumor cells isolated using a lateral magnetophoretic microseparator. *Biomed. Microdevices* **2016**, *18*, 91. [CrossRef] [PubMed]

24. Cho, H.; Kim, J.; Jeon, C.-W.; Han, K.-H. A disposable microfluidic device with a reusable magnetophoretic functional substrate for isolation of circulating tumor cells. *Lab Chip* **2017**, *17*, 4113–4123. [CrossRef] [PubMed]

25. Thomas, T.; Clarke, E. Antibody Compositions for Preparing Enriched Mesenchymal Progenitor Preparations. US 6645727B2, 11 November 2003.

26. Tsongalis, G.J.; Peterson, J.D.; De Abreu, F.B.; Tunkey, C.D.; Gallagher, T.L.; Strausbaugh, L.D.; Wells, W.A.; Amos, C.I. Routine use of the Ion Torrent AmpliSeq™ Cancer Hotspot Panel for identification of clinically actionable somatic mutations. *Clin. Chem. Lab. Med.* **2014**, *52*, 707–714. [CrossRef] [PubMed]

27. Chin, E.L.; Da Silva, C.; Hegde, M. Assessment of clinical analytical sensitivity and specificity of next-generation sequencing for detection of simple and complex mutations. *BMC Genet.* **2013**, *14*, 6. [CrossRef] [PubMed]

 *micromachines*

Review

Sorting of Particles Using Inertial Focusing and Laminar Vortex Technology: A Review

Annalisa Volpe [1,2,*], Caterina Gaudiuso [1,2] and Antonio Ancona [2]

[1] Physics Department, Università degli Studi di Bari 'Aldo Moro', Via G. Amendola 173, 70126 Bari, Italy
[2] Institute for Photonics and Nanotechnologies (IFN), National Research Council, Via Amendola 173, 70126 Bari, Italy
* Correspondence: annalisa.volpe@uniba.it; Tel.: +39-080-544-3235

Received: 1 August 2019; Accepted: 7 September 2019; Published: 10 September 2019

Abstract: The capability of isolating and sorting specific types of cells is crucial in life science, particularly for the early diagnosis of lethal diseases and monitoring of medical treatments. Among all the micro-fluidics techniques for cell sorting, inertial focusing combined with the laminar vortex technology is a powerful method to isolate cells from flowing samples in an efficient manner. This label-free method does not require any external force to be applied, and allows high throughput and continuous sample separation, thus offering a high filtration efficiency over a wide range of particle sizes. Although rather recent, this technology and its applications are rapidly growing, thanks to the development of new chip designs, the employment of new materials and microfabrication technologies. In this review, a comprehensive overview is provided on the most relevant works which employ inertial focusing and laminar vortex technology to sort particles. After briefly summarizing the other cells sorting techniques, highlighting their limitations, the physical mechanisms involved in particle trapping and sorting are described. Then, the materials and microfabrication methods used to implement this technology on miniaturized devices are illustrated. The most relevant evolution steps in the chips design are discussed, and their performances critically analyzed to suggest future developments of this technology.

Keywords: inertial micro-fluidics; inertial focusing; lab on a chip; micro-fluidics; vortex technology

1. Introduction

Many research studies in the biomedical field need to analyze individually specific cell types or micro-particles, contained in heterogeneous samples. For this purpose, the target components have to be isolated and sorted from the original mixture, in order to allow cell analysis. Stem cells studies, tissue and organs regeneration [1–4], lethal disease diagnosis and therapy [5,6] and drug screening [7] are made possible, thanks to the development of isolation, sorting and separation techniques for specific target cells or micro-particles. As an example, isolating rare circulation tumor cells (CTCs) from blood allows early an diagnosis and prognosis of cancer disease and its progression [5,8–11].

Conventional techniques like fluorescence-activated cell sorting (FACS), magnetically-activated cell sorting (MACS) and centrifugation methods have been proven successful to separate and extract the targeted cells. Nevertheless, they are affected by some technical limits, since they need labeled cells, long processing times, high operating costs and skilled personnel, despite having a low efficiency [12].

Significant advantages have been introduced by micro-fluidics and lab-on-a-chip (LoC) technology, which promotes novel techniques for cells sorting in a downsized scale. In fact, thanks to their reduced dimensions, LoCs ensure a good portability, paving the way for building a miniaturized platform for medical analysis, thus making this technology suitable for rapid diagnostics and therapy.

Active and passive methods have been implemented in micro-fluidic devices for cell sorting purposes: In the first case, external field forces are applied to the fluid sample to sort cells. On the other hand, passive methods do not employ external forces, but do rely upon the inherent characteristics of the cells, such as shape, dimensions, deformability [13–15].

Among the active methods, acoustophoresis (acoustic levitation) is based upon applying an acoustic force, by means of continuous ultrasound waves [16]. Here, particles suspended in a liquid are exposed to an acoustic standing wave, which causes them to move in the sound field. The magnitude of the exerted force depends on the features of particles, such as size, density and deformation. Therefore, cells separation can be obtained according to these characteristics [17–20]. However, care should be taken due to the large amount of ultrasound, causing temperature increase which can damage the cells, thus affecting their viability. Nevertheless, CTCs separation with a recovery rate of 83% at a throughput of 10^3 cells has been reported in [21]. Dielectrophoresis (DEP) uses an electric field as an external excitation source and relies on the dielectric characteristics of most of the biologicals cells in suspension. The magnitude and the direction of the exerted force is determined by the polarizability of the cells, besides the applied electric field. Therefore, a continuous electric field affects the cells' motion, and is capable of moving particles in different equilibrium positions. High purity (84%) and efficiency (92%) have been reported in [22] the case of stem cells. Magnetophoresis, instead, is based on the use of an external magnetic field to separate particles. In this case, magnetic beads attached to the target cells can be moved under the effect of a strong magnetic field, usually generated by a permanent magnet or electromagnets. This method has been reported for example for *Escherichia coli* (*E. coli*) separation from bovine blood [23], and for removing blood cells infected by malaria [24]. However, label-free magnetic methods have been also reported, for separating red blood cells (RCBs) and white blood cells (WBCs) [25,26], exploiting their native magnetic properties.

Optical methods have also been demonstrated very successful in cell sorting. Here, the optical force depends upon the refractive index and the size of the particles, and can be used to separate cells according to these features [27]. In particular, its magnitude decreases as the refractive index of the particle compared to the host fluid and its volume reduces. Therefore, within the photon description of light, each photon travels according to a geometrical optics ray description, and transfers momentum to the particles in the sample fluid by mean of collisions (elastic or inelastic) [28]. One of the most common optical sorting methods is the optical tweezer, where only one highly focused laser beam is used to exert a force on particles [29]. The possibility of manipulating particles of different dimensions relies on the capability of focusing a Gaussian beam on a very small area, i.e., the beam waist, where a strong electric field gradient is generated which attracts particles and makes them change their motion. Despite the Gaussian beams are the most used for such an application, being the fundamental selectable laser mode, also other transverse modes have been used, e.g., the doughnut beam [30].

Thanks to its simplicity, microfiltration sets as the most known passive method for cells sorting. Using membranes [31,32], pillars [33,34] and weirs [35], this microfiltration allows cell separation depending on their size. Among the advantages of these methods, the simple structures, the high separation efficiency and the precise control achievable with this kind of device, can be highlighted, despite suffering from clogging, which strongly affects their durability.

Hydrodynamic forces and channel shape are the key elements of the deterministic lateral displacement methods (DLD) [36], capable of separating cells depending on their sizes. This method uses the laminar flow hypothesis for the fluid sample motion. Therefore, the target particles suspended in the fluid move according to parallel streamlines, which define the parallel adjacent layers of the fluid. Thus, by placing on their trajectories arrays of obstacles in different directions with respect to the sample flow, the streamlines are bent, and the particles separated. A critical diameter determined by the size of the obstacles, their mutual distance and the distance at which they are separated after each row, defines the performances of the device. In fact, particles smaller than this critical diameter can be separated from bigger ones.

In this case, diffusion can limit the efficiency of these devices, leading particles to move laterally with respect to the streamlines, towards other rows, thus causing a mix between particles and the reduction of selectivity. However, a recovery of 99% and a cell viability higher than 87% [37] have been reported in the case of CTCs clusters isolation.

Recently, the possibility to exploit inertial effects for cell sorting has also been investigated, and several inertial effects are beginning to find a utilization in micro-fluidic systems. The predominantly used effects for inertial sorting are: (i) Inertial migration of particles in a straight channel and (ii) secondary flows in curved channels. As we will better describe in Section 2.1, inertial focusing in straight channels consists of particles' migration across streamlines, due to the onset of inertial forces whose intensity depends upon the particles size [14,38–43]. This phenomenon has been extensively studied for Reynold numbers between 1 and 1000 [44–46]. In this regime, the inertial effects become predominant on the viscous one, but a laminar flow is always possible.

Inertial focusing in a straight micro-channel has found various applications thanks to its ease of fabrication and use. High aspect ratio straight rectangular channels have been used for example by Hur et al. [47] for separating the murine adrenal cortical progenitor from the adrenal gland digest. In particular, they have shown that the inertial focusing depends on the particles' dimensions, finding that bigger cells were collected close to the channel center and separated from the smaller ones flowing in positions closer to the channel walls. Di Carlo et al. [48] have reported about the counting and differentiation process of red and white blood cells, with a device consisting of 256 high aspect ratio straight rectangular channels. A novel channel geometry has been introduced by Mach et al. to separate pathogen bacteria cells from blood [49]. They proposed to gradually expand the channel section. In particular, their device consisted of 40 single straight channels, each of them having three segments with different cross-sections. The parallel channels allowed a high throughput sorting process, with a >80% removal of pathogenic bacteria from blood after two passes of the single channel system. Zhou et al. [50] have in turn proposed a device designed on a two-stages inertial migration principle [51]. First, the sample fluid flows through a channel characterized by a high aspect ratio. Here, the particles are focused near the side walls, at half of the channel height. The second channel has a lower aspect ratio, thus focusing the particles at the center. The size-based differentiation between particles is achieved by the higher speed of the bigger particles, which allows them to relocate very quickly, keeping the smaller ones almost unaltered. Rectangular channels with innovative design have also been successfully exploited for other types of particles besides the cells. For example, Li et al. [14] have realized a device for the inertial focusing of a single-celled alga, *Euglena gracilis*, by using a straight channel followed by a gradually expanding chamber and five outlets.

Though particle migration in straight channels has been deeply studied, devices based on this principle are characterized by long channels, which leads to high resistance to the flow and limits the possibilities of miniaturization. That is why devices based on the secondary flow effect, known as "Dean flow", and originated by channel curvature changes, have been introduced.

The secondary flow effect strongly depends on the particle size; therefore, when two different particles are injected into a curved channel, they can be separated based on their different equilibrium positions. On this basis, Di Carlo et al. have demonstrated the focusing and sorting of cells under laminar flow using a serpentine pattern [12].

Kuntaegowdanahalli et al. [52] demonstrated, for the first time, the size-dependent focusing of particles at distinct equilibrium positions across the micro-channel cross-section in a spiral LoC. The device enabled them to separate neuroblastoma and glioma cells with 80% efficiency and high relative viability (>90%). The isolation and retrieval of circulating tumor cells using centrifugal forces in a spiral channel has been demonstrated by Hou et al. [53]. They obtained 85% cancer cell recovery from blood using an approach named Dean flow fractionation. Similarly, the study of Nivedita et al. [54] reports on an optimized Archimedean spiral device which, in few centimeter squares, reached a separation efficiency of about 95%, and high throughput with rates up to 1×10^6 cells per minute.

Warkiani et al. [55] successfully demonstrated CTC isolation and detection in blood samples of patients with a metastatic tumor, achieving ≥85% recovery of spiked cells across multiple cancer cell lines, and 99.99% depletion of white blood cells in whole blood. Further studies have been conducted by Guan [56] et al. and Warkiani et al. [57] to enhance the sorting performance of the spiral device, optimizing the channel cross section.

A comprehensive study on spiral inertial micro-fluidics for cell separation and biomedical applications can be found in [58].

Despite all the existing works on inertial particles sorting, the capability of significantly concentrating the target population into a smaller volume after operation is still unsatisfactory [44,59]. This issue has been addressed by combining inertial focusing in a straight channel, together with microscale laminar vortices, as first theoretically proposed in [60,61]. The combination of inertial focusing and this vortex trapping technique presents several advantages with respect to other sorting methods in performance, fabrication and operation. It does not need any filtering systems [62], obstacle structure or mechanical or electrical parts. Furthermore, the micro-fluidic system can be easily operated by only one syringe pump without the use of sheath flow, and aims to be a high throughput method by appropriately designing the devices.

Considering the growing interest in sorting methods based on inertial focusing and the continuous improvements implemented by groups all over the world, a review summarizing the evolution, achievements and indicating the perspectives of this method would be beneficial for the scientific community working in this field. Different reviews con be found in literature dealing with inertial effect in micro-fluidics. One of the most cited and complete one is by Di Carlo [41], which however does not include all the most recent results. Also devoted to inertial micro-fluidics is the review of Zhang et al. [40], where the perspectives of employing fluid inertia in micro-fluidics for particle manipulation are also discussed. Here, the inertial focusing with vortex technology is only partially treated.

The reviews of Sajeesh et al. [63] and Shields et al. [15] are specifically focused on micro-fluidics for particle sorting. Both of them provide to the reader a very comprehensive panorama of the developments in LoC cell sorting, reserving one section to the passive and label-free methods, but without mentioning the vortex technology, which has been only recently developed.

More recent is the Dalili et al. work [64]. Here, inertial-based devices specifically devoted to the biomedical application are treated. Unfortunately, most of the latest results exploiting vortex technology are missing, and the evolution of this method is not deeply discussed.

This review article aims to give a comprehensive overview on the development and diversity of this technology, which is still missing in literature, and in addition, to provide perspectives for future directions. We focus on the inertial sorting combined with microscale laminar vortices technology, by first explaining the physical mechanisms underlying inertial focusing. Successively, the particle trapping obtainable by laminar vortex is described, by highlighting the capability of such an approach for cell enrichment. Materials and methods usually employed for the LoC fabrication and testing are then illustrated.

What follows is a description and critical analysis of the most relevant evolution steps in the chips design. In this section, only the works strictly inherent to inertial focusing in combination with laminar vortices are treated, in order to provide to the reader a linear and clear picture of this technique, serving as a basis for future improvements.

2. Inertial Focusing and Laminar Vortex Technology

2.1. Working Principle

Segrè and Silberberg were the first in 1961 to observe that macroscopic, neutrally-buoyant, spherical particles, diluted in a liquid passing a laminar flow through a straight channel, collect in a thin, annular region at a position 0.6 times the radius of the pipe [65,66]. They experimented that the effect was proportional to the mean velocity of the flow U_m, and to the fourth power of the ratio of particle radius r to the tube radius R, and independent of the overall particle concentration at least up to the largest volume fraction 0.4% they used. Furthermore, the results were symmetrical around the axis (Figure 1a) of the tube, and were independent of the shape of the mouth of the tube.

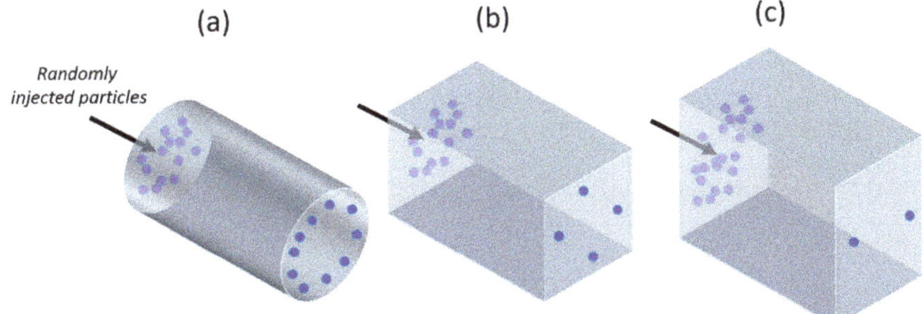

Figure 1. A schematic of inertial particle focusing in a (**a**) tubular, (**b**) square, and (**c**) rectangular channel.

Further works have demonstrated the same particles' focusing effect also in straight, non-circular cross-section micro-channels, e.g., square (Figure 1b) [67], rectangular (Figure 1c) [68], triangular [69], non-conventional shape [70] and gradually changing geometrical structures [38], providing also numerical methods for the fast prediction of the focusing patterns of particles along the channels' cross section [71,72]. Several experimental results on curved [73], serpentine [74] and spiral [55] micro-channels are also reported.

The ease of fabrication and the reduced number of equilibrium positions (only two located near the widest faces of the channel) have made the straight micro-fluidic channels with rectangular cross sections one of the mot studied geometries for inertial focusing [75].

Regardless of the geometry, different studies have tried to explain the origin of the focusing effect [76,77], bringing it back to the fluid inertia and drawing the particles as spheres. Two types of forces exert on neutrally buoyant particles randomly suspended in a laminar fluid flowing through a micro-channel, i.e., viscous drag forces and inertial lift forces [76]. The firsts, due to the viscous nature of the carried fluid, allow the particles to entrain along the flow streamlines. The parabolic nature of the laminar velocity profile in Poiseuille flow produces a shear-induced inertial lift force (F_{LS}) [59], pushing the particles away from the micro-channel center, or rather leading them towards the walls. Here, f_L, a, and W_c are the dimensionless lift coefficient, the particle diameter and the channel width (i.e., shorter face of the rectangular channel), respectively.

$$F_{LS} = \frac{f_L \rho U_m^2 a^3}{W_c} \tag{1}$$

On the other hand, particles experience a wall-effect-induced lift force (F_{LW}) [59]

$$F_{LW} = \frac{f_L \rho U_m^2 a^6}{W_c}, \tag{2}$$

which becomes more prominent as particles migrate closer to the channels' walls in two different ways. First, the wall generates an extra drag force making the particles lag behind the fluid. The bounded sphere undergoes a slower relative fluid velocity on the wall side than on the centerline side, thus causing an increased pressure upon the wall side, which pushes the sphere away. Second, the presence of the adjacent wall alters the streamlines around the sphere causing a dissymmetry of the wake vorticity distribution generated around its surface. This phenomenon induces a higher pressure on the wall side, repulsing the sphere away from the wall [78].

The equilibrium position of the lateral migration is then achieved by the balance of the two opposite lift forces F_{LS} and F_{LW} acting bi-directionally on a sphere (Figure 2a) [44,75].

The effect of the inertial forces increases proportionally to the flow velocity and to the particle size relative to the channel dimension. The particles' Reynolds number Re_p gives an estimation of this fluid dynamic phenomenon. This number represents the particle motion in a channel flow, and can be expressed as [75].

$$Re_p = Re_c \frac{a^2}{D_h^2} = \frac{U_m a^2}{v D_h},\tag{3}$$

where Re_c is the channel Reynolds number, defined as the ratio of inertial force (density x velocity) to viscous force (viscosity/length-scale) ($Re_c = U_m D_h / v$); D_h is the hydraulic diameter of the channel, U_m is the mean flow velocity and v is the kinematic viscosity of the fluid, while a is the particle diameter. The particle Reynolds number is additionally dependent on the particle size. The inertial lift forces become dominant to drive the particles in the equilibrium position when $Re_{ep} > 1$, conversely when $Re_{ep} << 1$ the drag force promotes the longitudinal particle migration.

In case of a channel expansion, the wall effect lift force becomes negligible due to the longer distance of the particles from the sidewall. As a consequence, the particles undergo a non-equilibrium situation, being pushed by the shear lift force in the expansion volume (b). Since the shear-gradient lift force scales with a^3, larger particles experience a stronger lateral lift force than the smaller ones. Furthermore, assuming that F_{LS} is balanced by the viscous drag force ($F_D = 3\pi\mu a U_L$), the lateral migration velocity of particles scales with particles size as $U_L \propto U_m^2 a^2$, consequently the larger particles migrate across the streamlines faster than the smaller ones.

When the Reynolds number is sufficiently high, multiple microscale laminar vortices (i.e., Moffatt's corner eddy flow [79]) are created in the chamber. In such situations, particles with different size are subjected to different lateral migration, which drives them across streamlines crossing the boundary between the main flow and the vortex (separatrix), where they remain isolated, orbiting in it. Therefore, size-selective trapping is achieved. Indeed, particles do not migrate at a sufficient rate to pass the separatrix if their size is below a cutoff a_c, and thus they remain in focused streams, flowing out of the device. Theoretical analyses have shown that the lateral equilibrium position of focused particles shifts toward the wall as Re_c increases. This effect brings the focused streams of flowing particles closer to the boundary streamline at the reservoir and enhances the capturing efficiency proportionally to their size [77].

2.2. Fabrication and Testing Methods

The core of the sorting device based on inertial focusing and laminar vortex technology usually consists of a rectangular cross-section micro-channel with an expansion (micro-chamber). Different channel and chamber dimensions have been investigated, but usually a high-aspect ratio channel of height H around 70–100 µm and width W_c of 40–50 µm are used. Indeed, high aspect ratio channels restrict the particles' equilibrium positions to two, in order to achieve an efficient cell sorting.

The chamber has the same height of the channel and width W_R of about 400–600 µm. The micrometric dimensions of the channels require techniques capable of achieving micrometric resolution and a high quality of the edge.

The most used technique to rapid prototyping lab on a chip is soft lithography [80], due to its high resolution and simplicity. Accordingly, the devices are made of polydimethylsiloxane (PDMS), an elastomeric polymer often used in LoC fabrication due to its optical transparency, thermal stability, low cost, biocompatibility, gas permeability and its ease of fabrication. The conventional procedure to fabricate micro-fluidic patterns using a standard soft-lithography technique requires the prototyping of a master template, PDMS replica molding and plasma oxidation [81,82]. The master template is made using a mask in contact photolithography.

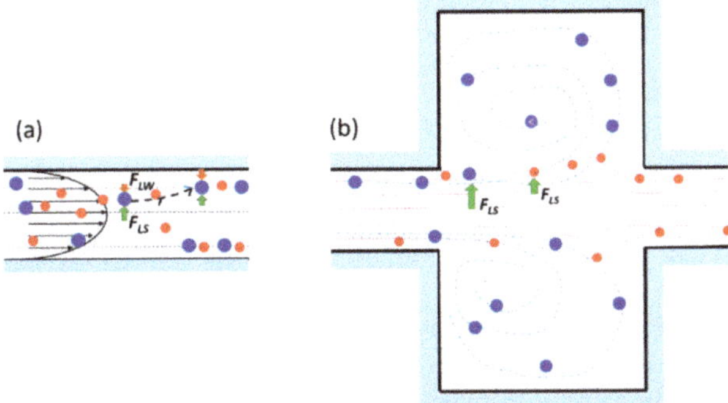

Figure 2. (**a**) The particles randomly flowing in a straight channel undergo two opposite lift forces: The wall effect lift force F_{LW} and the shear-gradient lift force F_{LS}, resulting in lateral equilibrium positions depending on the channel section. (**b**) At the entrance of the reservoir, the particles experience F_{LS}, which increases with their size: The larger blue particles are pushed toward the vortex center and trapped, meanwhile the smaller red ones flow in the main stream. Adapted from [83].

In order to avoid the expensive and time-consuming production of masks and the necessity of a clean room, femtosecond-laser micromachining technology (FLM) for the rapid prototyping of inertial sorting LoCs has been developed [83]. Thanks to the short timescale which characterizes the energy deposition during femtosecond-laser irradiation, material is removed with a negligible heat-affected zone, thus allowing micro-milled features with high precision and resolution [84] to be created. Moreover, being a non-clean room process, FLM provides an economical and flexible way to fabricate micro-fluidic patterns by simply varying the laser parameters. Moving the beam, designs are directly transferred from a computer file to the device, in a one-step fabrication protocol.

The fabricated and sealed micro-fluidic chips can be firstly tested by injecting dry polystyrene microspheres (density ~1.05 g/cm^3) into the micro-fluidic channels using a syringe pump at a specific flow rate, usually from a few μL/min to a few mL/min, in order to work in a laminar regime where the net inertial forces are strong enough to affect the particle motion [44–46]. The microspheres have diameters ranging from 1 to 15 μm, depending on the target application [81], and are suspended in an aqueous solution. The beads' density is close to the density of the liquid in which they are suspended in order to consider them neutrally buoyant. In the case of devices without lateral exit channels [62,85], the beads captured in the vortices are removed from the device (releasing step) by a final flush at a lower flow rate.

As soon as the device is validated with synthetic beads, it can be tested for the specific application with cells. For example, Hur et al. successfully validated such inertial devices separating breast cancer cells and cervical cancer cells spiked in blood [62]. Sollier et al. [85] extracted and enumerated CTCs from the blood of patients with breast and lung cancer.

2.3. Evolution of the Devices Design

The inertial lift forces combined with turbulent secondary flows generated in a topographically-patterned micro-channel was exploited in a new and simple way by Park et al. [81]. Their novel micro-fluidic method for focusing micro-particles exploited only the hydrodynamic forces exerted on suspended micro-particles as they passed through a series of contracting and expanding micro-channels. The influence of the micro-channel geometry was investigated on the particle focusing performance. In particular, two geometries were developed: (i) A straight micro-channel (Figure 3a), (ii) a multi-orifice micro-channel (Figure 3b).

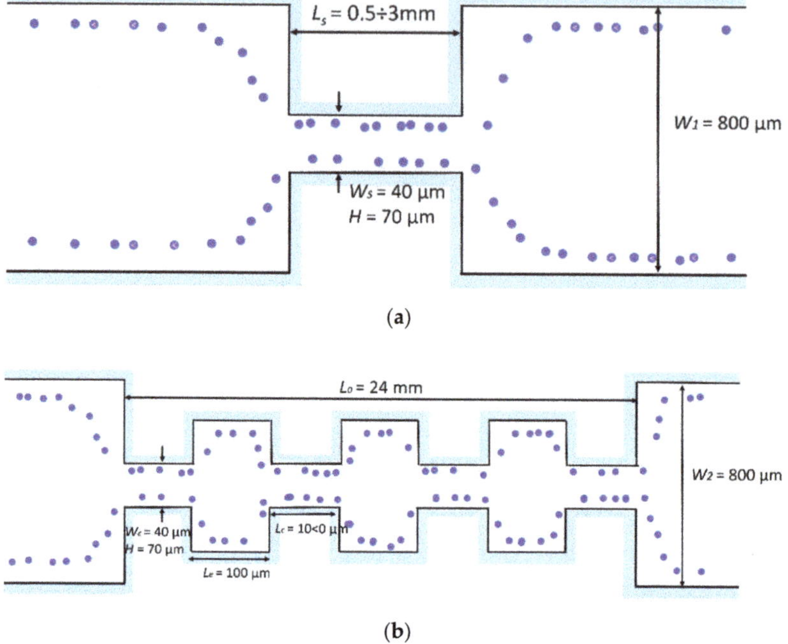

Figure 3. Design of the micro-channels used by Park et al. [81]. (**a**) Straight square channel. For the experiments, the micro-channel length was varied between 0.5 and 3 mm. (**b**) Multi orifice channel. All the structures are 70 μm high (*H*). Adapted from [81].

The authors observed how these innovative geometries generated secondary flows, i.e., vortices, with a low Reynolds number at the corners of the expansion chambers. This vortex formation depended on several factors, including flow velocity, surface roughness and the cross sectional area of the chamber. Computation fluid dynamic simulations helped to study vortex shape as a function of these parameters. The experimental tests, in agreement with the simulations, showed analogous vortex geometry as a function of the channel Reynolds number. The vortex zone increased with the Re_c, covering almost all of the cavity area when Re_c is over 78. The particle distribution as a function of the Re_p, i.e., of the flow rate, has been extensively studied. For a Re_p of 0.78–2.33, two distinguished distributions near the chamber expansions were observed, which disappeared as the Re_p increased due to the high-speed vortex flow generated in the expansion chamber. This phenomenon was ascribed to the vortex developed in the expansion at a high flow rate, generating a pressure in the main stream, which can substitute the wall-effect lift force. Since all of the mentioned tests have been carried out with a fixed particles' size, the description of the relationship between the particle Reynolds number and the focusing process was not possible from the experimental data. However, the authors suggested that continuous, non-intrusive, and high throughput techniques usable with real bio-samples can be built by exploiting the hydrodynamic focusing method in multi-orifice channels. Hur at al. presented a passive, continuous micro-fluidic device that isolated larger cancer cells from the smaller blood cells [62]. The core of the device was essentially a single straight high-aspect ratio channel (W_c = 50 μm, H = 70 μm and L = 4.5 cm), consisting of one inlet with coarse filters, two reservoirs/micro-chambers (Wr_1 = Wr_2 = 400 μm), and one outlet (Figure 4). The high-aspect ratio straight channel was implemented so that flowing particle/cells were inertially focused to two distinct lateral focusing positions closer to the channel walls. To increase the device efficiency, the tests were conducted on a device with eight parallel channels, each with ten trapping reservoirs.

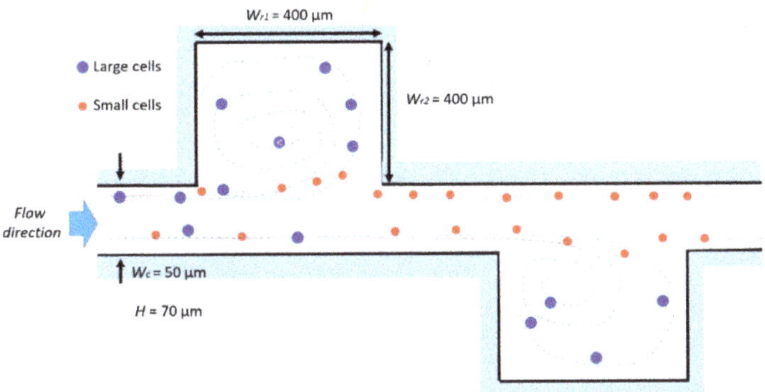

Figure 4. Design and working principle of the device of Hur et al. [56]. The larger cells remain trapped in the lateral chamber, while the smaller ones flow within the main channel, due to the dependence of the lift force on the particles' dimension. Wr_1 and Wr_2 indicate the chamber dimensions, and Wc is the chamber width. All of the structures are 70 μm high (H). The dashed paths indicate the fluid streamlines followed by the large (blue) and small (red) particles. Adapted from [62].

As found by [81], the vortex geometry highly depended upon the flow rate, i.e., the vortices gradually increased with an increasing flow rate, up to $Re_c = 215$, where the vortices occupied the entire reservoir. In addition, the authors empirically found that particles/cells with a diameter greater than a critical value a_c experience a shear-gradient lift force F_{LS}, which was strong enough to direct those particles into the reservoir, while the smaller particles remained in the main stream. The critical diameter depended on the Re_c, i.e., decreased when Re_c increased, and it is even greater for living cells with respect to synthetic beads. This outcome suggests that deformation-induced lift forces co-act in superposition with inertial lift forces, thus leading to modified lateral equilibrium positions, which result, dependent upon particle deformability. Consequently, deformability-induced target cell enrichment can be aimed based on the different lateral equilibrium positions among cell types and addressing the entrained target cells to distinct designated outlets. The maximum number of particles that each reservoir could trap and sustain was found to be 25, achieving an enrichment of cancer cells to a factor of up to 7.1, with a capturing efficiency of 23%. The authors suggested that these values can be further enhanced by simply parallelizing more channels or optimizing the reservoir geometry.

Sollier et al. [85] further improved this design using a vortex chip composed of 64 reservoirs, with eight paths in parallel and eight reservoirs per path (Figure 5).

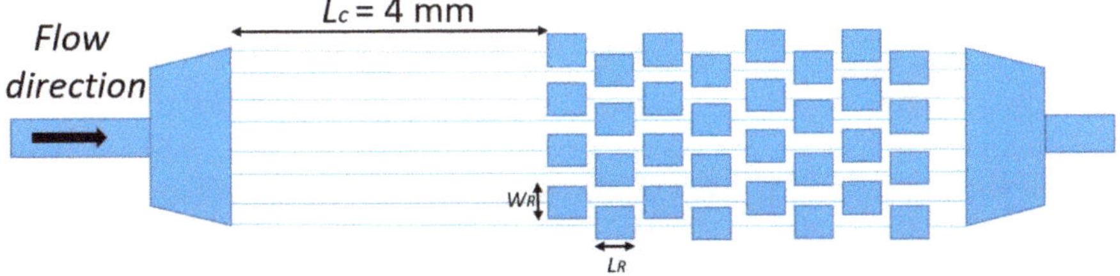

Figure 5. Vortex chip design of Sollier et al. [83]. The chip consists of eight channels (40 um wide, 80–85 um high) in parallel, with eight reservoirs (W_R = 480 um, L_R = 720 um) for each of them. The input channels of length L_C allow the focus of the randomly injected particles. Adapted from [85].

To identify the optimal design of the device for particle trapping, different geometric parameters were systematically compared with a set of experiments. In particular, the channel length before the first reservoir and the channel aspect ratio were fabricated and tested at different flow rates (from 2 to 11 mL/min), in order to study how these parameters influenced the inertial focusing and vortex flow patterns. They found that as the length before the reservoir was increased, the particles were more accurately focused, closer to the channel wall, increasing their chance to cross the separatrix and enter into the vortices, thus enhancing capture capability and reproducibility. Furthermore, higher aspect ratio AR rectangular micro-channels have been demonstrated to provide better capture efficiency, confirming previous results [41]. For high AR, particles are focused at two positions centered near the long channel walls. Under the right flow conditions, these particles can pass through the separatrix and enter in the recirculating vortex regions. In contrast, for square cross-section channels, the particles are distributed along four focusing positions (top, bottom and both side positions). Consequently, only the lateral particle streams can be trapped by the vortices, while the top and bottom ones flow undisturbed. In addition, the flow rate was also found to affect the capture efficiency, which is maximum for an optimal range. As previously found by [62], the vortices evolution depended on the flow rate, finally occupying the entire reservoir. Increasing the flow rate, the vortex center was progressively shifted towards the back of the reservoir.

Then the capture efficiency decreased, due to the larger particles, which hit the back wall leading to unpredictable trajectories, causing particles to re-enter into the main flow, rather than to move them laterally into the vortex region. Therefore, this design was particularly appropriate for capturing rare cells, and is more suitable for applications where there is the need to capture as many cells as possible. A dependence of the particle trapping efficiency on the reservoir position was also detected. This was probably due to the deformation of PDMS at high pressures, which modified the channel aspect ratio as well as the focusing position [86]. Due to the low elastic modulus of PDMS, at higher flow rates the channels cross-section undergoes deformation, losing the focusing.

Exploiting the size difference between cancer cells and blood cells, the authors demonstrated a high-throughput and label-free capture of cancer cells spiked into healthy human blood, independently of marker expression, suggesting that this approach can be used for the extraction of a large range of tumor cell types. A purity of the extracted sample of 80–100% has been demonstrated, in spite of a lower capture efficiency compared to other techniques, which can be compensated by processing a larger blood volume.

New reservoir layouts have also been investigated by Paiè et al. [87], in order to increase the trapping efficiency of the chip in Figure 5. Indeed, the billions of blood cells, in their flow through the device, can still interfere and interact with the target cells trapped in the vortices, thus affecting their trapping stability. In order to limit the particle–particle interactions, side channels were added to the standard reservoirs of [85], as shown in Figure 6.

These channels, by splitting the fluid stream lines, influence the path of the cells trapped in the vortex and force some of them to flow in the lateral fluidic circuit, where they remain trapped over many cycles of the main vortex. Through simulations and experimental validations, the authors have been able to properly choose the lateral fluidic circuit dimensions (namely L and W in Figure 6), in order to increase the maximum number of particles that can be stably trapped in each reservoir, and consequently, the total efficiency of the chip.

A trapping efficiency 19% higher than the value obtained by the standard Vortex chip [85] has been demonstrated using fluorescent beads. Also in this case, the trapping efficiency was found to be highly dependent on the reservoir position in the chip. This phenomenon was ascribed to the PDMS deformation effects, suggesting that a higher efficiency could potentially be reached with a rigid substrate which prevents from the flow-induced deformation at high flow rates.

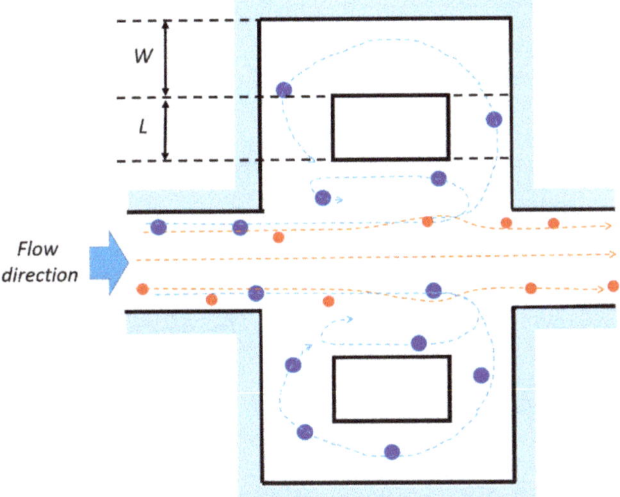

Figure 6. Schematic layout of the reservoir geometry tested by Paiè et al. [80]. Lateral channels of length *L* were added to the standard reservoirs [85] and joined together by a connection channel *W* wide. The dashed paths indicate the fluid streamlines followed by the big (blue) and small (red) particles. Adapted from [87].

A micro-fluidic chip based on inertial migration that achieves an efficient multimodal separation of mixtures of three types of micro-particles with small differences in size was demonstrated by Wang et el. [88] on PDMS. Their approach improved the designs described above, based on a straight micro-channel to pre-focus micro-particles, followed downstream by a channel expansion (micro-chamber), adding lateral siphoning outlets (Figure 7), thus offering a continuous, washing step-free operation device without any vortex saturation effect. Their numerical and experimental results showed that the position of the boundary streamline (i.e., the separatrix) could be precisely varied by tuning the input flow velocity U_m and engineering the channel fluidic resistance ratio r/R, where r is the resistance of a side outlet channels, and R is the resistance of the main outlet. Figure 8 shows the position of the separatrix (white dashed line) at increasing values of the r/R ratio. At a low r/R, the initial equilibrium position of all particles is within the flow exiting from the lateral outlets, leading to a non-selective extraction of all particles. As the r/R ratio increases, the boundary streamline moves towards the channel walls, leading to a possible size-based separation. In particular, the authors found that the cutoff diameter a_c increases with r/R, following the relationship $a_c \propto \left(\frac{r}{R}\right)^{1/2}$. Furthermore, the experimental results with synthetic micro-particles suggest that a_c scales as the first order of the input flow velocity $a_c \propto U_m$. This offers the possibility of tuning the cutoff diameter through the input flow velocity and the resistance modification of the outlet channels. These results were also supported by the work of Volpe et al. [89], where 3D computational fluid dynamics (CFD) simulations based on the lattice Boltzmann method were exploited to evaluate the performance of a micro-fluidic device, specifically designed to trap and extract particles by inertial focusing and microscale vortices, and whose scheme is similar to that in Figure 7. The simulations allowed them to characterize the flow properties of the micro-fluidic device as a function of the Reynolds number (Re), the input flow velocity, the chamber dimensions and the outlet channels resistance ratio r/R. In particular, they found a shift of the separatrix towards to the channels walls when r/R increases, with a consequent increase of the cutoff diameter, as also confirmed by the experimental results.

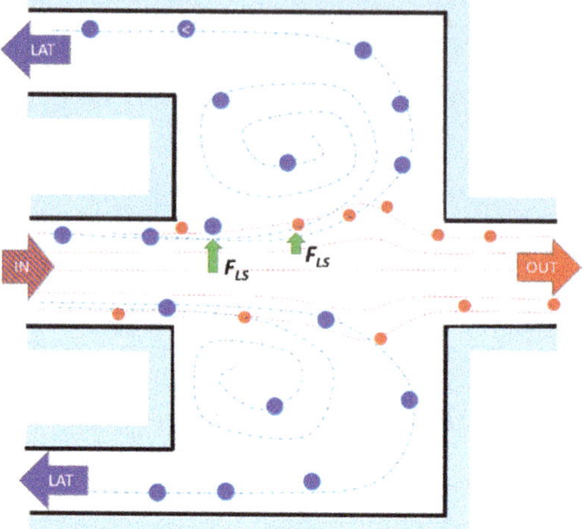

Figure 7. Scheme of the size-selective separation in micro-chambers with three outlets proposed by Wang and Papautsky [88]. At the chamber entrance, the particles are arranged in a two-focus position near the channel walls. The large particles (blue) undergo a larger shear-gradient induced lift force, being pushed in the chamber and exiting through the two later outlets (LAT), while the smaller particles (red) remain in the main channel exiting through the main outlet (OUT). The dashed paths indicate the fluid streamlines followed by the big (blue) and small (red) particles. Adapted from [88].

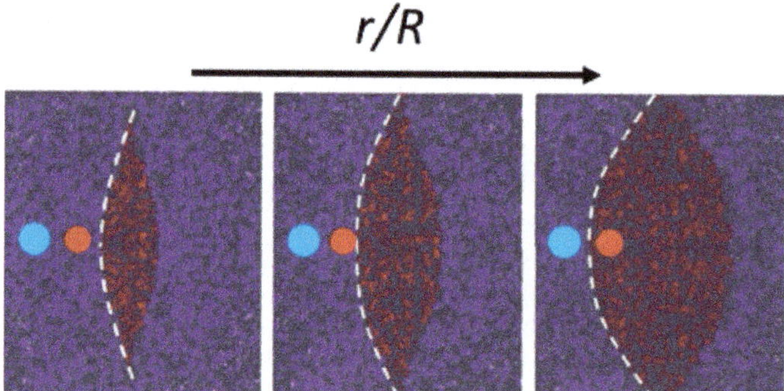

Figure 8. Cross-sectional view of the boundary surface (white dashed line) at the entrance of the micro-chamber increasing the r/R ratio. The red area represents the main flow exiting through the main outlet, while the blue area indicates the flow exiting through the lateral outlets. Increasing the r/R ratio, the boundary shifts toward the channel walls, allowing the sorting of particles of different dimensions. Adapted from [89].

To achieve multimodal separation, in [88] a combination of two (Figure 9) chambers in series has been developed. The resistance ratio was modulated in order to finely tune the cut-off diameter in each chamber. The capability of separating a heterogeneous sample into three distinct distributions was demonstrated by injecting a mixture of micro-particles with diameters continuously distributed in the range from 10 μm to 27 μm.

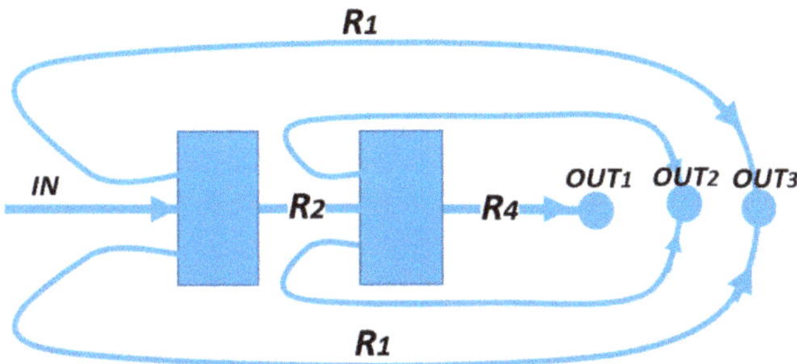

Figure 9. Inertial micro-fluidic multimodal separation designed by Wang et al. [81]. The micro-fluidic resistance network comprises two sorting chambers and three exits for particles, relying on their dimensions. Adapted from [88].

A similar continuous vortex-aided sorting device, integrating two units by connecting the outlets of the first unit into the inlet of the second unit, has also been tested (Figure 10) [90].

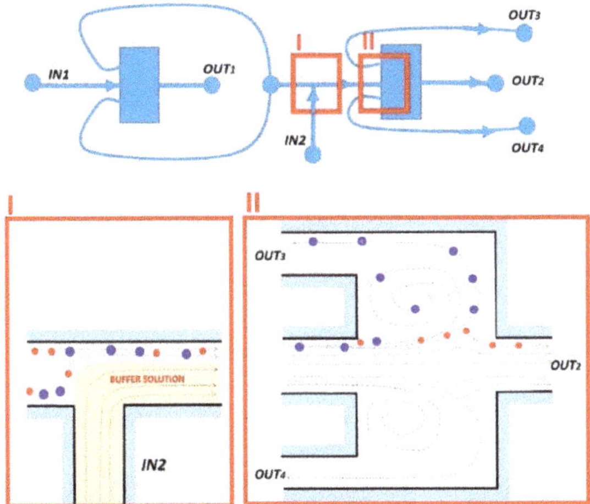

Figure 10. Schematic picture of two vortex sorting units integrated into a device for double sorting and purification. In the insets, a schematic detail of the device operation (I) at the second inlet stage IN2, and (II) in the second chamber. Adapted from [90].

The device operation can be summarized as follows: The first sorting unit splits the larger cells from the smaller background cells. To get further purification, the sorting product flows downstream from the side outlets of the first unit into the second unit, where a buffer solution is introduced from inlet 2 to accelerate the flow. By properly designing the fluidic resistance of the network, the condition for sorting the target cells larger than a cutoff diameter are created. After a double-sorting process, the target cells are thus extracted from outlet 3 (O3). The remaining small background cells flow through outlet 2 (O2). The feasibility of rare cells sorting, as well as the efficient removal of a large number of background cells, has been demonstrated for the continuous sorting of spiked human cancer stem-like cells from human blood with >90% efficiency and enhanced purity, as well as for the removal of red blood cells with ~99.97% efficiency [90]. Also in this work, the authors reported a channel deformation during the device operation due to the poor rigidity of PDMS. To this phenomenon could be ascribed the changes in channel resistance strongly affecting the cutoff diameter, which, nevertheless, could be compensated by slightly tuning the flow rate. The main advantage of this vortex-based device compared to the conventional ones was the continuous nature of the device operation, which led to high throughput and high volume applications. Conversely, owing to the presence of the lateral outlet channels, this approach prevents the possibility of packing many reservoirs in a compact footprint that would allow us to parallelize the sorting process and decrease the sample processing time.

The work of Volpe et al. [83] aimed at improving the throughput of previous implementations, while keeping the sorting efficiency high. In particular, the micro-fluidic devices presented in [83] consisted of micro-channels with expansion chambers in series. Partially inspired from the work of Wang and Papautsky [88], the chambers were supplied with siphoning outlets exiting from the backside of the chip. Relying on the simulation results presented in [89], the dimensions and number of chambers were varied, and different collecting geometries investigated at different flow rates. The layout with maximized trapping efficiency and enhanced throughput (Figure 11) comprised two-branch with two chambers for each of them, one inlet IN, one main outlet OUT and eight lateral outlets LAT.

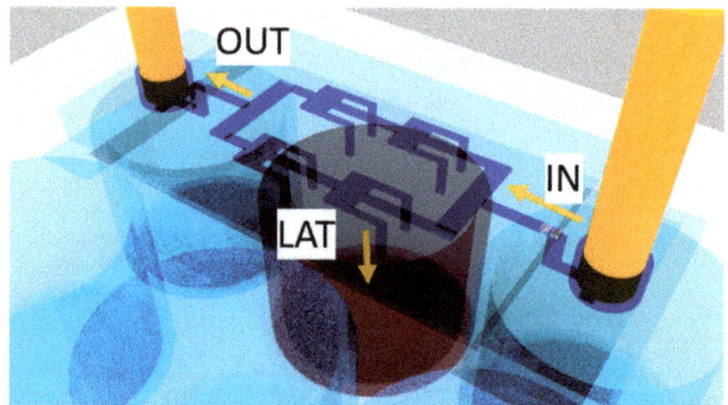

Figure 11. 3D sorting Lab on a Chip developed by Volpe et al. [75] on PMMA through fs-laser technology. The particles exiting through the lateral outlets (LAT) are collected in a multi-well plate, while the others flow through the main exit OUT. The ratio between the resistance of the LAT and the OUT is 20. The channels are 60 μm high and 50 μm wide. The chamber's dimensions are 480 μm × 540 μm. Adapted from [83].

The channels were 60 μm high and 50 μm wide, meanwhile the chambers had a width of 540 μm and a length of 480 μm. The drops coming out from the lateral outlets were directly collected in the multi-well plate with no tubes, thus increasing the simplicity of the experiments. The device characterizations were performed using separately 6- and 15-μm fluorescent beads, whose dimensions mimic the red blood cells and the CTC ones, respectively. About the 84% of the injected bigger particles were collected by lateral outlets, while just 10% of the smaller ones exited through the main one, thus preserving the purity of the sample.

This quasi-3D design demonstrates the possibility of multiplying the throughput of the device, by simply adding parallel micro-chambers. Furthermore, this LoC is fully inertial, reducing the complexity of other previously-reported, high-throughput devices [91,92]. It is worth mentioning that in the latter work, the fs-laser micro-machining technology was employed to fabricate the device. This technology offers the possibility to choose other polymers than PDMS as substrate material. In this work, PMMA was used, which is a biocompatible, transparent and much stiffer polymer than PDMS, thus withstanding higher flowing pressures without suffering from deformations.

3. Conclusions and Perspectives

In this work, a review on the evolution of the inertial particle sorting combined with microscale laminar vortices technology has been presented.

Recent works on micro-fluidic sorting techniques based on the combination of inertial focusing in a straight micro-channel and vortex technology have been critically presented, evidencing the limits and opportunities of each design. As emerged by this excursion, the main drawbacks of the first prototypes, i.e., the limited number of particles trapped per vortex and the deformation of the chamber at high pressure, have been gradually overcome, adding syphoning outlet and changing the fabrication technology from soft lithography to ultrashort laser micro-machining, and consequently, the material from PDMS to much stiffer polymers as PMMA. Furthermore, the possibility for increasing the throughput, without affecting the purity of the sample, has been demonstrated through a parallelization of the extraction process.

Inertial sorting combined with vortex trapping presents several pros in terms of performance, fabrication and operation with respect to other sorting techniques. In fact, it requires neither further mechanical nor electric parts, nor obstacle structures for filtering particles. Moreover, it is label-free, and employs only a simple single-layered micro-fluidic device, offering the possibility to parallelize the sorting process for reaching high purity and throughput. Furthermore, this technique exploits just passive fluid physics, without any complex high-resolution features, instrumentation or surface chemistry.

Thus, the per-sample cost for one chip provided of this pressure control system to run samples is expected to be much lower than ~$500 per test with respect to other traditional systems. Furthermore, this technology can be applied to various samples and cancer types, without adversely damaging the cells and ensuring high-purity. All of these advantages make the combination of inertial focusing in a straight micro-channel and vortex technology a successful technology in a clinical setting. The next step for realizing this goal will be the automation of the device operation, which is especially important for clinical adoption even by the non-specialized user.

Author Contributions: A.V. and C.G. performed bibliographic research and wrote the paper. A.A. coordinated and revised the study.

Funding: This research received no external funding.

Acknowledgments: The authors gratefully acknowledge Giuseppe Pascazio, Roberto Osellame and Petra Paiè for the useful discussions on this topic.

References

1. Lee, L.M.; Rosano, J.M.; Wang, Y.; Klarmann, G.J.; Garson, C.J.; Prabhakarpandian, B.; Pant, K.; Alvarez, L.M.; Lai, E. Label-free mesenchymal stem cell enrichment from bone marrow samples by inertial micro-fluidics. *Anal. Methods* **2018**, *10*, 713–721. [CrossRef]

2. Wylot, B.; Konarzewska, K.; Bugajski, L.; Piwocka, K.; Zawadzka, M. Isolation of vascular endothelial cells from intact and injured murine brain cortex-technical issues and pitfalls in FACS analysis of the nervous tissue. *Cytom. Part A* **2015**, *87*, 908–920. [CrossRef] [PubMed]

3. Chen, W.C.W.; Péault, B.; Huard, J. Regenerative translation of human blood-vessel-derived MSC precursors. *Stem Cells Int.* **2015**, *2015*, 1–11. [CrossRef] [PubMed]

4. Xue, K.; Zhang, X.; Qi, L.; Zhou, J.; Liu, K. Isolation, identification, and comparison of cartilage stem progenitor/cells from auricular cartilage and perichondrium. *Am. J. Transl. Res.* **2016**, *8*, 732–741. [PubMed]

5. Xu, H.; Dong, B.; Xu, S.; Xu, S.; Sun, X.; Sun, J.; Yang, Y.; Xu, L.; Bai, X.; Zhang, S.; et al. High purity micro-fluidic sorting and in situ inactivation of circulating tumor cells based on multifunctional magnetic composites. *Biomaterials* **2017**, *138*, 69–79. [CrossRef]

6. Jeanbart, L.; Swartz, M.A. Engineering opportunities in cancer immunotherapy. *Proc. Natl. Acad. Sci. USA* **2015**, *112*, 14467–14472. [CrossRef] [PubMed]

7. Haverty, P.M.; Lin, E.; Tan, J.; Yu, Y.; Lam, B.; Lianoglou, S.; Neve, R.M.; Martin, S.; Settleman, J.; Yauch, R.L.; et al. Reproducible pharmacogenomic profiling of cancer cell line panels. *Nature* **2016**, *533*, 333–337. [CrossRef]

8. Di Meo, A.; Bartlett, J.; Cheng, Y.; Pasic, M.D.; Yousef, G.M. Liquid biopsy: A step forward towards precision medicine in urologic malignancies. *Mol. Cancer* **2017**, *16*, 158. [CrossRef]
9. Varillas, J.I.; Zhang, J.; Chen, K.; Barnes, I.I.; Liu, C.; George, T.J.; Fan, Z.H. Micro-fluidic isolation of circulating tumor cells and cancer stem-like cells from patients with pancreatic ductal adenocarcinoma. *Theranostics* **2019**, *9*, 1417–1425. [CrossRef]
10. Kim, T.H.; Yoon, H.J.; Fouladdel, S.; Wang, Y.; Kozminsky, M.; Burness, M.L.; Paoletti, C.; Zhao, L.; Azizi, E.; Wicha, M.S. Characterizing circulating tumor cells isolated from metastatic breast cancer patients using graphene oxide based micro-fluidic assay. *Adv. Biosyst.* **2019**, *3*, 1–10. [CrossRef]
11. Kim, S.H.; Ito, H.; Kozuka, M.; Takagi, H.; Hirai, M.; Fujii, T. Cancer marker-free enrichment and direct mutation detection in rare cancer cells by combining multi-property isolation and micro-fluidic concentration. *Lab Chip* **2019**, *19*, 757–766. [CrossRef] [PubMed]
12. Lee, W.; Tseng, P.; di Carlo, D. Micro-fluidic Cell Sorting and Separation Technology. In *Microtechnology for Cell Manipulation and Sorting*; Lee, W., Tseng, P., di Carlo, D., Eds.; Springer International Publishing: Cham, Switzerland, 2017; pp. 1–14.
13. Chu, W.; Tan, Y.; Wang, P.; Xu, J.; Li, W.; Qi, J.; Cheng, Y. Centimeter-height 3d printing with femtosecond laser two-photon polymerization. *Adv. Mater. Technol.* **2018**, *3*, 1–6. [CrossRef]
14. Li, M.; Muñoz, H.E.; Goda, K.; di Carlo, D. Shape-based separation of microalga Euglena gracilis using inertial micro-fluidics. *Sci. Rep.* **2017**, *7*, 1–8.
15. Iv, C.W.S.; Reyes, C.D.; López, G.P. Micro-fluidic cell sorting: A review of the advances in the separation of cells from debulking to rare cell isolation. *Lab Chip* **2015**, *15*, 1230–1249.
16. Olm, F.; Urbansky, A.; Dykes, J.H.; Laurell, T.; Scheding, S. Label-free neuroblastoma cell separation from hematopoietic progenitor cell products using acoustophoresis—Towards cell processing of complex biological samples. *Sci. Rep.* **2019**, *9*, 8777. [CrossRef] [PubMed]
17. Shi, J.; Huang, H.; Stratton, Z.; Huang, Y.; Huang, T.J. Continuous particle separation in a micro-fluidic channel via standing surface acoustic waves (SSAW). *Lab Chip* **2009**, *9*, 3354–3359. [CrossRef] [PubMed]
18. Chen, Y.; Wu, M.; Ren, L.; Liu, J.; Whitley, P.H.; Wang, L.; Huang, T.J. High-throughput acoustic separation of platelets from whole blood. *Lab Chip* **2016**, *16*, 3466–3472. [CrossRef] [PubMed]
19. Nam, J.; Lim, H.; Kim, C.; Kang, J.Y.; Shin, S. Density-dependent separation of encapsulated cells in a micro-fluidic channel by using a standing surface acoustic wave. *Biomicro Fluid.* **2012**, *6*, 1–10.
20. Gupta, S.; Feke, D.L.; Manas-Zloczower, I. Fractionation of mixed particulate solids according to compressibility using ultrasonic standing wave fields. *Chem. Eng. Sci.* **1995**, *50*, 3275–3284. [CrossRef]
21. Li, P.; Mao, Z.; Peng, Z.; Zhou, L.; Chen, Y.; Huang, P.-H.; Truica, C.I.; Drabick, J.J.; El-Deiry, W.S.; Dao, M.; et al. Acoustic separation of circulating tumor cells. *Proc. Natl. Acad. Sci. USA* **2015**, *112*, 4970–4975. [CrossRef]
22. Rosano, J.M.; Garson, C.J.; Klarmann, G.J.; Perantoni, A.; Alvarez, L.M.; Lai, E.; Song, H.; Wang, Y.; Prabhakarpandian, B.; Pant, K. Continuous-flow sorting of stem cells and differentiation products based on dielectrophoresis. *Lab Chip* **2015**, *15*, 1320–1328.
23. Lee, J.J.; Jeong, K.J.; Hashimoto, M.; Kwon, A.H.; Rwei, A.; Shankarappa, S.A.; Tsui, J.H.; Kohane, D.S. Synthetic ligand-coated magnetic nanoparticles for micro-fluidic bacterial separation from blood. *Nano Lett.* **2014**, *14*, 1–5. [CrossRef] [PubMed]
24. Nam, J.; Huang, H.; Lim, H.; Lim, C.; Shin, S. Magnetic separation of malaria-infected red blood cells in various developmental stages. *Anal. Chem.* **2013**, *85*, 7316–7323. [CrossRef] [PubMed]
25. Han, K.H.; Frazier, A.B. Continuous magnetophoretic separation of blood cells in microdevlce format. *J. Appl. Phys.* **2004**, *96*, 5797–5802. [CrossRef]
26. Jung, J.; Han, K.-H. Lateral-driven continuous magnetophoretic separation of blood cells. *Appl. Phys. Lett.* **2008**, *93*, 223902. [CrossRef]
27. Landenberger, B.; Höfemann, H.; Wadle, S.; Rohrbach, A. Micro-fluidic sorting of arbitrary cells with dynamic optical tweezers. *Lab Chip* **2012**, *12*, 3177–3183. [CrossRef]
28. Yang, A.H.J.; Moore, S.D.; Schmidt, B.S.; Klug, M.; Lipson, M.; Erickson, D. Optical manipulation of nanoparticles and biomolecules in sub-wavelength slot waveguides. *Nature* **2009**, *457*, 71–75. [CrossRef]
29. Ashkin, A.; Dziedzic, J.M.; Chu, S. Observation of a single-beam gradient-force optical trap for dielectric particles in air. *Opt. Lett.* **1986**, *22*, 288–290. [CrossRef]
30. Kuga, T.; Torii, Y.; Shiokawa, N.; Hirano, T.; Shimizu, Y.; Sasada, H. Novel optical trap of atoms with a doughnut beam. *Phys. Rev. Lett.* **1997**, *78*, 4713–4716. [CrossRef]

31. Li, X.; Chen, W.; Liu, G.; Lu, W.; Fu, J. Continuous-flow micro-fluidic blood cell sorting for unprocessed whole blood using surface-micromachined microfiltration membranes. *Lab Chip* **2014**, *14*, 2565–2575. [CrossRef]

32. Kang, Y.-T.; Doh, I.; Byun, J.; Chang, H.J.; Cho, Y.-H. Label-free rapid viable enrichment of circulating tumor cell by photosensitive polymer-based microfilter device. *Theranostics* **2017**, *7*, 3179–3191. [CrossRef] [PubMed]

33. Doh, I.; Yoo, H.-I.; Cho, Y.-H.; Lee, J.; Kim, H.K.; Kim, J. Viable capture and release of cancer cells in human whole blood. *Appl. Phys. Lett.* **2012**, *101*, 43701. [CrossRef]

34. Alvankarian, J.; Bahadorimehr, A.; Majlis, B.Y. A pillar-based microfilter for isolation of white blood cells on elastomeric substrate. *Biomicro Fluid.* **2013**, *7*, 14102. [CrossRef] [PubMed]

35. Wilding, P.; Kricka, L.J.; Cheng, J.; Hvichia, G.; Shoffner, M.A.; Fortina, P. Integrated cell isolation and polymerase chain reaction analysis using silicon microfilter chambers. *Anal. Biochem.* **1998**, *257*, 95–100. [CrossRef]

36. McGrath, J.; Jimenez, M.; Bridle, H. Deterministic lateral displacement for particle separation: A review. *Lab Chip* **2014**, *14*, 4139–4158. [CrossRef] [PubMed]

37. Au, S.H.; Edd, J.; Stoddard, A.E.; Wong, K.H.; Fachin, F.; Maheswaran, S.; Haber, D.A.; Stott, S.L.; Kapur, R.; Toner, M. Micro-fluidic isolation of circulating tumor cell clusters by size and asymmetry. *Sci. Rep.* **2017**, *7*, 2433. [CrossRef] [PubMed]

38. Fan, L.L.; Yan, Q.; Guo, J.; Zhao, H.; Zhao, L.; Zhe, J. Inertial particle focusing in micro-channels with gradually changing geometrical structures. *J. Micromech. Microeng.* **2017**, *27*, 2017. [CrossRef]

39. Amini, H.; Lee, W.; di Carlo, D. Inertial micro-fluidic physics. *Lab Chip* **2014**, *14*, 2739–2761. [CrossRef] [PubMed]

40. Zhang, J.; Yan, S.; Yuan, D.; Alici, G.; Nguyen, N.T.; Warkiani, M.E.; Li, W. Fundamentals and applications of inertial micro-fluidics: A review. *Lab Chip* **2016**, *16*, 10–34. [CrossRef]

41. Di Carlo, D. Inertial micro-fluidics. *Lab Chip* **2009**, *9*, 3038–3046. [CrossRef]

42. Hur, S.C.; Henderson-Maclennan, N.K.; McCabe, E.R.B.; di Carlo, D. Deformability-based cell classification and enrichment using inertial micro-fluidics. *Lab Chip* **2011**, *11*, 912–920. [CrossRef] [PubMed]

43. Choi, Y.S.; Seo, K.W.; Lee, S.J. Lateral and cross-lateral focusing of spherical particles in a square micro-channel. *Lab Chip* **2011**, *11*, 460–465. [CrossRef] [PubMed]

44. Di Carlo, D.; Irimia, D.; Tompkins, R.G.; Toner, M. Continuous inertial focusing, ordering, and separation of particles in micro-channels. *Proc. Natl. Acad. Sci. USA* **2007**, *104*, 18892–18897. [CrossRef] [PubMed]

45. Masaeli, M.; Sollier, E.; Amini, H.; Mao, W.; Camacho, K.; Doshi, N.; Mitragotri, S.; Alexeev, A.; Di Carlo, D. Continuous inertial focusing and separation of particles by shape. *Phys. Rev. X* **2012**, *2*, 031017. [CrossRef]

46. Martel, J.M.; Toner, M. Inertial focusing in micro-fluidics. *Annu. Rev. Biomed. Eng.* **2014**, *16*, 371–396. [CrossRef] [PubMed]

47. Hur, S.C.; Brinckerhoff, T.Z.; Walthers, C.M.; Dunn, J.C.; di Carlo, D. Label-free enrichment of adrenal cortical progenitor cells using inertial micro-fluidics. *PLoS ONE* **2012**, *7*, e46550. [CrossRef] [PubMed]

48. Hur, S.C.; Tse, H.T.K.; Di Carlo, D. Sheathless inertial cell ordering for extreme throughput flow cytometry. *Lab Chip* **2010**, *10*, 274–280. [CrossRef]

49. Mach, A.J.; di Carlo, D. Continuous scalable blood filtration device using inertial micro-fluidics. *Biotechnol. Bioeng.* **2010**, *107*, 302–311. [CrossRef]

50. Zhou, J.; Giridhar, P.V.; Kasper, S.; Papautsky, I. Modulation of aspect ratio for complete separation in an inertial micro-fluidic channel. *Lab Chip* **2013**, *13*, 1919–1929. [CrossRef]

51. Zhou, J.; Papautsky, I. Fundamentals of inertial focusing in micro-channels. *Lab Chip* **2013**, *13*, 1121–1132. [CrossRef]

52. Kuntaegowdanahalli, S.S.; Bhagat, A.A.S.; Kumar, G.; Papautsky, I. Inertial micro-fluidics for continuous particle separation in spiral micro-channels. *Lab Chip* **2009**, *20*, 2973–2978. [CrossRef] [PubMed]

53. Hou, H.W.; Warkiani, M.E.; Khoo, B.L.; Li, Z.R.; Soo, R.A.; Tan, D.S.-W.; Lim, W.-T.; Han, J.; Bhagat, A.A.S.; Lim, C.T. Isolation and retrieval of circulating tumor cells using centrifugal forces. *Sci. Rep.* **2013**, *3*, 1259. [CrossRef]

54. Nivedita, N.; Papautsky, I. Continuous separation of blood cells in spiral micro-fluidic devices. *Biomicrofluidics* **2013**, *7*, 54101. [CrossRef] [PubMed]

55. Warkiani, M.E.; Khoo, B.L.; Wu, L.; Tay, A.K.P.; Bhagat, A.A.S.; Han, J.; Lim, C.T. Ultra-fast, label-free isolation of circulating tumor cells from blood using spiral micro-fluidics. *Nat. Protoc.* **2016**, *11*, 134–148. [CrossRef] [PubMed]

56. Guan, G.; Wu, L.; Bhagat, A.A.; Li, Z.; Chen, P.C.; Chao, S.; Ong, C.J.; Han, J. Spiral microchannel with rectangular and trapezoidal cross-sections for size based particle separation. *Sci. Rep.* **2013**, *3*, 1475. [CrossRef] [PubMed]

57. Warkiani, M.E.; Guan, G.; Luan, K.B.; Lee, W.C.; Bhagat, A.A.S.; Chaudhuri, P.K.; Tan, D.S.W.; Lim, W.T.; Lee, S.C.; Chen, P.C.; et al. Slanted spiral micro-fluidics for the ultra-fast, label-free isolation of circulating tumor cells. *Lab Chip* **2014**, *14*, 128–137. [CrossRef] [PubMed]

58. Liu, N.; Petchakup, C.; Tay, H.M.; Li, K.H.H.; Hou, H.W. Spiral Inertial Micro-fluidics for Cell Separation and Biomedical Applications. In *Applications of Micro-Fluidic Systems in Biology and Medicine*; Tokeshi, M., Ed.; Springer: Berlin/Heidelberg, Germany, 2019; pp. 99–150.

59. Di Carlo, D.; Edd, J.F.; Humphry, K.J.; Stone, H.A.; Toner, M. particle segregation and dynamics in confined flows. *Phys. Rev. Lett.* **2009**, *102*, 094503. [CrossRef] [PubMed]

60. Moffatt, H.K. Viscous and resistive eddies near a sharp corner. *J. Fluid Mech.* **1964**, *18*, 1–18. [CrossRef]

61. Cherdron, W.; Durst, F.; Whitelaw, J.H. Asymmetric flows and instabilities in symmetric ducts with sudden expansions. *J. Fluid Mech.* **1978**, *84*, 13. [CrossRef]

62. Hur, S.C.; Mach, A.J.; Di Carlo, D. High-throughput size-based rare cell enrichment using microscale vortices. *Biomicrofluidics* **2011**, *5*, 022206. [CrossRef]

63. Sajeesh, P.; Sen, A.K. Particle separation and sorting in micro-fluidic devices: A review. *Microfluid. Nanofluidics* **2014**, *17*, 1–52. [CrossRef]

64. Dalili, A.; Samiei, E.; Hoorfar, M. A review of sorting, separation and isolation of cells and microbeads for biomedical applications: Micro-fluidic approaches. *Analyst* **2019**, *144*, 87–113. [CrossRef] [PubMed]

65. Segré, G.; Silberberg, A.; Segr, A.S.G. Radial particle displacements in poiseuille flow of suspensions. *Nature* **1961**, *189*, 209–210. [CrossRef]

66. Segré, G.; Silberberg, A. Behaviour of macroscopic rigid spheres in Poiseuille flow. *J. Fluid Mech.* **1962**, *14*, 115. [CrossRef]

67. Chun, B.; Ladd, A.J.C. Inertial migration of neutrally buoyant particles in a square duct: An investigation of multiple equilibrium positions. *Phys. Fluids* **2006**, *18*, 31704. [CrossRef]

68. Liu, C.; Hu, G.; Jiang, X.; Sun, J. Inertial focusing of spherical particles in rectangular micro-channels over a wide range of Reynolds numbers. *Lab Chip* **2015**, *15*, 1168–1177. [CrossRef] [PubMed]

69. Kim, J.; Lee, J.; Wu, C.; Nam, S.; di Carlo, D.; Lee, W. Inertial focusing in non-rectangular cross-section micro-channels and manipulation of accessible focusing positions. *Lab Chip* **2016**, *16*, 992–1001. [CrossRef]

70. Tang, W.; Fan, N.; Yang, J.; Li, Z.; Zhu, L.; Jiang, D.; Shi, J.; Xiang, N. Elasto-inertial particle focusing in 3D-printed micro-channels with unconventional cross sections. *Microfluid. Nanofluidics* **2019**, *23*, 42. [CrossRef]

71. Mashhadian, A.; Shamloo, A. Inertial micro-fluidics: A method for fast prediction of focusing pattern of particles in the cross section of the channel. *Anal. Chim. Acta* **2019**, *1083*, 137–149. [CrossRef]

72. Chupin, L.; Cindea, N. *Numerical Method for Inertial Migration of Particles in 3D Channels*; To Cite This Version: Hal-01971999; CCSD: Las Vegas, VN, USA, 2019.

73. Bayat, P.; Rezai, P. Microfluidic curved-channel centrifuge for solution exchange of target? Microparticles and their simultaneous separation from bacteria. *Soft Matte.* **2018**, *14*, 5356–5363. [CrossRef]

74. Wang, L.; Dandy, D.S. High-throughput inertial focusing of micrometer- and sub-micrometer-sized particles separation. *Adv. Sci.* **2017**, *4*, 1700153. [CrossRef] [PubMed]

75. Bhagat, A.A.S.; Kuntaegowdanahalli, S.S.; Papautsky, I. Inertial micro-fluidics for continuous particle filtration and extraction. *Microfluid. Nanofluidics* **2009**, *7*, 217–226. [CrossRef]

76. Saffman, P.G. The lift force on a small shpere in a slow shear flow. *J. Fluid Mech.* **1965**, *22*, 385–400. [CrossRef]

77. Asmolov, E.S. The inertial lift on a spherical particle in a plane poiseuille flow at large channel Reynolds number. *J. Fluid Mech.* **1999**, *381*, 63–87. [CrossRef]

78. Matas, J.P.; Morris, J.F.; Guazzelli, É. Lateral forces on a sphere solid/liquid dispersions in drilling and production fluides chargés en forage et production pétrolière. *Oil Gas Sci. Technol. IFP* **2004**, *59*, 59–70. [CrossRef]

79. Moffatt, H.K. Viscous eddies near a sharp corner. *Fluid Dyn. Trans.* **2013**, 217–224. [CrossRef]

80. Kim, P.; Kwon, K.W.K.; Park, M.M.C.; Lee, S.H.S.; Kim, S.S.M.; Suh, K.Y.K. Soft lithography for micro-fluidics: A review. *Biochip J.* **2008**, *2*, 1–11.

81. Park, S.; Song, S.H.; Jung, H.I. Continuous focusing of micro-particles using inertial lift force and vorticity via multi-orifice micro-fluidic channels. *Lab Chip* **2009**, *9*, 939–948. [CrossRef] [PubMed]

82. Duffy, D.C.; Mcdonald, J.C.; Schueller, O.J.A.; Whitesides, G.M. Rapid prototyping of micro-fluidic systems in poly (dimethylsiloxane). *Anal. Chem.* **1998**, *70*, 4974–4984. [CrossRef] [PubMed]

83. Volpe, A.; Paiè, P.; Ancona, A.; Osellame, R. Polymeric fully inertial lab-on-a-chip with enhanced-throughput sorting capabilities. *Microfluid. Nanofluidics* **2019**, *23*, 37. [CrossRef]

84. Sima, F.; Sugioka, K.; Vázquez, R.M.; Osellame, R.; Kelemen, L. Review article Three-dimensional femtosecond laser processing for lab-on-a-chip applications. *Nanophotonics* **2018**, *7*, 1–22. [CrossRef]

85. Sollier, E.; Go, D.E.; Che, J.; Gossett, D.R.; O'Byrne, S.; Weaver, W.M.; Kummer, N.; Rettig, M.; Goldman, J.; Nickols, N.; et al. Size-selective collection of circulating tumor cells using Vortex technology. *Lab Chip* **2014**, *14*, 63–77. [CrossRef] [PubMed]

86. Sollier, E.; Murray, C.; Maoddi, P.; di Carlo, D. Rapid prototyping polymers for micro-fluidic devices and high pressure injections. *Lab Chip* **2011**, *11*, 3752–3765. [CrossRef] [PubMed]

87. Paiè, P.; Che, J.; Di Carlo, D. Effect of reservoir geometry on vortex trapping of cancer cells. *Microfluid. Nanofluidics* **2017**, *21*, 1–11. [CrossRef]

88. Wang, X.; Papautsky, I. Size-based micro-fluidic multimodal micro-particle sorter. *Lab Chip* **2015**, *15*, 1350–1359. [CrossRef]

89. Volpe, A.; Paiè, P.; Ancona, A.; Osellame, R.; Lugarà, P.M.; Pascazio, G. A computational approach to the characterization of a micro-fluidic device for continuous size-based inertial sorting. *J. Phys. D Appl. Phys.* **2017**, *50*, 2017. [CrossRef]

90. Wang, X.; Yang, X.; Papautsky, I. An integrated inertial micro-fluidic vortex sorter for tunable sorting and purification of cells. *Technology* **2016**, *4*, 88–97. [CrossRef]

91. Bhagat, A.A.S.; Kuntaegowdanahalli, S.S.; Kaval, N.; Seliskar, C.J.; Papautsky, I. Inertial micro-fluidics for sheath-less high-throughput flow cytometry. *Biomed. Microdevices* **2010**, *12*, 187–195. [CrossRef]

92. Nivedita, N.; Garg, N.; Lee, A.P.; Papautsky, I. A high throughput micro-fluidic platform for size-selective enrichment of cell populations in tissue and blood samples. *Analyst* **2017**, *142*, 2558–2569. [CrossRef]

Review

Fiber Optofluidic Technology Based on Optical Force and Photothermal Effects

Chenlin Zhang [1], Bingjie Xu [1,*], Chaoyang Gong [2], Jingtang Luo [3], Quanming Zhang [3] and Yuan Gong [2,*]

1   Science and Technology on Security Communication Laboratory, Institute of Southwestern Communication, Chengdu 610041, China
2   Key Laboratory of Optical Fiber Sensing and Communications (Ministry of Education), School of Information and Communication Engineering, University of Electronic Science and Technology of China, Chengdu 611731, China
3   State Grid Sichuan Economic Research Institute, Chengdu 610041, China
*   Correspondence: xbjpku@163.com (B.X.); ygong@uestc.edu.cn (Y.G.)

Received: 25 May 2019; Accepted: 19 July 2019; Published: 26 July 2019

Abstract: Optofluidics is an exciting new area of study resulting from the fusion of microfluidics and photonics. It broadens the application and extends the functionality of microfluidics and has been extensively investigated in biocontrol, molecular diagnosis, material synthesis, and drug delivery. When light interacts with a microfluidic system, optical force and/or photothermal effects may occur due to the strong interaction between light and liquid. Such opto-physical effects can be used for optical manipulation and sensing due to their unique advantages over conventional microfluidics and photonics, including their simple fabrication process, flexible manipulation capability, compact configuration, and low cost. In this review, we summarize the latest progress in fiber optofluidic (FOF) technology based on optical force and photothermal effects in manipulation and sensing applications. Optical force can be used for optofluidic manipulation and sensing in two categories: stable single optical traps and stable combined optical traps. The photothermal effect can be applied to optofluidics based on two major structures: optical microfibers and optical fiber tips. The advantages and disadvantages of each FOF technology are also discussed.

Keywords: optofluidics; optical force; photothermal effect; optical manipulation; optical fiber sensors

---

1. Introduction

Optofluidics is a reconfigurable, sensitive, and portable technology that combines microfluidic systems and the optical systems [1]. Microfluidic technology makes an optofluidic system more reconfigurable because the liquid of microfluidics has a unique flexibility for solid materials. Optical technology can enhance the sensitivity of optofluidics by introducing new functionality and opto-physical effects and can work at very small sizes in microfluidic channels. As a result, with the integration of microfluidics and an optical system, optofluidics has become suitable for multiple applications, such as medical diagnosis and treatment, environmental analysis, and substance analysis [2].

Fiber optofluidic (FOF) technology is an important branch of optofluidics. As shown in Figure 1, FOF has four major types of structures, the fiber-optic interferometer, fiber grating, microstructured optical fibers (MOFs), and optical micro/nano fibers. In practice, FOFs show several notable superiorities [3,4]. FOF devices are inexpensive and simple thanks to the mature manufacturing technology of optical fibers. An FOF system can easily couple light into a chip with channels at a submillimeter scale because of the tiny cross-section of optical fibers and can accurately interact with liquid samples based on its low loss transmission. Additionally, an FOF system can improve performance with the notable fabricability of its optical fibers. For example, optical micro/nano fibers fabricated with commercial fibers can enhance the sensitivity of the environment through the evanescent field. Microstructured optical fibers (MOFs) can serve as both light transmission and microfluidic channels, due to their specific structures, such as a suspended core, a hollow core, and air cladding [5].

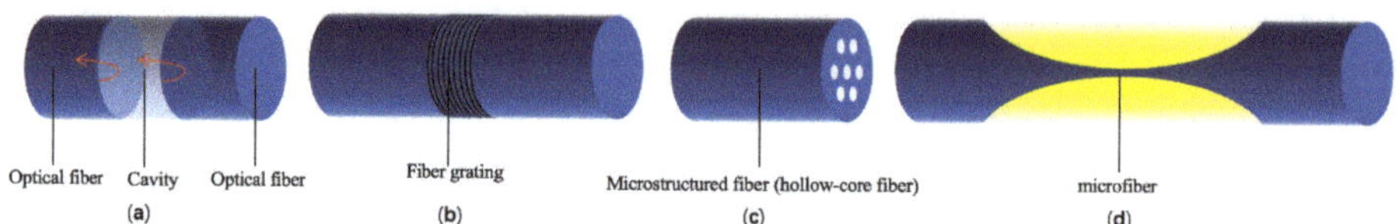

Figure 1. Fiber optofluidic (FOF) studies based on the structures of (**a**) the fiber-optic interferometer, (**b**) fiber grating, (**c**) microstructured optical fibers, and (**d**) optical microfibers.

Fiber-optic interferometers can be divided into several major categories according to their structures, including a Fabry–Perot interferometer (FPI), whose interference is based on single arm incident laser oscillated in a cavity; and Mach-Zehnder, Michelson, and Sagnac, whose interference occurs via a two-arm optical laser split from the incident laser [6]. The typical structures of fiber-optic interferometers are summarized in Table 1.

A Fabry–Perot interferometer (FPI) can be used for optofluidic sensing based on either an extrinsic or intrinsic cavity, which is usually fabricated with two parallel mirrors, as shown in Figure 1a. The FPI can be designed for sensing with high performance by filling the special material or optimizing the mirrors of the cavity. A high-visibility in-line with the optofluidic Fabry–Perot cavity was demonstrated by splicing a silica capillary tube into two single mode fiber (SMFs) and polishing the latter optical fiber of the FPI [7]. Refractive index detection with a high sensitivity of 1148.93 nm/RIU was achieved thanks to the smooth end faces of the SMFs. Another sensing probe based on FPI was fabricated by sandwiching a cavity between the single mode fiber (SMF) facet and a Nafion film [8]. The results indicated that Nafion could be used for temperature and humidity sensing.

Traditionally, two-arm interferometric structures, i.e., a Mach–Zehnder interferometer (MZI), a Michelson interferometer (MI), and a Sagnac interferometer (SI), are composed of two separate fibers [6]. Recent studies show that these two-arm interferometric structures can also be implemented as an in-line fiber optic core-cladding-mode interferometer, and thus have the advantage of compactness, simplicity, easy alignment, high coupling efficiency, high stability, and low-cost. Two-arm interferometric structures are promising for many sensing applications, such as label-free biosensing [9], humidity sensing [10,11], temperature-immune refractive index (RI) sensing [12,13], temperature sensing [14], and pressure sensing [15,16].

Table 1. A summary of sensing application based on fiber-optic interferometer.

Device Types	Structures	Applications	Ref.
FPI	Extrinsic/intrinsic cavity	RI sensing	[7]
		Temperature and humidity sensor	[8]
MZI/MI	Waist-enlarged fiber taper	Label-free biosensing	[9]
		Temperature-immune humidity sensing	[10,11]
		Temperature-immune RI sensing	[12]
MZI/MI	Core-offset	Temperature-immune RI sensing	[13]
MZI	Two micro-cavities	RI-immune temperature sensing	[14]
SI	Fiber loop mirror	Temperature-immune pressure sensing, High-temperature sensing	[15,16]

FPI is Fabry-Perot interferometer, MZI is Mach-Zehnder interferometer, MI is Michelson interferometer, SI is Sagnac interferometer, and RI is refractive index.

The optical fiber grating (OFG) is inscribed in the fiber core to form a sensing probe with the grating on the fiber tip or along the fiber axis. OFG can detect the changes of parameters in microfluids by mode coupling between the forward and backward transmission modes in the fiber core, or between the transmission mode in the core and in the cladding [17,18]. A sensing probe based on fiber Bragg grating inscribed in the photonic crystal fiber (PCF) was proposed for DNA detection, which is the first direct measurement of genomic DNA without a polymerase chain reaction (PCR) or other amplification reactions [17]. An in-line fiber optofluidic RI sensor was also proposed based on long-period fiber grating inscribed in a side-channel photonic crystal fiber [18]. A linear response and a sensitivity of 1145 nm/RIU was demonstrated.

Microstructured optical fibers (MOFs) are an excellent structure for optofluidic sensing due to their effective sample delivery and optical transmission. In [19], a side-channel photonic crystal fiber (PCF) was designed as a compact and ultrasensitive all-in-fiber optofluidic sensing platform. The large channel on one side of the fiber core enables a strong light-matter interaction and easy lateral access of liquid samples. It offers promising applications in chemical and biological analysis for monitoring the environment or biological/medical diagnosis. The working principle of optical micro/nano fiber sensors is based on an evanescent field. The electromagnetic field will partially penetrate into the cladding region to form an evanescent field, when the photon beam propagates through the core [20].

For the micro/nano fiber with a subwavelength diameter, the evanescent field is improved and interacts with the microfluid for sensing, as shown in Figure 1c. A microfiber, for which the diameter of the narrowed region is 7.8 μm, can achieve high humidity sensing [21]. A three-dimensional (3D) graphene network coated on the cladding of the microfiber can enhance the interaction between the moisture molecules and the three-dimensional graphene network (3-DGN) cladding. The relative humidity (RH) sensor displayed a fast response time of 4.0 s and an ultrahigh sensitivity of −4.118 dB/%RH in a relative humidity range from 79.5% RH to 85.0% RH.

Optofluidics not only structurally combine optics and microfluidics, but also induce new opto-physical effects from the energy transfer between light and microfluid and thus allows new schemes and functions for manipulation and sensing. However, few papers have comprehensively summarized the state-of-art fiber optofluidic (FOF) technologies in the literature. Recently, Vaiano et al. published an excellent review article about fiber optofluidics, which for the first time summarized the developments of the "lab on fiber (LOF)" concept for biological sensing applications [22]. They introduced the LOF technologies in three classes: lab on tip, lab around fiber, and lab in fiber, according to the integration location of functional unit. They compare the LOF technologies in terms of principle of operation, fabrication method, versatility in the design, and performance. They mainly focused on the functional material and structure of fibers in a microfluidic environment. The interaction between optics and microfluidics deserves another review to reflect its progress.

In this review, we summarize the latest progresses in fiber optofluidic (FOF) technology and analyze the correlation between fiber optics and microfluidics. Specifically, we mainly consider the optical force and photothermal effect, because they have been studied extensively, and many applications of manipulation and sensing in optofluidics have been demonstrated. Other photo-physical effects, such as the photoacoustic effect, are beyond the scope of this review.

The optical force can be generated by radiation pressure. It was first reported in 1970s, when Ashkin observed particle acceleration by a single laser beam [23]. Optical trapping with single laser beam was achieved in 1986 [24]. After that, researchers realized that optical force might be an effective tool for manipulating objects at a microscale [25–28]. Optical manipulation is a noncontact and nondestructive method and thus has great potential in the biological and healthcare sectors [17,25,26]. Additionally, optical force can be used for optofluidic sensing because its characters would be influenced by the trapped object or the microfluidic environment [29–32].

The photothermal effect may cause a thermal rise from laser beam, when liquid or objects in the microfluid absorb the laser energy. More specifically, the status of the target particle and/or the solvent may vary when the laser beam irradiates. The variations of charge carriers, molecular orientation, electrostriction, and radiation pressure in materials influence the conversion process of light energy to thermal energy. The changes in thermal energy further affect the temperature, refractive index, and the volume of the components in optofluidics, including liquid, objects, and the optical structure. In 1880, Bell and his coworkers reported the photothermal phenomenon in their paper on the photoacoustic effect [33], and the mechanism was reported by Terazima [34]. The photothermal effect has several unique advantages and can be used in optofluidics for multiple applications. First of all, the photothermal effect is applicable to a mix of solid, gas, and liquid matters states. Therefore, this effect is helpful to investigate the energy transfer between these states. Secondly, the photothermal effect has the potential for optofluidic multi-parameter detection and control. Lastly, the photothermal effect provides a non-contact method, which reduces the risk of mechanical damage in optofluidics.

The rest of this review is organized as follows. Section 2 introduces FOF technology based on optical force. The applications of optical trapping, manipulation, and sensing can be achieved with the stable single optical trap (SSOT) and stable combined optical trap (SCOT). The SSOT is formed only with optical force, while the SCOT is formed with the help of microfluidic flow force. Section 3 introduces FOF technology based on the photothermal effect. This device is mainly performed with two structures: the optical microfiber and optical fiber tip.

2. FOF Technology Based on Optical Force

Conventionally, optical force was often generated based on light beams in free space focused by a high numerical aperture (NA) microscope objective, which is known as optical tweezers [24,35]. This bulky device makes optical tweezers difficult to use and expensive. The optical force based on the optical fiber structure offers many advantages over conventional methods, such as its low cost, compact configuration, easy integration, flexibility, and long transparent distance. It can serve as a versatile tool for optical manipulation and sensing. However, it is challenging to generate a stable trap for the optical fiber structure due to the low NA of the fiber. To resolve this problem, two main categories of optical traps (i.e., the stable single optical trap (SSOT) and stable combined optical trap (SCOT)) have been developed. The SSOT is formed only with the optical force, which is mainly based on the structures of the lensed fiber or fiber taper, dual-beam fiber trap, and special constructions. SCOT is formed with a balance of optical force and microfluidic flow force, which can provide more flexible manipulation with a longer distance.

2.1. Stable Single Optical Trap with Optical Force

A stable single optical trap (SSOT) can be formed only with optical force to achieve trapping and manipulation. Since the optical force generated by a flat fiber tip usually pushes the object away, SSOT either utilizes a single optical force formed by a lensed fiber or fiber taper to enhance convergence (see Figure 2a) or utilizes the dual-optical forces formed by dual beams from different directions to maintain balance (see Figure 2b).

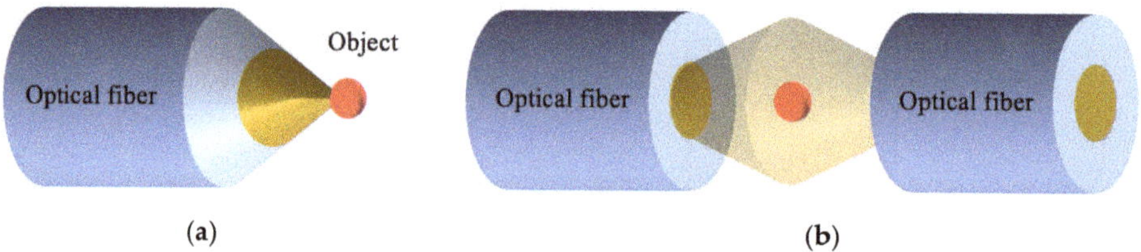

Figure 2. The schematic principle for stable single optical trap generation by (**a**) a fiber taper and (**b**) a dual-beam fiber trap.

2.1.1. SSOT Based on Single Optical Force

A fiber taper is a straightforward way to form an SSOT by enhancing the trapping efficiency of a single optical fiber. The fiber taper can strongly focus the laser beam with a lens-like structure, allowing a 3D SSOT to be achieved [36]. This structure is easy to fabricate with a commercial optical fiber by chemical etching [37–39], polishing [40], or heating-and-drawing [36]. However, it has the limitation of a short working distance and fixed SSOT due to the firm, sharp structure of the fiber taper. The fiber taper can trap an object, but it hardly changes the object's position without moving itself.

Yuan et al. reported a series of pioneering studies and demonstrated several new methods to solve this problem. In 2013, they first demonstrated controllable SSOT without moving the fiber [37]. A yeast cell can be manipulated for a distance of approximately 3 μm with the power ratio of a fundamental mode beam (LP01) and the low-order mode beam (LP11) generated in a normal single-core fiber taper. Then, they achieved multidimensional manipulation by using the LP11 mode beam excited with a special fiber taper [38]. In 2015, they achieved optical trapping and launching based on dual-wavelength single fiber optical tweezers [39]. As shown in Figure 3, a 980 nm laser was used to trap the object towards the fiber taper, and a 1480 nm laser was used to launch the object with a certain velocity. The trapping force and the launching force can be controlled independently with different laser powers at different wavelengths.

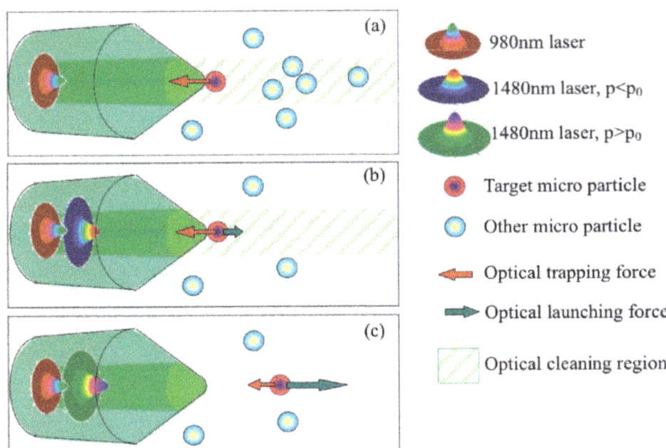

Figure 3. Schematic diagram of the dual-wavelength single fiber optical tweezers [39]. (**a**) Optical trapping force generated by the 980 nm laser beam. (**b**) Optical launching force generated by the 1480 nm laser beam, where the power of the 1480 nm laser beam is smaller than the power of the 980 nm laser beam, and thus the target is trapped while other objects are blown away. (**c**) If the power of the 1480 nm laser beam is larger, the launching force can launch the target away with a certain velocity.

An SSOT based on microfibers or nanofibers can achieve long-range manipulation due to its surface evanescent fields. Li and his coworker have done much research in this field and produced many theoretical and experimental results [40–42]. A nanofiber can trap the objects on its surface by the optical gradient force and propel trapped objects along the surface by the optical scattering force [42]. Further, as the evanescent field around the nanofiber can interact with the surrounding atoms, SSOT based on nanofibers can be used as an effective tool for atom manipulation and detection [43–47]. Recently, Li et al. developed a special nanofiber by embedding a silver nanowire (AgNW) into a polymethyl methacrylate (PMMA) nanofiber [40]. This nanofiber achieved trapping with a low laser power with a broad wavelength range because the AgNW enhanced the optical gradient force.

An SSOT based on a single optical force can trap a single object on or close to the fiber taper in a 3D fashion with a strongly focused beam. A manipulation length of several micrometers can be achieved with optical mode multiplexing and wavelength multiplexing, since the structure of the fiber taper is critical for functional performance, which also causes defects, such as the poor reconfigurability of devices and a short manipulation length [48–50].

2.1.2. SSOT Based on Dual-Optical Force

A dual-beam fiber trap (DFT). This DFT has the special advantage of stable and flexible manipulation between two laser beams based on its fiber. The typical structures of DFTs are shown in Figure 4. As shown in Figure 4a, the primary DFT can achieve long-range optical manipulation between two aligned flat or lensed fiber tips with the help of two optical forces. The aligned fiber tips provide two coaxial and opposite optical forces for the object, and the magnitude of optical force is determined by laser power and the distance between the fiber tip and the object. In this case, the position of the object can be controlled by the balance point of the optical forces, which can be easily adjusted by the laser power from each fiber [30]. A misaligned DFT further increases the manipulation dimension to implement precise object rotation [31]. In addition to the structure based on two strictly collimated fibers, hollow-core photonic crystal fiber (HC-PCF) [32] (Figure 4b) and inclined fibers [33] (Figure 4c) can also form a dual-beam trap with more functional advantages.

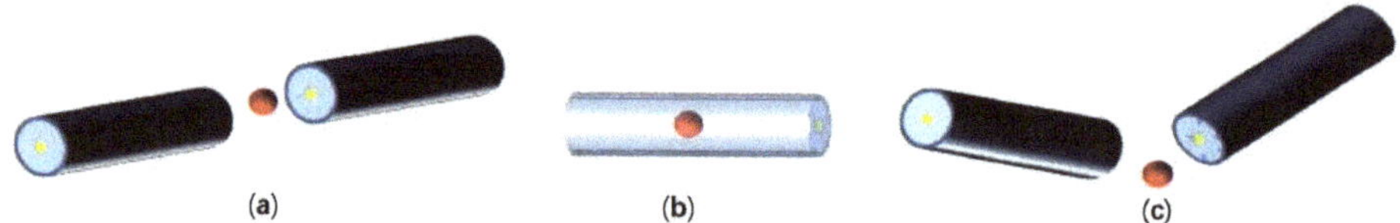

(a)   (b)   (c)

Figure 4. Optical trapping micro-objects based on a dual-beam fiber trap. (**a**) Aligned fibers; (**b**) Hollow-core photonic crystal fiber; (**c**) Inclined fibers.

In [51], a DFT was used for multiple object trapping and manipulation, as shown in Figure 5. Two aligned fibers with flat facets were separated by 160 μm, and each fiber emitted 100 mW of laser with two 980 nm laser diodes, which created a platform for optical trapping and manipulation. Figure 5a shows multiple polystyrene microspheres (PSMs), with 1 μm diameters, stably trapped between two fiber ends. The PSMs form linear arrays by themselves and get closer with more PSMs. When the number of PSMs is 13 or more, they start self-sustained oscillations with a range of 20 μm for a period of 0.5 s, as shown in Figure 5b. The DFT with an offset of about 5 μm can keep PSMs oscillating as a loop. The trajectory of the outermost PSM for eight-PSM oscillation loops is shown in Figure 5c.

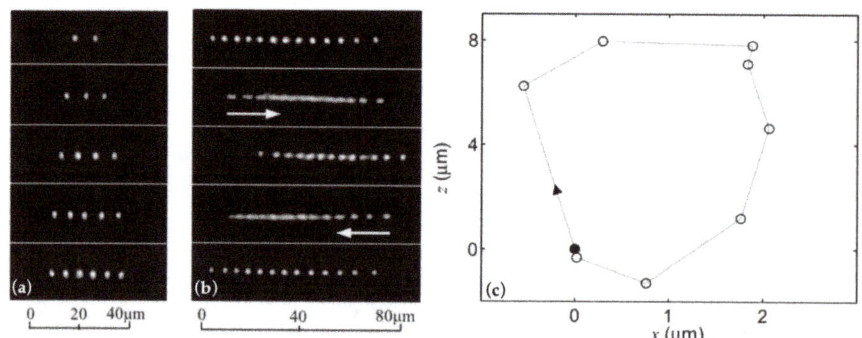

Figure 5. Multiple trapping and manipulation with the dual-beam fiber trap [51]. (**a**) Optical trapping of multiple objects. (**b**) Optical oscillating of a linear array of 13 objects. (**c**) Trajectory of the outermost polystyrene microsphere (PSM) for eight PSMs. The laser power of each fiber is 100 mW at 980 nm, and the objects are PSMs of with 1 μm diameters.

In addition to optical trapping and manipulation, DFT may also deform a soft object, such as a living cell. Guck and coworkers achieved measurement of cell membrane elasticity with a unique DFT [25]. They stably trapped a single cell in a DFT formed by two SMFs. Then, they stretched it along the optical axis because the force on each side of the cell membrane may be about several hundred pN. Additionally, the unfocused laser beams can avoid thermal damage to the living cell, even if the laser power is several watts. This has potential use in biological and medical research.

DFT can also be used for sensing. Force sensing was achieved by DFT with an inclination angle (θ) [52]. The inclination angle (θ) between the two fibers was used for function selection. For object trapping, θ should be $\leq 45°$, and for object lifting and force sensing, θ should be $\geq 50°$. A temperature sensor was developed with a trapped microparticle in the DFT [30]. The DFT was formed by two aligned fibers with concave tips and sealed in a quartz capillary. For temperature sensing, they used a 980 nm laser to adjust the position of the microparticle and a 1550 nm laser to form the interference spectra for sensing. Li and his coworkers also demonstrated an inclined DFT with two optical fiber tapers for cell regulation and analysis [53,54]. One fiber taper was used for trapping the cell or forming cell chain. The other fiber taper was used to manipulate the targeted cell. This method has the potential for investigation of cell growth, the intercellular singling pathway, and pathogenic processes.

As fabrication technology improves, SSOTs based on special constructions, such as multi-core fibers, PCFs, and nanofibers, are proposed with better performance. Multi-core fibers can increase the manipulation dimension combined with a special fiber taper [55]. Yuan and his coworkers used a dual-core fiber [53], four-core fiber [56], and coaxial core optical fiber [57] to achieve controllable optical manipulation, oscillation, and object shooting. Cristiani et al. proposed an SSOT based on a multicore optical fiber [58]. The cores were shaped with the proper angles to reflect the laser beams into a tight focus, which is the SSOT. A strong gradient optical force generates the SSOT for 3D trapping. This SSOT can trap and manipulate microparticles over a relatively long distance with better flexibility than a DFT. The hollow-core photonic crystal fiber (HC-PCF) is an excellent carrier for optofluidics, as it is a combined channel for both laser and microfluid. A centimeter-scale long distance optical manipulation was achieved by HC-PCF [59]. A focused laser beam vertical to the fiber was used to trap the object in front of the core of the HC-PCF, and then a horizontal laser beam was used to push the object in the HC-PCF. After that, the object could be manipulated along the HC-PCF with the laser power from each end of the HC-PCF. Additionally, this research could lead to a promising new approach for biomechanical detection, because it can achieve cell deformation with the help of shear force. SSOT formed by HC-PCF can also be used for sensing. The multiple-parameter sensing of temperature, transverse mechanical vibration, and electric/magnetic fields was achieved with the help of a trapped object in the HC-PCF [32]. The object was trapped and adjusted to the sensing area with the counter-propagating laser in the HC-PCF. The change of optofluidics was reflected by a back-scattered light.

SSOT based on a single optical force is easy to integrate and move but has the disadvantage of a short and fixed trapping/manipulation range. An SSOT based on dual-optical force uses two counter-propagating laser beams to effectively extend its trapping/manipulation range. Moreover, this method reduces the requirements of the fiber facet, thus makes the system simpler and reconfigurable for fabrication and also harmless to the trapped object. However, SSOT based on fiber inherently has an inflexible manipulation range as it uses optical force only to control its trapping position. Introducing the flow force in microfluidics is helpful to achieve a flexible and controllable scheme. To be specific, by adjusting both the optical force and flow force, the object can be manipulated along the optical axis, as discussed below.

2.2. Stable Combined Optical Trap with Optical Force and Microfluidic Flow Force

In this sub-section, we will introduce a stable combined optical trap (SCOT), which is formed with the combination of optical force from a single optical fiber and the flow force from the microfluid. Optofluidic applications based on SCOT will be introduced in two categories according to the type of optical force. One is a SCOT based on the optical scattering force from the fiber tip, and the other is a SCOT based on the optical gradient force from optical microfibers.

2.2.1. SCOT Based on Optical Scattering Force

The principle of the SCOT near the optical scattering force is shown in Figure 6. The optical manipulation, along with the optical axis, is based on the force balance on the object between the axial optical force, F_{ao}, and the microfluidic flow force, F_v. The flow force can be calculated by Stokes law,

$$F_v = k_1 v. \tag{1}$$

Here, $k_1 = 6\pi\eta a$, where η is the coefficient of viscosity of water, and a is the radius of the microparticle. As the direction of F_v is the same with the microfluidic flow, F_v is directed toward the fiber tip. In contrast, F_{ao}, consisting of the scattering force, F_{as}, and the axial gradient force, F_{ag}, forms a counter force to push the object away from the fiber end. The F_{ag} is negligible compared to F_{as}, due to low acceleration of light intensity generated by the optical fiber tip. F_{ao} is directly proportional to the laser power and inversely proportional to manipulation length, L_m, which is the vertical dimension between the center of the microparticle to the fiber tip. The object can be trapped at a certain L_m, corresponding to the position of the SCOT, because the total force on the object is zero. This process can be described as

$$F_v(v) = F_{ao}(L_m) \tag{2}$$

Henceforth, L_m can be controlled by both the flow rate (v) and the laser power (P), and v can also be calibrated by P and L_m.

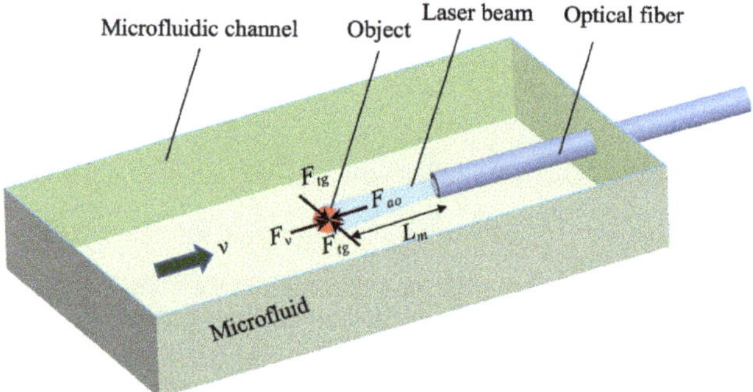

Figure 6. The principle of the stable combined optical trap based on a single optical fiber tip.

In comparison to the SSOT on the fiber taper, the SCOT over the fiber tip possesses the advantage of being easy to fabricate, flexible to manipulate, and compactable to be integrated. Moreover, it greatly extended the manipulation length based on the balance between the optical force and fluid flow and shows the potential for truly 3D optical manipulation.

The SCOT over the fiber tip can achieve controllable manipulation of single object with a long range along the optical laser beam. Gong and his coworkers made much progress in this respect. They achieved long range optical manipulation with the graded-index fiber (GIF) due to its periodic focusing effect [29]. In [60], a controllable manipulation length of over 177 μm was achieved by integrating a GIF taper with a microcavity. In this work, the manipulation length L_m was directly controlled by adjusting the laser power, the flow rate, or the length of the air cavity (L_c), where the air cavity was formed by the two flat fiber ends of the GIF and SMF.

Figure 7 shows the principle of manipulation based on the air cavity (L_c). In brief, the L_c affects the incident angle and the coupling intensity of the laser beam from the SMF to GIF. Due to the periodic focusing effect of the GIF, the different incident angle and the coupling intensity produced a different light distribution from the GIF taper. The optical force was controllable, and the object could be manipulated according to the newly balanced SCOT.

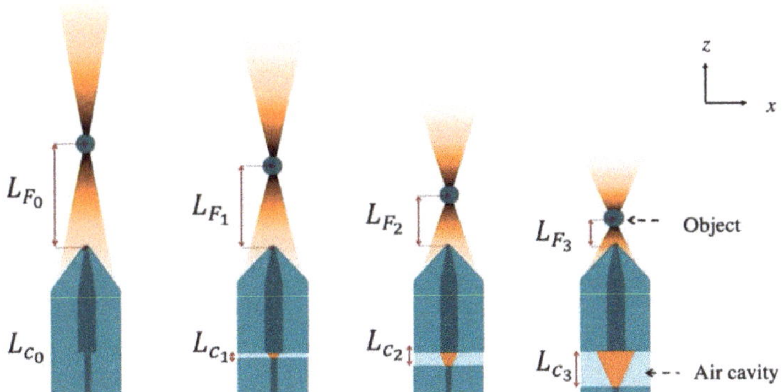

Figure 7. Principle of controllable optical manipulation based on air cavity length.

In 2016, a strain controllable optical manipulation was proposed with a longer range up to 1314.1 µm [61]. This method manipulates the object by directly stretching the GIF. As the light beam converges and diverges periodically in the GIF, the change of fiber length can be used to control the distribution of the emergent field. Figure 8a shows a sequence of microscopic images of stretching the 52.5 cm GIF with a step of 50 µm. Compared to the optical manipulation of the SCOT with an optical fiber taper, the method in [61] consists of a simpler fabrication process with high repeatability and more stable performance. Figure 8b showed that the manipulation length (L_m) changes by controlling the strain on the GIF.

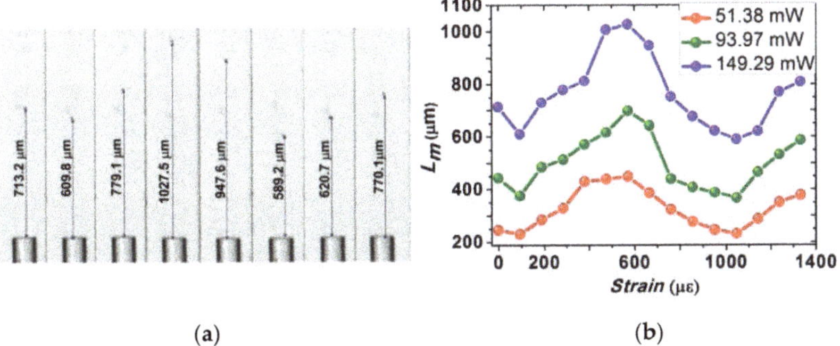

(a)   (b)

Figure 8. (a) Microscopic images of strain controllable optical manipulation by increasing the strain to 0, 95 µε, 286 µε, 571 µε, 667 µε, 1048 µε, 1143 µε, and 1238 µε, respectively. (b) Manipulation length versus strain at flow rate of v = 150 nL/min [61].

Although the GIF provides a new optical manipulation method based on the periodic focusing effect, it is difficult to mass-produce it with consistent performance by precisely controlling the GIF length. The distribution of the emergent field is sensitive to the length of the GIF, so the performance of optical manipulation may vary substantially even with very slight difference between each GIF.

The introduced flow force reduces the requirement of the light convergence so that the SCOT can be achieved with a flat SMF. This is the simplest scheme, with the advantage of being easy-to-fabricate and use, having high uniformity and availability for mass production, and low cost. Optofluidic flow rate detection was also achieved with a structure similar to that in Figure 6. According to Equation (2), the flow rate can be calculated by laser power or manipulation length [61–63].

The performance of the flow rate sensing with SCOT based on the cleaved SMF is shown in Figure 9 [62]. By coordinating the proper laser power, this device can detect the flow rate in a large dynamic range from 20 nL/min to 22 µL/min and can manipulate the object from 3 µm to 715 µm. The method calculating the flow rate with the laser power is named the open-loop mode, which is particularly useful for detecting a low flow rate but limited for detecting a high flow rate, because the manipulation length, L_m, is reversely proportional to the flow rate, v. In this case, the dual-mode detection induced a closed-loop mode method and enlarged the dynamic-range by four orders of magnitude, from 10 nL/min to 100,000 nL/min [63]. The sensing performance of the dual-mode flowmeter is shown in Figure 10. The mode switching threshold was set as 5000 nL/min with an initial laser power of 23.5 mW.

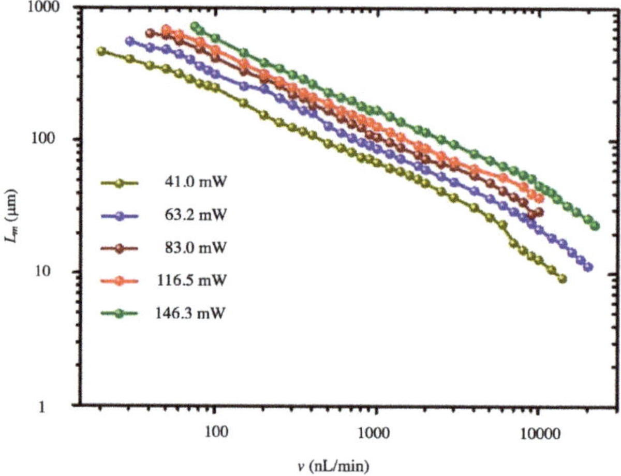

Figure 9. The performance of the flow rate sensing with the stable combined optical trap (SCOT) based on the cleaved single mode fiber (SMF) [62].

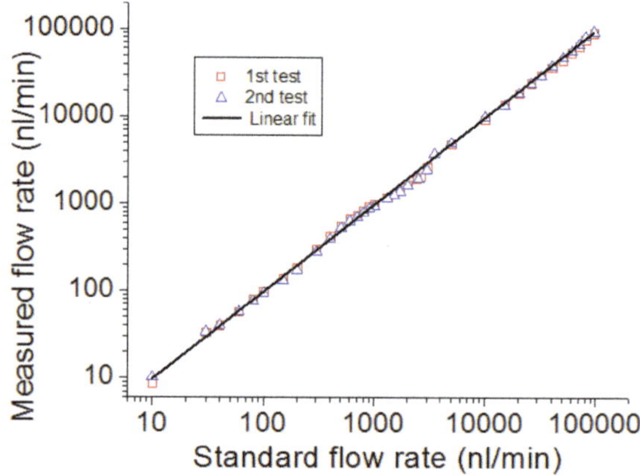

Figure 10. Sensing performance of the dual-mode flowmeter [63].

Optical trapping and orientation were achieved with an abruptly tapered SMF [64]. Microfluid delivered the *Escherichia coli* cell to the fiber tip with a velocity of 16 µm/s. A single *E. coli* cell was trapped with a laser power of 30 mW at a 980 nm wavelength, as shown in Figure 11a. The cell with an arbitrary azimuthal angle θ (Inset I of Figure 11b) was trapped with a fixed orientation in the final stable state. Figure 11b shows the calculated restoring torque on the object, with a θ at the central axis of the fiber tip. As shown in inset II of Figure 11b, the most stable orientation for trapping occurred at $\theta = 0$, as the torque is 0. Figure 11c reflected the trapping ability of the abrupt tapered SMF. The cell was trapped at $\theta = 0$ (Inset I of Figure 11c) with the energy density distribution simulated as Inset II of Figure 11c. It can be seen that the cell can be manipulated with a range of less than 8 µm and with a trapped optical force of more than 2 pN when the velocity of microfluid is 16 µm/s.

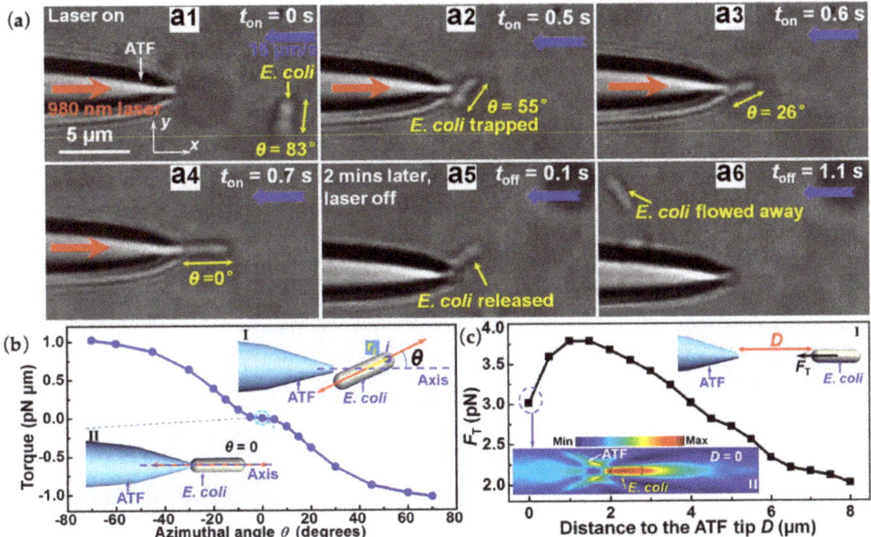

Figure 11. Optical trapping of a single *E. coli* cell [64]. (**a**) Optical microscope images of the trapping and orientation process of a single *E. coli* cell. The blue, red, and yellow arrows indicate flow velocity (16 mm/s), the input laser with an optical power of 30 mW, and the orientation of the *E. coli*, respectively. (**b**) The calculated torque acting on an *E. coli* cell as a function of azimuthal angle. (**c**) Calculated trapping force (F_T) exerted on a single *E. coli* cell as a function of distance (D) between the cell and the fiber tip.

The SCOT over the fiber tip can also achieve the organization and transport of multiple objects. In 2013, Li and his coworkers reported an optofluidic method for realizing and retaining stable cell–cell contact and controlling the trapped cells number using an abrupt tapered fiber (ATF) [64]. As shown in Figure 12, an optical power of 30 mW at a 980 nm wavelength was launched into the ATF. Cells delivered by the microfluid with a flow velocity of 3 µm/s were trapped onto the fiber tip one after another, thereby forming a highly organized cell chain. All the trapped and connected cells were aligned with the same orientation. In 2017, they achieved controllable organization of the cell chain with a large-tapered-angle fiber probe, and demonstrated the performance with *E. coli* cells, yeast cells, and human red blood cells [65]. The cell chain can be moved by a change of laser power and flow rate.

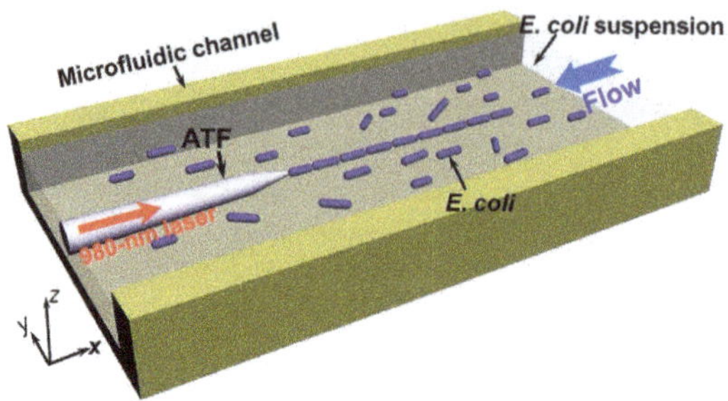

Figure 12. Schematic of cell–cell contact realization and retaining process. A laser at 980 nm was launched into an abrupt tapered fiber (ATF), which was placed in the microfluidic channel with a flowing suspension of *E. coli* cells. Multiple cells were trapped and connected in order at the tip of the ATF [64].

Microstructured optical fibers have customized structures for some unique applications. A hollow annular-core fiber taper (HAFC) was used to manipulate and transport living cells [66]. A schematic diagram of optical manipulation based on the HACF is shown in Figure 13. The hollow structure of the HACF helps to realize the sterile transport of particles in the optical fiber and provides a flow force by liquid viscous resistances (LVR). LVR is determined by the size of the object and the relative flow rate. Thus, the HACF tweezers were used in object selection and manipulation. Moreover, it is easy to clean the fiber probe and convenient for repeated use.

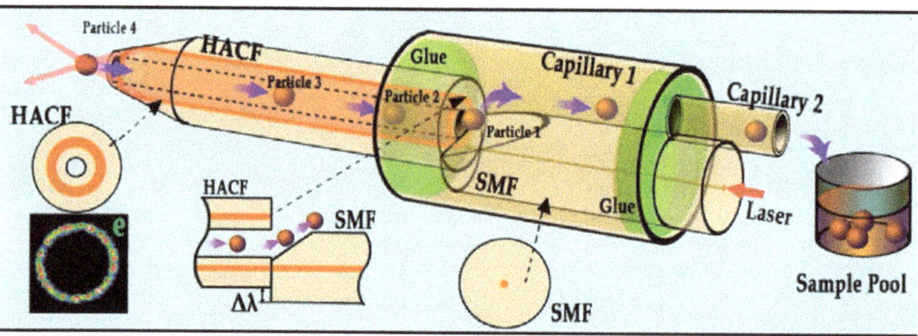

Figure 13. Schematic diagram of optical manipulation based on the hollow annular-core fiber taper (HACF) [66].

A SCOT based on optical scattering force can achieve controllable long-range optical manipulation and a sensitive flowmeter with an optical fiber probe. A larger optical scattering force is generated by laser irradiation from the fiber probe, which can be balanced with the flow force generated by the microfluid from the opposite direction of the laser irradiation. The optical scattering force can be adjusted by the laser power, and the flow force can be related to the flow rate. In this case, controllable long-range optical manipulation is achieved by adjusting the optical laser's power and flow rate, and the flow rate is calculated from the manipulation length or the laser power.

2.2.2. SCOT Based on Optical Gradient Force

A SCOT around the microfiber could achieve long-range manipulation along the fiber. The optical gradient force plays a crucial role in optical trapping based on SCOT. A microfiber with a subwavelength diameter enhances the evanescent field and exerts a large optical gradient force perpendicular to the fiber surface. When the microfluid flows against the optical gradient force, the object can be trapped with the combination of optical force generated by the light leaked from the optical fiber and the dragging force induced by the fluidic flow and then move along the surface via the optical scattering force that occurs in the direction of light propagation [67].

A SCOT around the microfiber can achieve optical transport along the fiber with the help of flow force, optical gradient force, and optical scattering force. Li et al. reported the backward optical transport of Polystyrene (PS) nanoparticles (713 nm in diameter) using an optical nanofiber with a diameter of 710 nm [68]. Figure 14 shows the schematic of the experiment. The optical forces, including gradient force (F_g) and scattering force (F_s), were generated and controlled using a diode laser with a 980 nm wavelength, and the flow force (F_d) induced by the microfluid was dependent on flow rate. When the laser was on, the evanescent field of the nanofiber applied F_g and F_s to the PS particle. F_g, directed towards the stronger optical intensity region, traps the particles to the surface of the nanofiber. The F_s with a direction parallel to the light propagation propels the particle to move along the nanofiber. By varying the laser power from 0 to 90 mW and flow velocity from 0 to −20 μm/s, the backward transport velocity exhibits a linear dependence. Furtherly, bidirectional optical transport can be achieved with two counter-propagating laser beams from each end of the optical nanofiber [42]. The transportation direction and velocity of the particles can be controlled by changing the difference between the laser power from each side of the nanofiber.

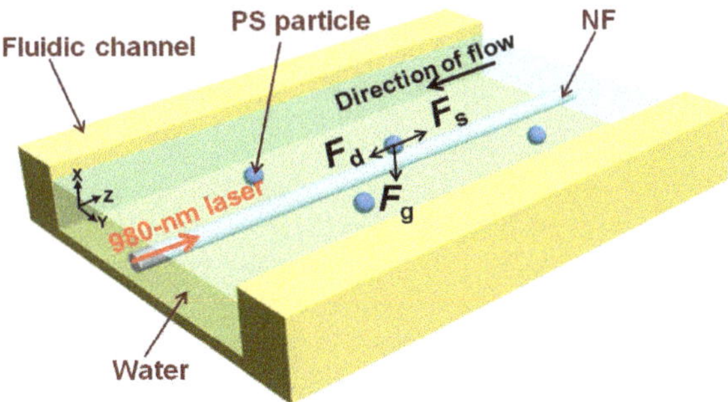

Figure 14. Schematic diagram of the experiment. F_d shows the drag force on the particle (in blue) induced by the fluidic flow. F_g and F_s denote the gradient and scattering forces, respectively, from the evanescent field [68].

Following this scheme, a particle separation method was demonstrated [69]. Figure 15 shows a schematic of the SCOT around the microfiber. A 1.55 μm laser was launched into the microfiber for particle separation. The microfiber with a 1.2 μm diameter was placed in a channel. The suspensions flowed into the channel for separation. Three types of particle mixtures were successfully separated, including 5/10 μm PMMA particles, 2.08 /5.65 μm SiO$_2$ particles, and 2.08 μm SiO$_2$/yeast cells.

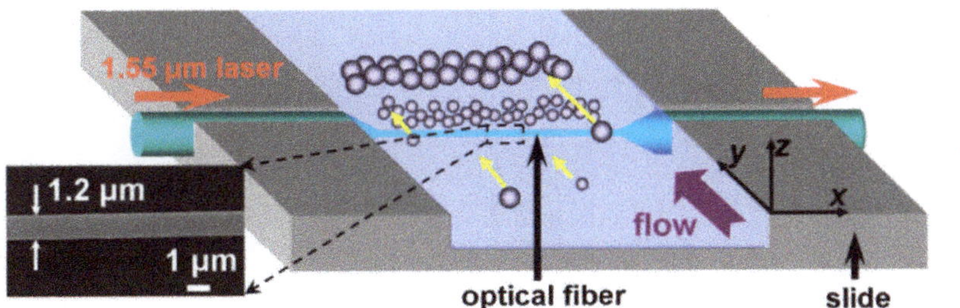

Figure 15. Schematic diagram for particle separation in fluidic flow by an optical fiber. The inset shows the scanning electron microscope image of a 1.2 μm optical fiber [69].

In conclusion, a SCOT based on optical scattering force (F_s) achieves force balance within a small cross-section along the optical axis. Thus, it can accurately trap or manipulate a single object, even far from the fiber probe. A SCOT based on optical gradient force (F_g) achieves force balance vertical to the microfiber surface. Thus, it can simultaneously trap or manipulate a larger number of objects.

The major difference between a SCOT and SSOT is that a SCOT utilizes the flow force in the microfluidic system as an extra control factor. By tuning the flow force, a SCOT can achieve more flexible and longer-range trapping or manipulation, without requiring sophisticated fiber structures as a SSOT does. In addition, a SCOT can calibrate the flow rate by measuring the laser power and the manipulation length.

3. FOF Technology Based on a Photothermal Effect

Photothermal effects can be used for optical manipulation and sensing in optofluidics. The photothermal effect is usually weak and needs to be enhanced by increasing laser radiation or absorption. In FOFs, laser radiation is usually increased by applying a microfiber, and laser absorption is usually increased by integrating special materials with different components of the optofluidic system.

3.1. FOF Technology Based on Photothermal Effect with Microfibers

In a standard optical fiber, the optical field is well confined to the fiber core, unable to interact with microfluid from the side [70]. Therefore, the key to generating a photothermal effect is improving the radiation of the optical laser from the fiber core to the microfluid. The optical microfiber provides an efficient solution for radiation enhancement, which has been extensively investigated.

Optical microfibers can easily be fabricated from commercial optical fibers by heating at a melting temperature and stretching to an appropriate size, enabling much higher flexibility and compatibility over conventional fiber-based systems [71,72]. Laser at a wavelength with high absorption to the solution is often launched into the microfiber to further enhance the photothermal effect. This can achieve optical manipulation and sensing for optofluidic applications.

The photothermal effect can achieve massive particle trapping and manipulation based on its derivative effects, i.e., the photophoresis effect and temperature gradient effect. The photophoresis effect is generated by an uneven heat distribution when the photothermal effect acts on the particles in the microfluid [73]. Uneven heat distribution will increase the movement of the surrounding water molecules, and eventually generate negative photophoresis or positive photophoresis to drive the particle towards or away from the light source, separately. The temperature gradient effect is based on the strong laser absorption of the liquid in the microfluid and can drive the particles to move to the colder region [74].

Li and his co-workers have been achieved massive photothermal trapping and manipulation with different structures, such as tapered optical fiber (TF) [75], subwavelength diameter optical fiber (SDF) [76], and optical fiber ring (FR) [77], as shown in Figure 16 Using a TF with a diameter of 3.1 μm for the taper (Figure 16a), plenty of particles were assembled into a spindle-shaped region, when the laser power was 170 mW at 1550 nm. There was a space of about 380 μm between the assembled particles and the TF after 15 min. Using an SDF (Figure 16b), particles were assembled around the SDF with a laser power of 200 mW, and reached saturation after 360 s. With an FR (Figure 16c), particles were trapped and assembled in the center of the FR with a power of 97 mW.

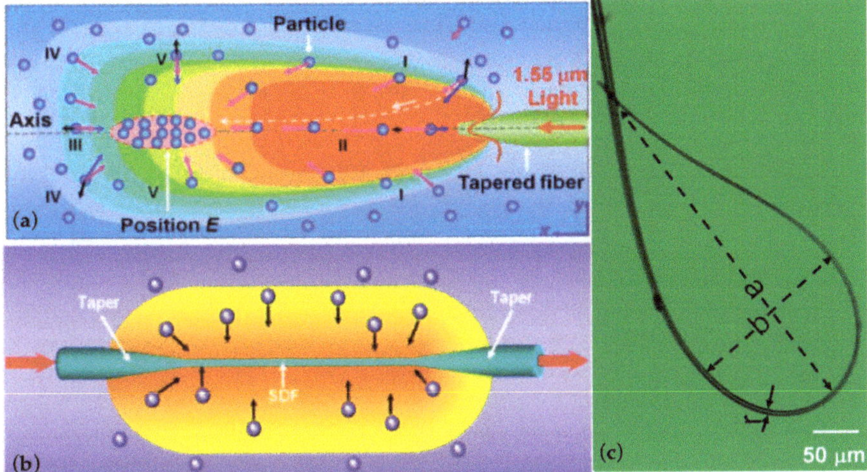

Figure 16. Massive photothermal trapping and assembly of particles using (**a**) tapered optical fiber [75], (**b**) subwavelength diameter optical fiber [76], and (**c**) optical fiber ring [77].

Photothermal effect can achieve optofluidic sensing based on the evanescent field around the microfiber during laser transmission. The microfiber with a diameter of several micrometers or less can enhance the evanescent field and is sensitive to the ambient temperature around it [78].

An optofluidic flow rate sensor based on the photothermal effect in a microfluid has been proposed by Gong and coworkers [79]. The side view and the cross section of the sensor are shown in Figure 17. A microfiber with a waist of approximately 3 μm was fabricated by heating and drawing a commercial SMF. A hollow round capillary acted as an optofluidic ring resonator perpendicular to the microfiber. A small fraction of the incident light of the microfiber was coupled into the capillary due to the evanescent field and kept circulating in the wall due to the total reflection of the smooth inside of the round capillary. A part of the reflection was coupled into microfiber and transmitted to the detector. The wavelength shift of the transmission spectrum can be used as a function of the flow rate. As the full width at half magnitude (FWHM) of the linewidth is narrow, this structure can achieve flow rate sensing with high sensitivity.

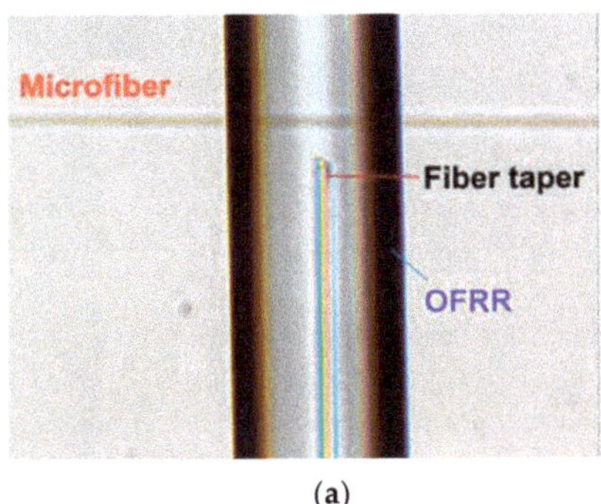

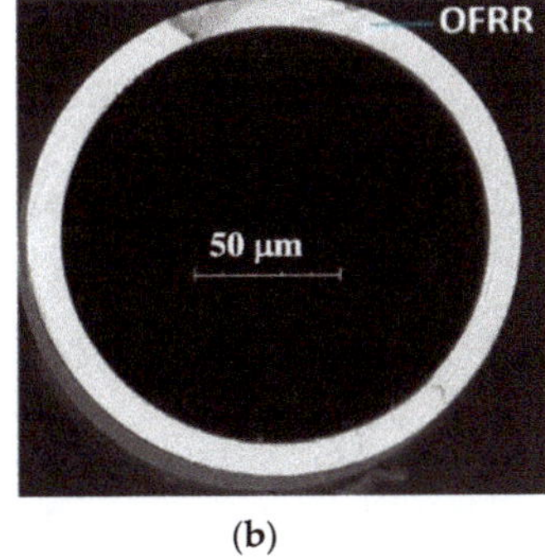

Figure 17. (**a**) The side view and (**b**) the cross section of the sensor of the optofluidic ring resonator [79].

A laser at 1480 nm coupled into the fiber taper was used to heat the liquid in the capillary for temperature change. The fiber taper can be easily fabricated with a commercial fiber by different methods, such as chemical etching, mechanical polishing, and flame heating. Since the fundamental mode of the evanescent field was powerful due to its larger radius near the capillary and the effective index (n_{eff}), it has often been chosen as the output for sensing. The microfiber was close to the outside of the round capillary. Therefore, while the fundamental mode can act on the capillary, it cannot pass through to change the parameters of the microfluid. As a result, the difference of temperature can be calibrated with the relative wavelength shift as

$$\frac{\Delta\lambda}{\lambda} = \left(\alpha + \frac{\kappa_{wall}}{n_{eff}}\frac{\partial n_{eff}}{\partial n_{wall}}\right)\Delta T \tag{3}$$

where α is the thermal expansion coefficient of the resonator, which can be calculated by $1/r\,(\partial r/\partial T)$, and $\kappa_{wall} = \partial n/\partial T$ is the photothermal effect coefficient of the capillary. The photothermal effect occurs near the fiber taper. First, the temperature of the microfluid increased near the fiber taper and then transferred to the round capillary. The wavelength shift is dependent on two factors in Equation (3): the thermal expansion of the capillary and its photothermal effects.

Optical microfibers can enhance the effective photothermal effect for manipulation and sensing applications. For optical manipulation, the photothermal effect enables massive objects manipulation with high flexibility. For optical sensing, the photothermal effect could be employed together with another microresonator for local detection of the microfluidic flow rate with high sensitivity by detecting the wavelength shift.

3.2. FOF Technology Based on a Photothermal Effect with Special Materials

Recently, materials with strong laser absorption, such as gold nanoparticles (Au NPs), graphene oxide (GO), and carbon nanotubes (CNTs), have been extensively investigated to improve the efficiency of photothermal conversion. These kinds of materials can enhance the photothermal effect for FOF to achieve applications of optical manipulation and sensing [80]. Photothermal materials are mostly integrated with optical fibers or microfluids, as introduced below.

3.2.1. Materials Integrated with Optical Fiber (MIFs)

It is difficult to generate an optimal photothermal effect with an untreated optical fiber due to its low light-thermal conversion efficiency. Integrating photothermal materials with the fiber (MIF) is a useful method to improve the photothermal effect. MIF can achieve optical trapping and manipulation based on the photothermal materials coated on the optical fiber tip or on the cylindrical surface.

Xing and co-workers investigated an optical manipulation method based on a graphene-coated microfiber probe (GCMP) [81]. As shown in Figure 18a, a 980 nm laser coupled into the GCMP can effectively trap erythrocytes based on photothermal effect induced thermophoresis and natural convection flow and can arrange the trapped erythrocytes over a long distance, combining with the optical scattering force. The MIF of the graphene oxide on the cylindrical surface of the fiber achieved mobile vortex arrays with high stability for the no-time-delay, non-contact delivery of massive trapped objects along the arbitrary direction [82], as shown in Figure 18b. When a 1070 nm laser was coupled into the coated fiber, a temperature gradient was generated and excited the oscillatory wave to trap and deliver the particles.

Optical sensing can also be achieved with MIF. Fiber optofluidic microbubble-on-tip (µBoT) sensors, featuring a flat fiber tip coated with carbon nanotube (CNT) film [83] or gold nanofilm [84], have been proposed. The process and sensing mechanism are mainly based on a reconfigurable microbubble. When the laser was irradiated on the coated fiber tip, a microbubble was generated and gradually expanded. The generation of the microbubble can be monitored using both the microscope and interference spectrum of the interferometer formed by the fiber tip and the surface of the microbubble, and the changes of parameters in microfluidics can be calibrated with the growth rate of the microbubble.

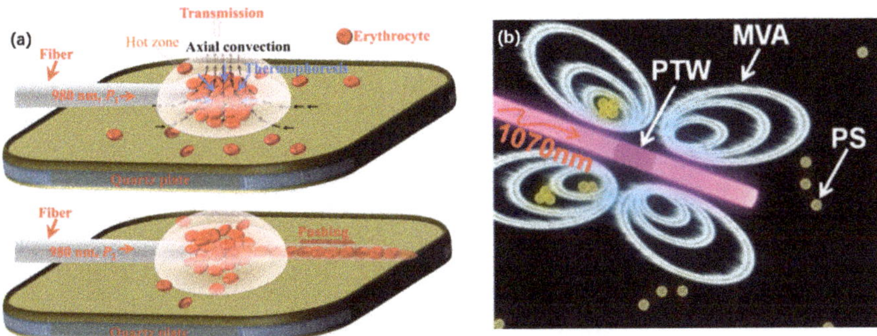

Figure 18. Schematic diagrams of optical trapping and manipulation based on the photothermal materials coated (**a**) on the optical fiber tip [81] or (**b**) on the cylindrical surface [82].

A CNT-coated µBoT sensor can detect the temperature and flow rate with the laser at 980 nm. The sensing signal was the free spectral range (FSR) of the microbubble heated with the same duration. As the diameter of the µBoT interferometer increases over time, the FSR decreases. For temperature sensing, the microbubble expands mainly based on gas generation from liquid vaporization around the fiber tip. The principle of flow rate sensing is mainly based on microbubble expansion with the dissolving gas in the flow fluid.

A gold-coated µBoT sensor can detect the concentration of the solution with a laser at 1550 nm. Sucrose and H_2O_2 were chosen as models to demonstrate sensing performance, which represents two different sensing mechanisms. One is based on the evaporation of liquid near the fiber tip, and the other is based on heat-induced chemical decomposition. For sucrose sensing, the µBoT sensor achieved a dynamic range of two orders of magnitude, from 0.5 wt% to 50.0 wt%. For H_2O_2 sensing, the µBoT sensor achieved a dynamic range of five orders of magnitude, from 10^{-5} M to 1 M, as shown in Figure 19. The microscopic images of the microbubbles generated with different concentrations of H_2O_2 were recorded at different heating times (Figure 19a). The imaging method was chosen due to its low cost, and Δd as a function is shown in the log-log scale in Figure 19b.

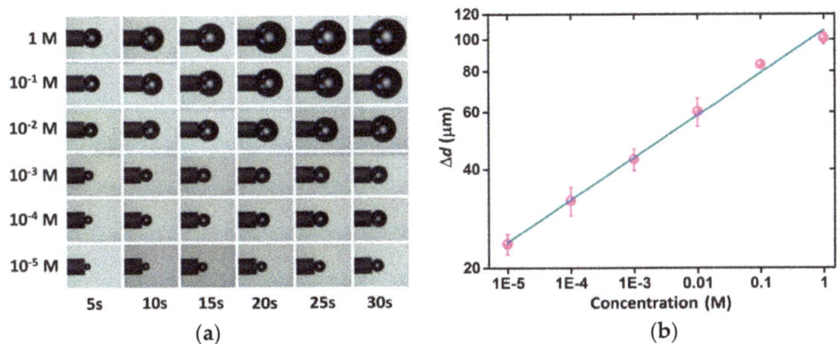

Figure 19. (**a**) Microscopic images of the microbubbles and (**b**) H_2O_2 concentration detection [84].

The result of sensing shows a large dynamic range and high sensitivity, which demonstrate the high performance of sensing based on this mechanism. This is the first report on concentration sensing based on a reconfigurable µBoT structure. This technique shows many advantages for optofluidic detection, such as flexibility, reconfigurability, low cost, ease of fabrication, and ease of use.

Besides coating the photothermal materials on the fiber tip, compact fiber-optic sensors also can be achieved by integrating materials into the optical fibers. A miniature, all-optical, fiber-optic sensor has been demonstrated for thermal conductivity measurements [85]. A vanadium doped fiber spliced with an SMF was used as a highly absorbent part and coated with a thin zirconia film to create a semi-reflective surface. Thus, the short section of vanadium doped fiber formed an all-fiber F–P interferometer for sensing. The all-silica design makes the sensor compatible with most chemical environments and has good potential for use at elevated temperatures and high pressures.

3.2.2. Materials Integrated with Microfluids

Materials with strong laser absorption abilities can also be directly integrated with microfluids, such as the channel of microfluidics or the objects in the microfluid. This method increases photothermal conversion efficiency and shortens the response time for manipulation or sensing.

Photothermal materials on the channel have been widely applied to handling liquids [86–88], manipulation in microfluid [89–91], and micromaching processes [92] because the photothermal effect on the channel can cause fluid dynamics, phase changing, interfacial action, and a strong vertical temperature gradient. A laser induced microbubble-based device was introduced as an example of a photothermal effect on the channel [93]. Graphene oxide (GO) was integrated with the channel, which could serve as a miniature heat source to generate a microbubble and control dynamic behaviors of flow by adjusting optical laser power at the micrometer scale. A microfiber was used to simulate the photothermal effect at the locality of the microfluid with a 1070 nm CW laser. This device can be used for optical manipulation based on the thermal convection around a microbubble, which can be controlled by an optical laser. A simulation of thermal convection is shown in Figure 20. Based on controllable thermal convection, i.e., vertical convection and Marangoni convection, the microfluidic flow around the microbubble can be controlled easily, and the massive objects around the microbubble can be manipulated.

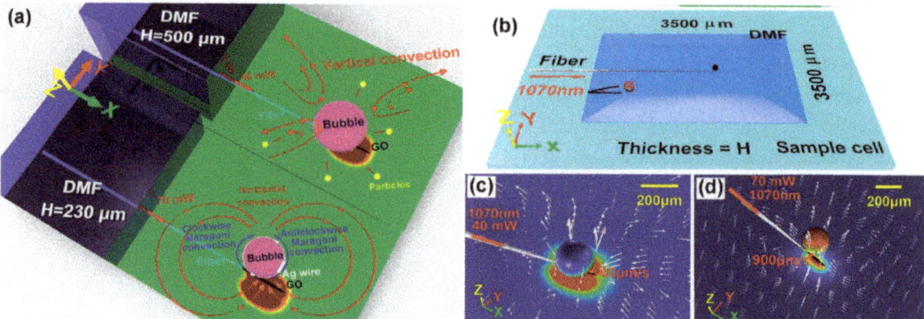

Figure 20. Simulation of thermal convection when the bubble is located at H = 230 μm and 500 μm and P = 40 mW and 70 mW [93]. (**a**) Schematic model of different functional flows. (**b**) Simulated model of the microfluidic system. (**c**) Vertical convection when a 160 μm diameter microbubble is located above the heater with T = 400 K at H = 500 μm and P = 40 mW. The longest arrow shows the maximum velocity of 40 μm/s. (**d**) Horizontal convection induced by the heater when the 160 μm diameter microbubble is located at the side of the graphene oxide (GO) heater with T = 450K at H = 230 μm and P = 70 mW. The longest arrow shows that the maximum velocity of 900 μm/s. Marangoni convection (green arrows) at the surface of the bubble would influence movement of particles when the thickness of the N, N-Dimethylformamide (DMF) was less than 300 μm. Flow-induced deformation of the bubble is not considered in these simulations.

Photothermal materials integrated with the object have received great attention because the materials can efficiently convert adsorbed photons into thermal energy, and this function can directly act on the target with minimally invasive effects [94,95]. Based on these characteristics, photothermal therapy and other bio-medical applications based on IT have been increasingly and widely investigated as facile oncological treatment methods [96]. In 2003, Halas and co-workers first demonstrated the selective destruction of breast carcinoma cells based on the target region integrated with gold-on-silica nanoshells [97]. After that, a variety of hybrid materials with different compositions or structures have been explored for photothermal therapy (PTT) [98,99]. Optical fiber used to induce laser power to a target are minimally invasive due to their flexibility, optical conductivity, and biocompatibility.

A hybrid material with components of zinc phthalocyanine (ZnPc), polyethylene glycol (PEG), core–shell nanoparticles (NPs), a gold (Au) core, and graphene oxide nanocolloid (GON) (ZnPc-PEG-Au@GON NPs), were successfully applied to the in vitro photothermal ablation of HeLa (Human cervical cancer cell line) cells [98]. A 660 nm fiber coupled laser was used to generate a effect on treated HeLa cells, and induced death of nearly all of them, as shown in Figure 21a. Figure 21b shows the contrast of the cell viability of HeLa cells treated with ZnPc and ZnPc-PEG-Au@GON NPs. The results clearly indicated that ZnPc-PEG-Au@GON NPs could enhance photothermal efficiency for medical treatment.

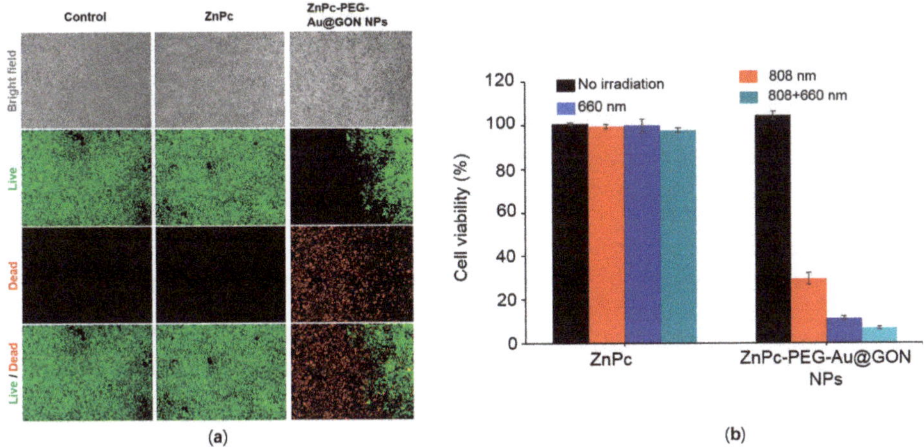

Figure 21. (**a**) Bright-field and fluorescence microscopy images of control, ZnPc, and ZnPc-PEG-Au@GON nanoparticle (NP) treated HeLa cells wgucg were exposed to a 660 nm fiber coupled laser with 67 mW/cm^2 for 15 min and subsequently stained by a lince/Dead assay reagent, green: live cells, red: dead cells. (**b**) Cell viability of HeLa cells treated with ZnPc, and ZnPc-PEG-Au@GON NPs and subsequent exposure to 808 nm with 0.67 W/cm^2 for 20 min, 660 nm with 0.2 W/cm^2 for 10 min, or both light sources with the same condition, sequentially [98].

In conclusion, FOF technology based on a photothermal effect can achieve massive object manipulation and microfluid sensing with microfibers or with special materials, which enhance laser radiation or absorption, respectively. However, because light energy needs a relatively long time to be converted to heat energy, FOF technology based on a photothermal effect usually has a long response time for manipulation or sensing. Therefore, photothermal materials are often used to further increase photothermal conversion efficiency. However, this method maybe unstable in some cases, as photothermal materials could be washed away by the microfluid.

4. Conclusions

This review mainly introduces the fiber optofluidic technology (FOF) based on two major opto-physical effects: optical force and the photothermal effect. Optical force is used for optofluidic manipulation and sensing with a stable single optical trap (SSOT) and a stable combined optical trap (SCOT), and the photothermal effect is used for various microfluidic control applications with an optical microfiber and special absorption materials.

SSOT and SCOT exploit different types of forces in optofluidic systems. SSOT is formed only by optical force and can be further subdivided into those based on a single optical force and those based on a dual-optical force. The former often uses fiber taper to generate a large optical force through a strong convergent laser beam, while the latter uses two beams to generate a pair of balanced optical force via two counter-propagating laser beams. SCOT is formed with a balance of optical force and flow force in the microfluid. Furthermore, according to the components of optical force balance with flow force, SCOTs can be subdivided into those based on optical scattering force (F_s) and those based on optical gradient force (F_g). The SCOT based on F_s is often generated by a fiber tip, whose end is against the fluid flow direction. This SCOT can consciously adjust the position of the trapped object along the optical axis with a large manipulation range. Further, it can also be used for flow rate sensing with excellent performance in dynamic range and sensitivity. The SCOT based on F_g is often generated by an optical microfiber, which is perpendicular to the fluid flow. This SCOT is an effective method for massive trapping and manipulation of objects at the micro/nano scale.

The photothermal effect is used for FOF technology in two major ways: by using the optical microfiber and by using special absorption materials. The optical microfiber is often used for enhancing laser radiation and can achieve massive object manipulation, as well as for flow rate sensing. Integration with photothermal materials, such as carbon nanotubes (CNTs), and gold (Au), is a common and effective method to enhance laser absorption. Photothermal materials can be flexibly integrated with any FOF components when required. FOF technology can be integrated with fibers and generate a laser-controlled thermal field to achieve optical manipulation and sensing. Materials integrated with a channel can be used for handling liquids, manipulation in microfluid, micromachining processes, and other applications of microfluidic control. Materials integrated with a target can directly act on the target with a certain position and less thermal energy loss. This method has broad prospects in biological research and medical treatment.

The main results of fiber optofluidic technology based on optical force and photothermal effects are summarized in Tables 2 and 3, respectively. Using optical force and photothermal effects, FOF technology present various advantages, including easy fabrication, miniaturization, low cost, high sensitivity, and a large dynamic range. We believe that many new applications will be explored for physical, chemical, and biological use based on FOF technology in the near future [100,101].

Table 2. A summary of fiber optofluidic technology based on optical force.

Device Types	Principle	Fabrication	Features
Stable Single Optical Trap based on single optical force	Strongly focused beam	Fiber taper	Single object trapping with short range [36–39]
	Surface evanescent fields	Micro/nanofiber	Massive object trapping on the fiber surface [40–47]
Stable Single Optical Trap based on dual optical force	Balance of two optical forces	Two aligned fiber probes	Object manipulation (~200 μm) [30]. Object deformation [25]
		Two misaligned fiber probes	Object rotation [31] Object oscillating, and moving around [51] Object lifting, and force sensing [52] Cell regulation and analysis [53,54]
		HC–PCF	Multiple parameter sensing [32] A centimeter-scale long distance optical manipulation, cell deformation [59]
Stable Combined Optical Trap based on optical scattering force	$F_s = F_v$	Fiber probe	Long range manipulation and large dynamic range flowmeter [60–63] Multiple object organization [64,65] Manipulation and transportation [66]
Stable Combined Optical Trap based on optical gradient force	$F_g = F_v$	Microfiber	Massive trapping, manipulation, and selection [42,67–69]

Table 3. A summary of fiber optofluidic technology based on a photothermal effect.

Device Types	Principle	Fabrication	Features
FOF technology based on a photothermal effect with a microfiber	Enhancing the laser radiation	Fiber taper	Massive particle trapping and manipulation [75]
		Micro/nanofiber	Massive object trapping on fiber surface [76] Flow rate sensing [80]
		optical fiber ring	Massive object trapping in center of ring [77]
FOF technology based on a photothermal effect with special materials	Enhancing the laser absorption	Materials integrated with the fiber	Massive object trapping and delivering [81,82] Multiple parameter sensing [83–85]
		Materials integrated with the channel	Microfluid control [86–88], massive object manipulation [89–91], micromaching processes [92]
		Materials integrated with the channel	Medical treatment with minimally invasive effects [94–99]

Author Contributions: Writing-Original Draft Preparation, C.Z.; Writing-Review and Editing, C.Z., J.L., Y.G., and C.G.; Supervision, B.X. and Q.Z.

Funding: We acknowledge the financial support from the National Natural Science Foundation of China (61575039, 61875034) and the 111 Project (B14039).

References

1. Schmidt, H.; Hawkins, A.R. The photonic integration of non-solid media using optofluidics. *Nat. Photonics* **2011**, *5*, 598–604. [CrossRef]

2. Fan, X.; White, I.M. Optofluidic microsystems for chemical and biological analysis. *Nat. Photonics* **2011**, *5*, 591–597. [CrossRef] [PubMed]

3. Sudirman, A.; Margulis, W. All-Fiber optofluidic component to combine light and fluid. *IEEE Photonics Technol. Lett.* **2014**, *26*, 1031–1033. [CrossRef]

4. Erickson, D.; Sinton, D.; Psaltis, D. Optofluidics for energy applications. *Nat. Photonics* **2011**, *5*, 583–590. [CrossRef]

5. Cubillas, A.M.; Unterkofler, S.T.; Euser, G.; Etzold, B.J.; Jones, A.C.; Sadler, P.J.; Wasserscheid, P.; Russell, P.S.J. Photonic crystal fibres for chemical sensing and photochemistry. *Chem. Soc. Rev.* **2013**, *42*, 8629–8648. [CrossRef] [PubMed]

6. Zhu, T.; Wu, D.; Liu, M.; Duan, D.W. In-line fiber optic interferometric sensors in single-mode fibers. *Sensors* **2012**, *12*, 10430–10449. [CrossRef]

7. Zhang, Q.; Hao, P.; Tian, X. High-visibility in-line fiber-optic optofluidic Fabry–Pérot cavity. *Appl. Phys. Lett.* **2017**, *111*, 191102. [CrossRef]

8. Liu, S.; Ji, Y.; Yang, J. Nafion film temperature/humidity sensing based on optical fiber Fabry-Perot interference. *Sens. Actuators A* **2018**, 313–321. [CrossRef]

9. Chen, L.H.; Chan, C.C.; Ni, K.; Hu, P.B.; Li, T.; Wong, W.C.; Poh, C.L. Label-free fiber-optic interferometric immunosensors based on waist-enlarged fusion taper. *Sens. Actuators B* **2013**, *178*, 176–184. [CrossRef]

10. Shao, M.; Qiao, X.; Fu, H. A Mach–Zehnder interferometric humidity sensor based on waist-enlarged tapers. *Opt. Laser Eng.* **2014**, 86–90. [CrossRef]

11. Hu, P.; Dong, X.; Ni, K. Sensitivity-enhanced Michelson interferometric humidity sensor with waist-enlarged fiber bitaper. *Sens. Actuators B* **2014**, *194*, 180–184. [CrossRef]

12. Zhang, S.; Zhang, W.; Geng, P.; Gao, S. Fiber Mach-Zehnder interferometer based on concatenated down-and up-tapers for refractive index sensing applications. *Opt. Commun.* **2013**, *288*, 47–51. [CrossRef]

13. Tian, Z.; Yam, S.S.; Loock, H.P. Single-mode fiber refractive index sensor based on core-offset attenuators. *IEEE Photonics Technol. Lett.* **2008**, *20*, 1387–1389. [CrossRef]

14. Jiang, L.; Yang, J.; Wang, S.; Li, B.; Wang, M. Fiber Mach-Zehnder interferometer based on micro cavities for high-temperature sensing with high sensitivity. *Opt. Lett.* **2011**, *36*, 3753–3755. [CrossRef] [PubMed]

15. Fu, H.Y.; Tam, H.Y.; Shao, L.Y.; Dong, X.; Wai, P.K.A.; Lu, C.; Khijwania, S.K. Pressure sensor realized with polarization-maintaining photonic crystal fiber-based Sagnac interferometer. *Appl. Opt.* **2008**, *47*, 2835–2839. [CrossRef] [PubMed]

16. Moon, D.S.; Kim, B.H.; Lin, A.; Sun, G.; Han, T.G.; Han, W.T.; Chung, Y. The temperature sensitivity of Sagnac loop interferometer based on polarization maintaining side-hole fiber. *Opt. Express* **2007**, *15*, 7962–7967. [CrossRef] [PubMed]

17. Bertucci, A.; Manicardi, A.; Candiani, A.; Giannetti, S.; Cucinotta, A.; Spoto, G.; Konstantaki, M.; Pissadakis, S.; Selleri, S.; Corradini, R. Detection of unamplified genomic DNA by a PNA-based microstructured optical fiber (MOF) Bragg-grating optofluidic system. *Biosens. Bioelectron.* **2015**, *63*, 248–254. [CrossRef]

18. Zhang, N.; Humbert, G.; Wu, Z.; Li, K.; Shum, P.P.; Zhang, N.M.Y.; Wei, L. In-line optofluidic refractive index sensing in a side-channel photonic crystal fiber. *Opt. Express* **2016**, *24*, 27674–27682. [CrossRef] [PubMed]

19. Zhang, N.; Li, K.; Cui, Y. Ultra-sensitive chemical and biological analysis via specialty fibers with built-in microstructured optofluidic channels. *Lab. Chip.* **2018**, *18*, 655–661. [CrossRef]

20. Rifat, A.A.; Ahmed, R.; Yetisen, A.K. Photonic crystal fiber based plasmonic sensors. *Sens. Actuators B* **2017**, 311–325. [CrossRef]

21. Xing, Z.; Zheng, Y.; Yan, Z. High-sensitivity humidity sensing of microfiber coated with three-dimensional graphene network. *Sens. Actuators B* **2019**, 953–959. [CrossRef]

22. Vaiano, P.; Carotenuto, B.; Pisco, M.; Ricciardi, A.; Quero, G.; Consales, M.; Cusano, A. Lab on Fiber Technology for biological sensing applications. *Laser Photonics Rev.* **2016**, *10*, 922–961. [CrossRef]

23. Ashkin, A. Acceleration and trapping of particles by radiation pressure. *Phys. Rev. Lett.* **1970**, *24*, 156. [CrossRef]

24. Ashkin, A.; Dziedzic, J.M.; Bjorkholm, J.; Chu, S. Observation of a single-beam gradient force optical trap for dielectric particles. *Opt. Lett.* **1986**, *11*, 288–290. [CrossRef] [PubMed]

25. Guck, J.; Ananthakrishnan, R.; Moon, T.; Cunningham, C.; Käs, J. Optical deformability of soft biological dielectrics. *Phys. Rev. Lett.* **2000**, *84*, 5451. [CrossRef] [PubMed]

26. Jess, P.; Garcés-Chávez, V.; Smith, D.; Mazilu, M.; Paterson, L.; Riches, A.; Herrington, C.; Sibbett, W.; Dholakia, K. Dual beam fibre trap for Raman microspectroscopy of single cells. *Opt. Express* **2006**, *14*, 5779–5791. [CrossRef] [PubMed]

27. Xu, X.; Cheng, C.; Xin, H.; Lei, H.; Li, B. Controllable orientation of single silver nanowire using two fiber probes. *Sci. Rep.* **2015**, *4*, 3989. [CrossRef]

28. Taguchi, K.; Atsuta, K.; Nakata, T.; Ikeda, R. Levitation of a microscopic object using plural optical fibers. *Opt. Commun.* **2000**, *176*, 43–47. [CrossRef]

29. Gong, Y.; Huang, W.; Liu, Q.F.; Wu, Y.; Rao, Y.; Peng, G.D.; Lang, J.; Zhang, K. Graded-index optical fiber tweezers with long manipulation length. *Opt. Express* **2014**, *22*, 25267–25276. [CrossRef]

30. Zhang, Y.; Liang, P.; Liu, Z.; Lei, J.; Yang, J.; Yuan, L. A novel temperature sensor based on optical trapping technology. *J. Lightwave Technol.* **2014**, *32*, 1394–1398. [CrossRef]

31. Xiao, G.; Yang, K.; Luo, H.; Chen, X.; Xiong, W. Orbital rotation of trapped particle in a transversely misaligned dual-fiber optical trap. *IEEE Photonics J.* **2016**, *8*, 1–8. [CrossRef]

32. Bykov, D.S.; Schmidt, O.A.; Euser, T.G.; Russell, P.S.J. Flying particle sensors in hollow-core photonic crystal fibre. *Nat. Photonics* **2015**, *9*, 461. [CrossRef]

33. Bell, A.G. On the production and reproduction of sound by light. *Am. J. Sci.* **1880**, 305–324. [CrossRef]

34. Terazima, M.; Hirota, N.; Braslavsky, S.E.; Mandelis, A.; Bialkowski, S.E.; Diebold, G.J.; Miller, R.; Fournier, D.; Palmer, R.A.; Tam, A. Quantities, terminology, and symbols in photothermal and related spectroscopies (IUPAC Recommendations 2004). *Pure Appl. Chem.* **2004**, *76*, 1083–1118. [CrossRef]

35. Ashkin, A. Forces of a single-beam gradient laser trap on a dielectric sphere in the ray optics regime. *Biophys. J.* **1992**, *61*, 569–582. [CrossRef]

36. Liu, Z.; Guo, C.; Yang, J.; Yuan, L. Tapered fiber optical tweezers for microscopic particle trapping: Fabrication and application. *Opt. Express* **2006**, *14*, 12510–12516. [CrossRef] [PubMed]

37. Liu, Z.; Wang, L.; Liang, P.; Yuan, L. Mode division multiplexing technology for single-fiber optical trapping axial-position adjustment. *Opt. Lett.* **2013**, *38*, 2617–2620. [CrossRef]

38. Zhang, Y.; Liang, P.; Lei, J. Multi-dimensional manipulation of yeast cells using a LP11 mode beam. *J. Lightwave Technol.* **2014**, *32*, 1098–1103. [CrossRef]

39. Liu, Z.; Liang, P.; Zhang, Y.; Zhang, Y.; Zhao, E.; Yang, J.; Yuan, L. Micro particle launcher/cleaner based on optical trapping technology. *Opt. Express* **2015**, *23*, 8650–8658. [CrossRef]

40. Cheng, C.; Xu, X.; Lei, H.; Li, B. Plasmon-assisted trapping of nanoparticles using a silver-nanowire-embedded PMMA nanofiber. *Sci. Rep.* **2016**, *6*, 20433. [CrossRef]

41. Li, Y.; Xin, H.; Xu, X.; Liu, X.; Li, B. Fibre-optic trapping and manipulation at the nanoscale. *Adv. Mater. Lett.* **2018**, *9*, 567–577. [CrossRef]

42. Lei, H.; Xu, C.; Zhang, Y.; Li, B. Bidirectional optical transportation and controllable positioning of nanoparticles using an optical nanofiber. *Nanoscale* **2012**, *4*, 6707–6709. [CrossRef] [PubMed]

43. Brambilla, G.; Murugan, G.; Wilkinson, J.; Richardson, D. Optical manipulation of microspheres along a subwavelength optical wire. *Opt. Lett.* **2007**, *32*, 3041–3043. [CrossRef] [PubMed]

44. Xu, L.; Li, Y.; Li, B. Size-dependent trapping and delivery of submicro-spheres using a submicrofibre. *New J. Phys.* **2012**, *14*, 033020. [CrossRef]

45. Zhang, Y.; Lei, H.; Li, B. Refractive-Index-Based Sorting of Colloidal Particles Using a Subwavelength Optical Fiber in a Static Fluid. *Appl. Phys. Express* **2013**, *6*, 072001. [CrossRef]

46. Zhang, Y.; Li, B. Particle sorting using a subwavelength optical fiber. *Laser Photonics Rev.* **2013**, *7*, 289–296. [CrossRef]

47. Harada, Y.; Asakura, T. Radiation forces on a dielectric sphere in the Rayleigh scattering regime. *Opt. Commun.* **1996**, *124*, 529–541. [CrossRef]

48. Ribeiro, R.S.R.; Soppera, O.; Oliva, A.G.; Guerreiro, A.; Jorge, P.A.S. New trends on optical fiber tweezers. *J. Lightwave Technol.* **2015**, *33*, 3394–3405. [CrossRef]

49. Gong, Y.; Ye, A.Y.; Wu, Y.; Rao, Y.J.; Yao, Y.; Xiao, S. Graded-index fiber tip optical tweezers: Numerical simulation and trapping experiment. *Opt. Express* **2013**, *21*, 16181–16190. [CrossRef]

50. Nylk, J.; Kristensen, M.V.G.; Mazilu, M.; Thayil, A.K.; Mitchell, C.A.; Campbell, E.C.; Dholakia, K. Development of a graded index microlens based fiber optical trap and its characterization using principal component analysis. *Biomed. Opt. Express* **2015**, *6*, 1512–1519. [CrossRef]

51. Gordon, R.; Kawano, M.; Blakely, J.T.; Sinton, D. Optohydro dynamic theory of particles in a dual-beam optical trap. *Phys. Rev. B* **2008**, *77*.

52. Liu, Y.; Yu, M. Investigation of inclined dual-fiber optical tweezers for 3D manipulation and force sensing. *Opt. Express* **2009**, *17*, 13624–13638. [CrossRef] [PubMed]

53. Huang, J.; Liu, X.; Zhang, Y.; Li, B. Optical trapping and orientation of Escherichia coli cells using two tapered fiber probes. *Photonics Res.* **2015**, *3*, 308–312. [CrossRef]

54. Liu, X.; Huang, J.; Zhang, Y.; Li, B. Optical regulation of cell chain. *Sci. Rep.* **2015**, *5*, 11578. [CrossRef] [PubMed]

55. Zhang, Y.; Liu, Z.; Yang, J. A non-contact single optical fiber multi-optical tweezers probe: Design and fabrication. *Opt. Commun.* **2012**, *285*, 4068–4071. [CrossRef]

56. Zhao, H.; Farrell, G.; Wang, P. Investigation of particle harmonic oscillation using four-core fiber integrated twin-tweezers. *IEEE Photonics Technol. Lett.* **2016**, *28*, 461–464. [CrossRef]

57. Deng, H.; Zhang, Y.; Yuan, T. Fiber-based optical gun for particle shooting. *ACS Photonics* **2017**, *4*, 642–648. [CrossRef]

58. Liberale, C.; Minzioni, P.; Bragheri, F.; Angelis, F.D.; Fabrizio, E.D.; Cristiani, I. Miniaturized all-fibre probe for three-dimensional optical trapping and manipulation. *Nat. Photonics* **2007**, *1*, 723–727. [CrossRef]

59. Unterkofler, S.; Garbos, M.K.; Euser, T.G.; Russell, P.S.J. Long-distance laser propulsion and deformation-monitoring of cells in optofluidic photonic crystal fiber. *J. Biophotonics* **2013**, *6*, 743–752. [CrossRef] [PubMed]

60. Gong, Y.; Zhang, C.; Liu, Q.F.; Wu, Y.; Wu, H.; Rao, Y.; Peng, G.D. Optofluidic tunable manipulation of microparticles by integrating graded-index fiber taper with a microcavity. *Opt. Express* **2015**, *23*, 3762–3769. [CrossRef] [PubMed]

61. Zhang, C.L.; Gong, Y.; Liu, Q.F.; Wu, Y.; Rao, Y.J.; Peng, G.D. Graded-index fiber enabled strain-controllable optofluidic manipulation. *IEEE Photonics Technol. Lett.* **2016**, *28*, 256–259. [CrossRef]

62. Gong, Y.; Liu, Q.F.; Zhang, C.L.; Wu, Y.; Rao, Y.J.; Peng, G.D. Microfluidic flow rate detection with a large dynamic range by optical manipulation. *IEEE Photonics Technol. Lett.* **2015**, *27*, 2508–2511. [CrossRef]

63. Gong, Y.; Qiu, L.; Zhang, C.; Wu, Y.; Rao, Y.J.; Peng, G.D. Dual-Mode fiber optofluidic flowmeter with a large dynamic range. *J. Lightwave Technol.* **2017**, *35*, 2156–2160. [CrossRef]

64. Xin, H.; Zhang, Y.; Lei, H.; Li, Y.; Zhang, H.; Li, B. Optofluidic realization and retaining of cell–cell contact using an abrupt tapered optical fibre. *Sci. Rep.* **2013**, *3*, 1993. [CrossRef] [PubMed]

65. Liu, X.; Huang, J.; Li, Y.; Zhang, Y.; Li, B. Optofluidic organization and transport of cell chain. *J. Biophotonics* **2017**, *10*, 1627–1635. [CrossRef] [PubMed]

66. Zhang, Y.; Li, Y.; Zhang, Y.; Hu, C.; Liu, Z.; Yang, X.; Yuan, L. HACF-based optical tweezers available for living cells manipulating and sterile transporting. *Opt. Commun.* **2018**, *427*, 563–566. [CrossRef]

67. Xin, H.; Li, B. Fiber-based optical trapping and manipulation. *Front. Optoelectron.* **2019**, *12*, 97–110. [CrossRef]

68. Xu, C.; Lei, H.; Zhang, Y.; Li, B. Backward transport of nanoparticles in fluidic flow. *Opt. Express* **2012**, *20*, 1930–1938. [CrossRef]

69. Lei, H.; Zhang, Y.; Li, B. Particle separation in fluidic flow by optical fiber. *Opt. Express* **2012**, *20*, 1292–1300. [CrossRef]

70. Gloge, D. Weakly guiding fibers. *Appl. Opt.* **1971**, *10*, 2252–8225. [CrossRef]

71. Li, K.; Liu, G.; Wu, Y.; Hao, P.; Zhou, W.; Zhang, Z. Gold nanoparticle amplified optical microfiber evanescent wave absorption biosensor for cancer biomarker detection in serum. *Talanta* **2014**, *120*, 419. [CrossRef] [PubMed]

72. Sun, D.; Guo, T.; Ran, Y.; Huang, Y.; Guan, B. In situ DNA hybridization detection with a reflective microfiber grating biosensor. *Biosens. Bioelectron.* **2014**, *61*, 541. [CrossRef] [PubMed]

73. Soong, C.; Li, W.; Liu, C.; Tzeng, P. Theoretical analysis for photophoresis of a microscale hydrophobic particle in liquids. *Opt. Express* **2010**, *18*, 2168–2182. [CrossRef] [PubMed]

74. Duhr, S.; Braun, D. Optothermal molecule trapping by opposing fluid flow with thermophoretic drift. *Phys. Rev. Lett.* **2006**, *97*, 038103. [CrossRef] [PubMed]

75. Xin, H.; Li, X.; Li, B. Massive photothermal trapping and migration of particles by a tapered optical fiber. *Opt. Express* **2011**, *19*, 17065–17074. [CrossRef] [PubMed]

76. Lei, H.; Zhang, Y.; Li, X.; Li, B. Photophoretic assembly and migration of dielectric particles and Escherichia coli in liquids using a subwavelength diameter optical fiber. *Lab. Chip* **2011**, *11*, 2241–2246. [CrossRef] [PubMed]

77. Xin, H.; Lei, H.; Zhang, Y.; Li, X.; Li, B. Photothermal trapping of dielectric particles by optical fiber-ring. *Opt. Express* **2011**, *19*, 2711–2719. [CrossRef] [PubMed]

78. Cadarso, V.J.; Llobera, A.; Puyol, M. Integrated photonic nanofences: Combining subwavelength waveguides with an enhanced evanescent field for sensing applications. *ACS Nano* **2015**, *10*, 778–785. [CrossRef] [PubMed]

79. Gong, Y.; Zhang, M.; Gong, C. Sensitive optofluidic flow rate sensor based on laser heating and microring resonator. *Microfluid. Nanofluid.* **2015**, *19*, 1497–1505. [CrossRef]

80. Liu, Z.; Lei, J.; Zhang, Y. A Method to Gather/Arrange Particles Based on Thermal Convection. In Proceedings of the 24th International Conference on Optical Fibre Sensors, Curitiba, Brazil, 28 September 2015; SPIE: Bellingham, WA, USA, 2015.

81. Li, Z.; Yang, J.; Liu, S.; Jiang, X.; Wang, H.; Hu, X.; Xing, X. High throughput trapping and arrangement of biological cells using self-assembled optical tweezer. *Opt. Express* **2018**, *26*, 34665–34674. [CrossRef]

82. Yang, J.; Li, Z.; Wang, H.; Zhu, D.; Cai, X.; Cheng, Y.; Xing, X. Optofluidic trapping and delivery of massive mesoscopic matters using mobile vortex array. *Appl. Phys. Lett.* **2017**, *111*, 191901. [CrossRef]

83. Zhang, C.L.; Gong, Y.; Zou, W.L.; Wu, Y.; Rao, Y.J.; Peng, G.D.; Fan, X. Microbubble-Based Fiber Optofluidic Interferometer for Sensing. *J. Lightwave Technol.* **2017**, *35*, 2514–2519. [CrossRef]

84. Zhang, C.L.; Gong, Y.; Wu, Y.; Rao, Y.; Peng, G.; Fan, X. Lab-on-tip based on photothermal microbubble generation for concentration detection. *Sens. Actuators B* **2018**, 2504–2509. [CrossRef]

85. Matjasec, Z.; Donlagic, D. All-optical, all-fiber, thermal conductivity sensor for identification and characterization of fluids. *Sens. Actuators B* **2017**, *242*, 577–585. [CrossRef]

86. Zhang, K.; Jian, A.; Zhang, X.; Wang, Y.; Li, Z.; Tam, H.Y. Laser-induced thermal bubbles for microfluidic applications. *Lab. Chip* **2011**, *11*, 1389–1395. [CrossRef]

87. Tovar, A.R.; Patel, M.V.; Lee, A.P. Lateral air cavities for microfluidic pumping with the use of acoustic energy. *Microfluid. Nanofluid.* **2011**, *10*, 1269–1278. [CrossRef]

88. Ahmed, D.; Chan, C.Y.; Lin, S.C.S.; Muddana, H.S.; Nama, N.; Benkovic, S.J.; Huang, T.J. Tunable pulsatile chemical gradient generation via acoustically driven oscillating bubbles. *Lab. Chip* **2013**, *13*, 328–331. [CrossRef]

89. Rogers, P.; Neild, A. Selective particle trapping using an oscillating microbubble. *Lab. Chip* **2011**, *11*, 3710–3715. [CrossRef]

90. Xu, Y.; Hashmi, A.; Yu, G.; Lu, X.; Kwon, H.J.; Chen, X.; Xu, J. Microbubble array for on-chip worm processing. *Appl. Phys. Lett.* **2013**, *102*, 023702. [CrossRef]

91. Fujii, S.; Kobayashi, K.; Kanaizuka, K.; Okamoto, T.; Toyabe, S.; Muneyuki, E.; Haga, M.A. Manipulation of single DNA using a micronanobubble formed by local laser heating on a Au-coated surface. *Chem. Lett.* **2009**, *39*, 92–93. [CrossRef]

92. Davim, J.P. *Nontraditional Machining Processes: Research Advances*; Springer: Berlin/Heidelberg, Germany, 2013; pp. 42–44. ISBN 978-3-662-45088-8.

93. Cheng, Y.; Yang, J.; Li, Z.; Zhu, D.; Cai, X.; Hu, X.; Xing, X. Microbubble-assisted optofluidic control using a photothermal waveguide. *Appl. Phys. Lett.* **2017**, *111*, 151903. [CrossRef]

94. Skirtach, A.G.; Munoz Javier, A.; Kreft, O.; Köhler, K.; Piera Alberola, A.; Möhwald, H.; Sukhorukov, G.B. Laser-induced release of encapsulated materials inside living cells. *Angew. Chem. Int. Ed.* **2006**, *45*, 4612–4617. [CrossRef] [PubMed]

95. Ochs, M.; Carregal-Romero, S.; Rejman, J.; Braeckmans, K.; De Smedt, S.C.; Parak, W.J. Light-Addressable Capsules as Caged Compound Matrix for Controlled Triggering of Cytosolic Reactions. *Angew. Chem. Int. Ed.* **2013**, *52*, 695–699. [CrossRef] [PubMed]

96. Dykman, L.; Khlebtsov, N. Gold nanoparticles in biomedical applications: Recent advances and perspectives. *Chem. Soc. Rev.* **2012**, *41*, 2256–2282. [CrossRef] [PubMed]

97. Loo, C.; Lowery, A.; Halas, N.; West, J.; Drezek, R. Immunotargeted nanoshells for integrated cancer imaging and therapy. *Nano Lett.* **2005**, *5*, 709–711. [CrossRef] [PubMed]

98. Kim, Y.K.; Na, H.K.; Kim, S.; Jang, H.; Chang, S.J.; Min, D.H. One-Pot Synthesis of Multifunctional Au@ Graphene Oxide Nanocolloid Core@ Shell Nanoparticles for Raman Bioimaging, Photothermal, and Photodynamic Therapy. *Small* **2015**, *11*, 2527–2535. [CrossRef] [PubMed]

99. Shao, J.; Xuan, M.; Dai, L.; Si, T.; Li, J.; He, Q. Near-Infrared-Activated Nanocalorifiers in Microcapsules: Vapor Bubble Generation for In Vivo Enhanced Cancer Therapy. *Angew. Chem. Int. Ed.* **2015**, *54*, 12782–12787. [CrossRef] [PubMed]

100. Gong, C.; Gong, Y.; Chen, Q.; Rao, Y.J.; Peng, G.D.; Fan, X. Reproducible fiber optofluidic laser for disposable and array applications. *Lab Chip* **2017**, *17*, 3431–3436. [CrossRef] [PubMed]

101. Yang, X.; Shu, W.; Wang, Y.; Gong, Y.; Gong, C.; Chen, Q.; Rao, Y.J. Turbidimetric inhibition immunoassay revisited to enhance its sensitivity via an optofluidic laser. *Biosens. Bioelectron.* **2019**, *131*, 60–66. [CrossRef] [PubMed]

MDPI

St. Alban-Anlage 66

4052 Basel

Switzerland

Tel. +41 61 683 77 34

Fax +41 61 302 89 18

www.mdpi.com

Micromachines Editorial Office

E-mail: micromachines@mdpi.com

www.mdpi.com/journal/micromachines